Climate Change Economics and Policy

An RFF Anthology

Edited by Michael A. Toman

Resources for the Future
Washington, DC

Printed in the United States of America

An RFF Press book
Published by Resources for the Future
1616 P Street, NW, Washington, DC 20036–1400
www.rff.org

Library of Congress Cataloging-in-Publication Data

Climate change economics and policy : an RFF anthology / edited by Michael Toman.
    p.   cm.
    Includes bibliographical references and index.
    ISBN 1–891853–04–X (alk. paper)
    1. Climatic changes—Economic aspects. 2. Climatic changes—Social aspects. 3. Climatic changes—Government policy. 4. Climatic changes—Environmental aspects. 5. Climatic changes—International cooperation. 6. Environmental policy. I. Toman, Michael A. II. Resources for the Future
QC981.8.C5 C511384   2001
363.738'747—dc21   00–054868

f e d c b a

This book was typeset in Futura Condensed and ITC Berkeley Oldstyle by Betsy Kulamer. It was copyedited by Pamela Angulo. The cover was designed by Debra Naylor Design.

About
Resources for the Future
and RFF Press

Founded in 1952, Resources for the Future (RFF) contributes to environmental and natural resource policymaking worldwide by performing independent social science research.

RFF pioneered the application of economics as a tool to develop more effective policy about the use and conservation of natural resources. Its scholars continue to employ social science methods to analyze critical issues concerning pollution control, energy policy, land and water use, hazardous waste, climate change, biodiversity, and the environmental challenges of developing countries.

RFF Press supports the mission of RFF by publishing book-length works that present a broad range of approaches to the study of natural resources and the environment. Its authors and editors include RFF staff, researchers from the larger academic and policy communities, and journalists. Audiences for RFF publications include all of the participants in the policymaking process—scholars, the media, advocacy groups, NGOs, professionals in business and government, and the general public.

Contents

Part 3: Policy Design and Implementation Issues

Part 4: International Considerations

Conclusion

Preface

Having risen from relative obscurity as few as 10 years ago, climate change now looms large among environmental policy issues. Its scope is global, the potential environmental and economic impacts are ubiquitous, the potential restrictions on human choices touch the most basic goals of people in all nations, and the sheer scope of the potential response—a significant shift away from using fossil fuels as the primary energy source in the modern economy—is daunting.

The magnitude of these changes has motivated experts worldwide to study the natural and socioeconomic effects of climate change as well as policy options for slowing climate change and reducing its risks. The various options serve as fodder for the ongoing, often testy negotiations within and among nations under the auspices of the 1992 U.N. Framework Convention on Climate Change (UNFCCC) and the 1997 Kyoto Protocol. These negotiations concern how and when to mitigate climate change, who should take action, and who should bear the costs.

In 1996, Resources for the Future (RFF) launched a program of extended research on and communication about these complex issues in an effort to reach various audiences. As part of this program, RFF began publishing a series of Climate Issue Briefs, which were meant to build on an extensive and growing body of research at RFF and to draw on the best research being carried out elsewhere. The intent of producing short papers that were less technical (than the usual Discussion Paper format) was to make the material accessible to a broad audience and to draw attention to key climate problems and possible solutions. The Climate Issue Briefs, which now number more than 20, have been widely disseminated via RFF's climate change website, *Weathervane*, as well as in print.

As the series grew, so did RFF's interest in producing an anthology that would be a useful and accessible guide to climate economics and policy for students of, participants in, and observers of the policy process. The collection compiled here combines most of the Climate Issue Briefs in the series with additional material, some of which has not been published previously. All of the "older" (from 1999 and earlier) Climate Issue Briefs were reviewed and updated by the authors—in some cases, substantially—for inclusion as chapters in this book.

Given our focus on economics and policy in the source material, this anthology is not a comprehensive treatment of climate issues. We have not attempted to review all of the many issues that surround the effects of climate change, and although we allude to the continuing scientific controversy that surrounds the issue, we do not explore these important impacts in any depth. Nevertheless, this volume provides a solid introduction to the importance of climate change, the history of the policy debate, and the ways in which the challenges of developing and implementing climate policy might be addressed, domestically and internationally.

As editor of the volume, I express my profound gratitude to my many colleagues at RFF and elsewhere who contributed the content for this volume. Every one of these authors took time out of a busy research schedule to produce well-written, accessible, yet rigorous material. It would have been impossible to produce this book without their help.

Production of the book has been aided hugely by the very competent research assistance of Jennifer Lee, as well as the earlier assistance of Marina Cazorla. The able editorial and production assistance of Pam Angulo, Felicia Day, Betsy Kulamer, and Katherine Murphy also was indispensable.

Last, but certainly not least, I thank the G. Unger Vetlesen Foundation for financial support of the RFF Climate Issue Brief series and of subsequent efforts to assemble this book.

MICHAEL A. TOMAN
Resources for the Future

Contributors

J.W. Anderson, Journalist-in-Residence, Resources for the Future

Allen Blackman, Fellow, Resources for the Future

Dallas Burtraw, Senior Fellow, Resources for the Future

Marina V. Cazorla, Environmental Specialist, California Coastal Commission; former Research Assistant, Resources for the Future

Sarah A. Cline, Research Assistant, Resources for the Future

Pierre Crosson, Senior Fellow, Resources for the Future

Joel Darmstadter, Senior Fellow, Resources for the Future

Carolyn Fischer, Fellow, Resources for the Future; former staff economist, Council of Economic Advisers, Executive Office of the President (1994–1995)

Kenneth D. Frederick, Senior Fellow, Resources for the Future

Lawrence H. Goulder, Associate Professor of Economics and Fellow, Institute for International Studies, Stanford University; University Fellow, Resources for the Future

Adam B. Jaffe, Professor of Economics, Brandeis University

Pamela Jagger, Research Analyst, International Food Policy Research Institute; former Research Assistant, Resources for the Future

Suzi Kerr, Director and Senior Fellow, Motu Economic and Public Policy Research, Wellington, New Zealand

Raymond J. Kopp, Vice President for Programs and Senior Fellow, Resources for the Future

Alan J. Krupnick, Senior Fellow and Division Director, Resources for the Future; former senior staff economist for climate change and other environmental issues, Council of Economic Advisers, Executive Office of the President (1993–1994)

Ramón López, Professor of Economics, Department of Agricultural and Resource Economics, University of Maryland, College Park

Richard G. Newell, Fellow, Resources for the Future

Karen L. Palmer, Senior Fellow, Resources for the Future; former visiting economist, Office of Economic Policy, Federal Energy Regulatory Commission (1996–1997)

Ian W.H. Parry, Fellow, Resources for the Future

William A. Pizer, Fellow, Resources for the Future

Roger A. Sedjo, Senior Fellow, Resources for the Future; convening lead author, Working Group 3, Third Assessment Report of the Intergovernmental Panel on Climate Change

Jason F. Shogren, Stroock Distinguished Professor of Natural Resource Conservation and Management, Department of Economics and Finance, University of Wyoming; former senior staff economist for climate change and other environmental issues, Council of Economic Advisers, Executive Office of the President (1997)

Brent Sohngen, Assistant Professor, Department of Agricultural, Environmental, and Development Economics, Ohio State University

Robert N. Stavins, Albert Pratt Professor of Business and Government, John F. Kennedy School of Government, Harvard University; University Fellow, Resources for the Future

Michael A. Toman, Senior Fellow and Division Director, Resources for the Future; former senior staff economist for climate change and other environmental issues, Council of Economic Advisers, Executive Office of the President (1994–1996)

Jonathan Baert Wiener, Associate Professor, Law School and Nicholas School of the Environment, Duke University; former senior staff economist for climate change and other environmental issues, Council of Economic Advisers, Executive Office of the President (1992–1993), and former senior aide on environmental policy issues at the White House Office of Science & Technology Policy and the Department of Justice (1991–1992)

Climate Change Economics and Policies
An Overview

Michael A. Toman

Life on Earth is possible partly because some gases, such as carbon dioxide (CO_2) and water vapor, which naturally occur in Earth's atmosphere, trap heat—like a greenhouse does. Humans are greatly adding to the presence of such gases, commonly referred to as greenhouse gases (GHGs), by burning fossil fuels and through other industrial activities as well as various kinds of land use, such as deforestation.

Anthropogenic GHG emissions (those that are produced, induced, or influenced by human activity) work against us when they trap too much sunlight and block outward radiation. Many scientists worry that the accumulation of these gases in the atmosphere has changed the climate and will continue to change it. Potential negative climate risks include more severe weather patterns; hobbled ecosystems, with less biodiversity; changes in patterns of drought and flood, with less safe drinking water; inundation of coastal areas from rising sea levels; and a greater spread of infectious diseases such as malaria, yellow fever, and cholera. On the positive side, climate change might benefit agriculture and forestry in various locations by increasing productivity as a result of longer growing seasons and increased fertilization.

A great deal of controversy surrounds the issue of climate change. Some participants in the debate say that climate change is one of the greatest threats fac-

ing humankind, one that calls for immediate and strong controls on GHGs, particularly CO_2 emissions from the burning of fossil fuel (by far the largest source of anthropogenic GHGs). Others say that the risks are weakly documented scientifically, that adaptation to a changing climate will substantially reduce human vulnerability, and consequently, that little action is warranted other than additional study and the development of future technological options. The same kinds of divisions arise in discussing policy options to reduce GHG emissions, in which some parties predict net benefits to the economy and others fear the loss of several percentage points of national income.

These disagreements surface in international negotiations under the 1992 U.N. Framework Convention on Climate Change (UNFCCC). Article 2 of the UNFCCC requires signatories to take actions to "prevent dangerous anthropogenic interference with the climate system" from GHG emissions (and other actions, such as deforestation). However, the term "dangerous" in Article 2 does not have an unambiguous, purely scientific definition; it is inherently a question of human values. Article 3 states that precautionary risk reduction should be guided by equity across time and wealth levels, as expressed in the concept of "common but differentiated responsibilities"; however, this phrase is not precisely defined. Article 4 states that nations should cooperate to improve human adaptation and mitigation of climate change through financial support and low-emission technologies. Articles 3 and 4 also refer to the use of cost-effective response measures. These articles provide a basis for an ongoing debate on policies to implement the goals of the convention.

More than 160 nations signaled their commitment to address the problem of climate change by initiating the Kyoto Protocol in December 1997. The protocol requires industrialized "Annex I" countries to reduce their total emissions of CO_2 and other GHGs by an average of roughly 5% compared with 1990 levels by 2008–12. Developing countries are not required to meet quantitative emission goals, but all signatories to the 1992 UNFCCC have certain obligations to measure and report emissions and to encourage climate-friendly activities. How the Kyoto Protocol targets will be met—in particular, the design of international "flexibility mechanisms" for low-cost GHG abatement—continues to be negotiated. However, larger questions remain in the United States and other countries about whether, when, and how the Kyoto Protocol will be ratified and implemented.

The authors of this book address many economic and policy issues related to climate change. Several points set the context for addressing these issues. First, some degree of climate change appears inevitable. Given current emissions trends and the inertia of the climate system, even if emissions were stabilized or substantially reduced, scientific models suggest that climatic changes and their consequences would continue. To stabilize atmospheric concentrations of GHGs (and thus their effects) would eventually require very large cuts in emissions from current levels, not to mention the future levels implied by continued economic growth under a "business as usual" scenario. For example, to stabilize CO_2 concentrations at something over twice preindustrial levels would require emissions ultimately to fall by more than 70% from their current level. The authors of this book allude to the scientific underpinnings of climate change and the controversies that surround them, but the emphasis is on economics and policy issues.

Second, the potential human consequences of climate change are key to analyzing economic and policy issues. The findings of climate scientists and studies of physical impacts from climate change alone do not suffice. This simple but important point often seems to be overlooked in debates about "what the science says."

Finally, the problem is global. Rich and poor countries alike argue over how the burden of GHG emissions reductions should be allocated. However, no solution can be effective in the long term unless it ultimately leads to reductions in total global emissions, not only emissions in selected countries.

The economics perspective reflected in this book suggests several basic points for evaluating climate change risks and response costs:

- **Think comprehensively about risks.** Climate change can have several possible impacts over space and time, and each of these impacts remains uncertain at present.

- **Address adaptation.** The risks of climate change depend not only on what happens to the natural world as a consequence of GHG accumulation in the atmosphere but also on how humans respond to the prospect of those impacts.

- **Consider the long term.** The impacts of climate change are driven strongly by the accumulation of GHGs in the atmosphere over decades and, to some extent, by the speed of climate change over decadal periods. This attribute makes climate change quite different from many other local and regional environmental problems.

- **Make the focus international.** The sources of GHGs, the risks posed by climate change, and the opportunities for mitigating risks all are global.

- **Keep in mind distributional issues.** Climate change will have very uneven effects over space and time, and the capacities to mitigate climate change risk also vary greatly.

- **Estimate control costs comprehensively and realistically.** It is important to remember the many subtle costs that economies can experience from attempts to limit GHGs, even while recognizing opportunities to reduce GHGs affordably through new technologies and the reduction of market distortions.

Several themes also recur in the analysis of what constitutes effective and efficient climate policies:

- **Incorporate economic incentives into emissions-reduction policy.** Years of research and an increasing body of international experience provide strong support for the cost-effectiveness of this approach in lieu of technology-oriented policy mandates that are less flexible. This approach also can be used to address distributional concerns that surround the impacts of climate policies.

- **Provide opportunities for emissions reductions wherever possible.** Because the sources of GHGs are global and the costs of GHG control vary by location, large joint benefits—for developing as well as developed countries—can result from seeking the cheapest abatement options.

- **Allow flexibility in the timing of cumulative emissions reductions to reduce overall costs.** This controversial proposition relates to how rapidly GHG emissions reductions goals should be implemented and ratcheted upward. Whereas a strong case can be made for current action, one also can be made for more flexible policy goals and for a more gradual phasing-in of significant emission limits.

- **Encourage the development of the climate change knowledge base and improved technology for emissions reduction.** Given the long-term nature of climate change, improved knowledge is critical for refining policy and for increasing public understanding and support.

- **Increase the emphasis on adaptation.** The focus of international negotiation has been almost exclusively on reducing GHG emissions. Yet, many opportunities for increasing resilience to the potential negative effects of climate change need to be pursued given the likely prospects of some negative effects. Increased resilience lowers the long-term cost of climate change, allows more gradual implementation of costly GHG mitigation measures, and in many cases pays immediate dividends in addressing other environmental and social problems.

Outline of the Book

This book is organized in four sections.

Part 1: Introduction

In Part 1, Anderson (Chapter 2) and Darmstadter (Chapter 3) begin by presenting the historical evolution of concern for climate change, the development of the existing international regime for climate policy, and the backdrop of energy use and economic activity that gives rise to GHG emissions. Shogren and Toman (Chapter 4) discuss how the benefits and costs of climate change, mitigation, and adaptation should be addressed. They address both the scale of climate change policy goals (how much GHG emissions reduction) and the timing, and provide an overview of how costs and benefits are assessed. Cline (Appendix A) provides more detail about cost assessment for mitigating GHG emissions through a review of many of the large-scale computer simulation models used for this task.

Part 2: Impacts of Greenhouse Gas Emissions

In Part 2, several possible impacts of climate change are addressed from an economic as well as a physical perspective. Crosson (Chapter 5), Frederick (Chapter 6), and Sedjo and Sohngen (Chapter 7) show how economic analysis combines scientific information about climate change impacts with social scientific information about human responses and valuations. These chapters strongly illustrate how adaptation can greatly reduce the expected negative impact of climate change. The authors also draw attention to the importance of policies that can increase resilience (for example, improved water and forest management and agricultural productivity) and the possible constraints on resilience measures (for example, poverty as an obstacle to improved agricultural technology and the challenges of balancing harvest and biodiversity values in forestry). Burtraw and Toman (Chapter 8) instead look at how reducing GHG emissions can have benefits beyond emissions reduction in the long-term risk of climate change. The authors particularly focus on how well-designed GHG policies can yield short-term dividends from improved local air quality and how these ancillary benefits of GHG policy can affect the desirability of some degree of GHG emissions control. Krupnick (Appendix B) summarizes a key concern about the impacts of climate change beyond natural resources, namely, the possibility of increased threats to human health. Again, the message is that the concern is real, but adaptation—here, in the context of various practices for improving public health and avoiding disease—lessen the risk, perhaps substantially.

Note that Part 2 does not address the whole panoply of concerns about climate change, which include the loss of current coastal areas from rising sea levels, threats to unique areas and other ecological resources, and unlikely but catastrophic changes such as the rerouting of ocean currents and massive storm surges. Even taking into account the possibilities for adaptation, the last word on the threat of climate change is not yet in. Given the information available, the initiation of long-term efforts to abate GHG emissions seems indicated; however, climate change is not a clear and present danger that requires a draconian short-term response.

Part 3: Policy Design and Implementation Issues

Part 3 addresses the questions that surround the development and implementation of climate policies as well as climate aspects of other policies that can have important implications for GHG emissions (such as electricity sector restructuring). Pizer (Chapter 9) lays out the economic trade-offs between using the kinds of policy tools embedded in the Kyoto Protocol framework—binding quantitative limits on GHG emissions—versus another class of policies designed to induce the reduction of GHG emissions through price incentives without a binding emission ceiling. A fossil energy tax on the basis of its carbon content is the standard example of this latter approach. Pizer shows that a tax-based approach could substantially increase the overall efficiency of GHG emissions policy. It highlights one of several sharp dichotomies between the Kyoto Protocol policy architecture and the implications of economic analysis. Pizer also shows how a hybrid policy can combine aspirational quantity targets with the cost-reducing flexibility of the price-based approach.

Fischer, Kerr, and Toman (Chapter 10); Parry (Chapter 11); and Goulder (Chapter 12) consider in more detail several elements in designing cost-effective quantity and price policies. Fischer and others review the many key components going into an effective domestic emissions trading program for GHG emissions control. This general approach involves issuing emission allowances that can be bought and sold, so that GHG sources have a built-in economic incentive to seek the lowest-cost opportunities for GHG emissions abatement (conservation, switching fuel, or otherwise) throughout the economy. An emissions trading program already has proved successful in cost-effectively reducing sulfur dioxide emissions from power plants in the United States.

Fischer and others (Chapter 10) argue that an effective domestic GHG emissions control program needs to regulate the sources of fossil fuels themselves rather than attempt to measure and regulate the myriad sources of actual emissions in the economy. They further argue that on both efficiency and fairness grounds, the emissions allowances should be largely auctioned by the government rather than distributed free of charge to fossil fuel suppliers or users. This set of issues is explored more extensively by Parry (Chapter 11) and Goulder (Chapter 12). Parry shows how revenue-raising policies that can be recycled in the economy through other tax cuts can have substantially lower overall cost than policies that raise no revenue. Both carbon taxes and auctioned permits are revenue-raising policies. Goulder shows how one of the key distributional concerns with GHG emissions policy—the negative impact on profit and net worth in the fossil fuel industry—could be ameliorated at a relatively low cost in terms of economic efficiency loss by allocating a fairly small portion of the total amount of GHG emissions permits free of charge to the industry. The value of these assets would offset the loss in value of the fossil fuel assets. This approach is likely to be more cost-effective than less targeted corporate tax relief.

Sedjo, Sohngen, and Jagger (Chapter 13) look at the complex set of economic and policy issues that surround carbon sinks (that is, biological storage of carbon in trees, plant roots, soils, and so forth). Land use changes—in particular, deforestation and reforestation—can have substantial effects on the balance between stored carbon and atmospheric CO_2. Sedjo and others note first that under the Kyoto Protocol, the status of carbon sinks as a policy tool for GHG emissions abatement is highly ambiguous. They also discuss some of the challenges encountered in designing carbon sink policies—for example, how does one deal with the possibility that increased forest preservation in one location could be offset by increased forest cutting elsewhere to satisfy global timber demand?

The next three chapters in Part 3 examine broader policy changes and their connections to policy for GHG emissions. Fischer and Toman (Chapter 14) examine prospects for obtaining environmental and economic efficiency improvements through reducing direct or indirect subsidies (the latter include special tax breaks, for example). There is no question in principle that such "win-win" possibilities exist. But with reference to tax policies in the United States, the environmental dividends of energy tax reform may be small—policies that benefit specific kinds of producers or fuel types without substantially encouraging energy consumption will have economic costs but little environmental impact. The authors also critique the use of "green subsidies" to provide environmental benefits, arguing that this approach is less cost-effective than other methods and can be directed at technologies with limited environmental as well as economic merit.

Palmer (Chapter 15) critically assesses how a restructuring of the electricity industry in the United States could affect GHG emissions in that sector. The net effect on emissions depends in a complicated way on how restructuring alters incentives to invest in new plants and retire old plants as well as how plants are used. Whereas some studies suggest that restructuring could reduce GHG emissions by increasing efficiency, Palmer suggests that increased emissions are a real possibility. She also reviews one of the key elements in the restructuring debate: a provision that would require a certain percentage of electricity generation to be undertaken with renewable energy sources.

Darmstadter (Chapter 16) addresses this factor and many others that influence the historical role and future potential of renewable energy sources. He emphasizes that in many cases, technical progress in the provision of renewable energy has been impressive, but actual market penetration has been hampered by offsetting reductions in the cost of using conventional energy (increased efficiency in both primary fuel production and generation fuel efficiency). These observations suggest that renewable energy sources could play an expanded role in the U.S. fuel mix, but unless historical patterns of technical progress accelerate greatly, it would occur only if policy increased the cost of using fossil fuel (as would be the case for example with a carbon tax).

Jaffe, Newell, and Stavins (Chapter 17) and Fischer (Chapter 18) look at complementary aspects of the relationship between GHG emissions policies and technical change. Jaffe and others discuss the long-standing debate between economists and technologists on the prospects for improved energy efficiency at low economic cost. The authors point out that whereas energy markets are not perfect, there are reasons to believe that technical and engineering estimates of efficiency potential overstate what can be achieved in practice at any particular economic cost. Jaffe and others also review analyses of what determines energy efficiency choices in the market and note that energy prices, information about energy efficiency, and technology performance standards all are ways to increase energy efficiency. But the ultimate economic costs of these policies are not the same; they argue that the most cost-effective way to increase the use of more energy-efficient technology over time is energy price signals.

Fischer considers the complementary problem of how different GHG emissions policies influence the development and diffusion of new technology that emits fewer GHGs. She notes that both emissions trading and carbon tax policies augment such incentives, but the strength of the incentives depends on several factors, including the extent to which the technology developer can reap the benefits of innovation. Various GHG emissions policies can create too much or too little incentive for innovation, and policies cannot be simply ranked on the basis of this criterion. Kopp (Appendix C) explores in more detail the underpinnings of the innovation process (discussed in Chapter 18) and how it is affected by economic incentives.

Parry and Toman (Chapter 19) examine a specific set of policy proposals for "early reductions" of GHG emissions, where early reductions refer to actions taken before the Kyoto Protocol limits would become binding in 2008 (assuming ratification and implementation of the protocol). Several early reduction proposals involve voluntary action by individual GHG emitters to reduce emissions compared with some counterfactual baseline, in exchange for preferred consideration if and when binding GHG emissions limits are imposed in the future. The authors argue that this approach is fraught with several inefficiencies compared with early establishment of the kind of formal emissions trading system discussed by Fischer and others in Chapter 10. They also predict substantial potential cost savings if low-cost early reductions could be banked for use after 2008. Under the Kyoto Protocol, such banking of domestic early reductions is not allowed.

Part 4: International Considerations

Part 4 contains discussions of various issues related to the establishment and maintenance of effective international agreements for GHG control. The first three chapters explore various aspects of the "flexibility mechanisms" that are part of the Kyoto Protocol. In effect, these mechanisms are different forms of emissions trading at an international level. One flexibility mechanism envisages the same kinds of formal emissions trading internationally that Fischer and others (Chapter 10) discuss in a domestic context. Other mechanisms involve project-specific emissions reductions relative to some baseline performance measure, which generate emission reduction credits that can be used or traded. The Kyoto Protocol is unprecedented in its embrace of these mechanisms, but many details on their operation remain to be settled.

Wiener (Chapter 20) provides a broad overview of the key economic and legal issues governing the functioning of these types of mechanisms and a comparison of these quantity-based approaches with price-based approaches. Whereas Pizer (Chapter 9) argues that price-based approaches are more cost-effective strategies for controlling GHG emissions given an enforcement system, Wiener argues that quantity-based approaches are more likely to generate effective international agreement for participation in GHG emissions control.

Toman (Chapter 21) explores in detail the Clean Development Mechanism (CDM), the project-based approach under the Kyoto Protocol that allows for voluntary participation by developing countries. The basic message is that the CDM can be an imperfect but effective tool in promoting both cost-effective GHG emissions control and tangible economic and environmental benefits for developing countries. To do this, however, the design of the CDM must allow a real market mechanism to emerge, and the scope of projects eligible for inclusion in the CDM must be broad enough to provide opportunities for a wide range of developing countries.

Kerr (Chapter 22) emphasizes the question of liability for the "integrity" of transactions in both formal emissions trading and the CDM. She argues that in a system where the Annex I targets are subject to some degree of enforcement penalty if not met, then sellers are in the best position to ensure that permits or credits sold reflect legitimate reductions in GHG emissions, and so sellers ought to be liable. However, with the CDM, the host developing country has no national obligation for GHG reductions control. In this case, the investor/partner in the project from the developed world—or the initial purchaser of the CDM credits—needs to be liable for the integrity of the credits to help ensure that CDM projects do not become loopholes for sham emissions reductions.

Cazorla and Toman (Chapter 23) step back from international policy instruments to consider the broader question of what will motivate developed and developing countries to join and adhere to international climate agreements. As one would expect, a decisive factor is how equitable the distribution of mitigation burdens appears. Given the large disparities between and among developed and developing countries in income, population, and current and historical GHG emissions, the search for equity principles to guide the burden distribution is a difficult one. No single workable principle can be identified.

A combination of criteria must be used, with dynamic adjustments that allow developing countries to "graduate" to additional obligations as their living standards improve. Cazorla and Toman suggest that a focus on shorter-term cooperation through the CDM may help to build a base for future negotiation of longer-term equity rules.

Blackman (Chapter 24) and López (Chapter 25) address more specific sets of issues related to the potential for developing countries to cost-effectively reduce GHGs. Blackman reviews the economics of technology diffusion as it applies to developing countries in particular, describing how several market distortions in developing countries can inhibit the diffusion and use of economically efficient low-GHG technologies. López provides further concrete illustrations of how energy and land use policy changes in developing countries could profitably reduce GHG emissions and how the CDM might be used to catalyze such activities.

In Chapter 26, I summarize some of my own views on productive future directions for climate policy. My points draw heavily on the content of previous chapters, but the opinions expressed in that chapter are mine alone.

Supplemental Reading

Council of Economic Advisers. 1998. *The Kyoto Protocol and the President's Policies to Address Climate Change: Administration Economic Analysis.* July. Washington, DC: Executive Office of the President.

EIA (Energy Information Administration). 1998. *Impacts of the Kyoto Protocol on U.S. Energy Markets and Economic Activity.* Washington, DC: EIA.

Grubb, Michael J., Christiaan Vrolijk, and Duncan Brack. 1999. *The Kyoto Protocol: A Guide and Assessment.* London, U.K.: Royal Institute of International Affairs.

IPCC (Intergovernmental Panel on Climate Change). 1996. *Climate Change 1995: The Science of Climate Change.* Contribution of Working Group I to the Second Assessment Report of the Intergovernmental Panel on Climate Change. New York: Cambridge University Press.

———. 1996. *Climate Change 1995: Impacts, Adaptations, and Mitigation of Climate Change: Scientific-Technical Analysis.* Contribution of Working Group II to the Second Assessment Report of the Intergovernmental Panel on Climate Change. New York: Cambridge University Press.

———. 1996. *Climate Change 1995: Economic and Social Dimensions of Climate Change,* edited by James P. Bruce, Horsang Lee, and Erik F. Haites. Contribution of Working Group III to the Second Assessment Report of the Intergovernmental Panel on Climate Change. New York: Cambridge University Press.

———. 1998. *The Regional Impacts of Climate Change: An Assessment of Vulnerability.* New York: Cambridge University Press.

IWG (Interlaboratory Working Group). 1997. *Scenarios of U.S. Carbon Reductions: Potential Impacts of Energy Technologies by 2010 and Beyond.* Report LBNL-40533 and ORNL-444. September. Berkeley, CA, and Oak Ridge, TN: Lawrence Berkeley National Laboratory and Oak Ridge National Laboratory.

Jacoby, Henry, Ronald Prinn, and Richard Schmalensee. 1998. *Kyoto's Unfinished Business.* Foreign Affairs 77(4): 54–66.

Jaffe, Adam B., and Robert N. Stavins. 1994. *The Energy-Efficiency Gap: What Does It Mean?* Energy Policy 22(1): 804–10.

Nordhaus, William (ed.). 1998. *Economics and Policy Issues in Climate Change.* Washington, DC: Resources for the Future.

NRC (National Research Council). 2000. *Reconciling Observations of Global Temperature Change.* Washington, DC: National Academy Press.

OTA (Office of Technology Assessment). 1991. *Changing by Degrees: Steps to Reduce Greenhouse Gases.* OTA-O-482. Washington, DC: U.S. Government Printing Office.

UNFCCC (United Nations Framework Convention on Climate Change). 1999a. *Convention on Climate Change.* UNEP/IUC/99/2. Geneva, Switzerland: Published for the Climate Change Secretariat by the UNEP's Information Unit for Conventions (IUC). http://www.unfccc.de (accessed October 18, 2000).

———. 1999b. *The Kyoto Protocol to the Convention on Climate Change.* UNEP/IUC/99/10. Paris, France: Published by the Climate Change Secretariat with the Support of UNEP's Information Unit for Conventions (IUC). http://www.unfccc.de (accessed October 18, 2000).

———. 1999c. *Guide to the Climate Change Negotiation Process.* http://www.unfcc.de/resource (accessed October 18, 2000).

Part 1

Introduction

2

How the Kyoto Protocol Developed
A Brief History

J.W. Anderson

With the Kyoto Protocol, the world's governments proposed to adopt legally binding commitments to slow global warming. Although this treaty is still hardly more than a draft and not yet in force, it constitutes an unprecedented attempt to organize international cooperation to protect the global environment. It would require fundamental changes in the ways in which the world produces and uses energy. But the protocol has become the center of great political controversy, for reasons that in some part arose from the way it developed.

In this chapter, I present the issues that rapidly evolving scientific knowledge first began to press on politicians and diplomats nearly half a century ago. I consider the Montreal Protocol to protect the stratospheric ozone layer—a tremendous success for international environmental cooperation, but an unsatisfactory model for the climate change agreement with its different circumstances. I then discuss the negotiation of the Kyoto text and conclude with a review of the current efforts by some but not all of the signatory governments to put the treaty into effect by 2002.

Beginnings

Like most environmental issues, questions about Earth's changing climate began with scientists. Scientists had known for many years that the concentration of carbon dioxide (CO_2) in the atmosphere could affect worldwide temperatures. They also knew that industrial growth and rising oil and coal consumption were increasing the amounts of CO_2 being emitted into the air enormously. But through the first half of this century, they had generally assumed that most of that CO_2 was being absorbed harmlessly by the oceans. It was only in 1958 that researchers began to test that assumption by taking measurements from the top of Mauna Loa, a Hawaiian volcano far from any smokestacks. The Mauna Loa data series soon showed that the amount of CO_2 in the atmosphere was steadily increasing.

For the next decade, the concern about the effects of human activity on the climate remained largely theoretical, because temperatures seemed stable. From the early 1900s until about 1940, the global average temperature rose. From then until 1970, for reasons that are still not entirely clear, it leveled off and even fell a little. But around 1970, it started to increase again, strongly and consistently.

When that happened, the possibility that human activity was changing the global climate began to attract serious attention among scientists and government officials with environmental responsibilities. Table 1 summarizes some of the key milestones in the international evolution of climate policy over the past 20 years. In the United States, the National Academy of Sciences addressed the possibility of

Table 1. Summary of Key Milestones in Climate Policy, 1979–99.

1979	First World Climate Conference
1985	A conference on greenhouse gases in Villach, Austria, demonstrates a growing scientific consensus that human activity is affecting the climate. Here, for the first time, the need for policies to respond to climate issues is discussed.
1988	The Toronto Conference, an unofficial gathering of scientists and politicians from 48 countries, ends with a call for a 20% reduction of carbon dioxide emissions from 1988 levels by 2005.
1990	First Assessment Report of the Intergovernmental Panel on Climate Change (IPCC) is published; initial evidence that human activities might be affecting climate, but significant uncertainty exists.
1990	Second World Climate Conference; agreement to negotiate a "framework treaty."
1992	The U.N. Framework Convention on Climate Change (UNFCCC) is established at the U.N. Conference on Environment and Development (also known as the Earth Summit) in Rio de Janeiro, Brazil. This treaty sets a nonbinding goal for the developed countries of reducing emissions of greenhouse gases. Annex I developed countries pledge to return emissions to 1990 levels by 2000. United States ratifies UNFCCC later in the year.
1993	Clinton administration publishes its Climate Change Action Plan, a collection of largely voluntary emission-reduction programs.
1995	IPCC Second Assessment Report completed (published in 1996); strong conviction expressed that human activities could be adversely affecting climate.
1995	Berlin Mandate developed at the first conference of the parties to the UNFCCC (COP-1) to the UNFCCC. Agreement to negotiate *legally binding* targets and timetables to limit emissions in Annex I countries.
1997	U.S. Senate passes Byrd–Hagel resolution, 95 to 0, stating that the United States should accept no climate agreement that did not demand comparable sacrifices of all participants and calling for the administration to justify any proposed ratification of the Kyoto Protocol with analysis of benefits and costs.
1997	COP-3 is held in Kyoto, Japan, leading to the Kyoto Protocol. The agreement would require the developed countries to reduce their emissions of greenhouse gases by an average of about 5% of 1990 levels by the 5-year period 2008-12.
	Annex I/Annex B countries agree to binding emission reductions averaging 5% below 1990 levels by 2008–12, with "flexibility mechanisms" (including emissions trading) for compliance; no commitments for emission limitation by developing countries.
1998	COP-4 is held in Buenos Aires, Argentina; emphasis on operationalizing the "flexibility mechanisms" of the Kyoto Protocol.
	IPCC Third Assessment begins.
1999	COP-5 is held in Bonn, Germany; continued emphasis on operationalizing the flexibility mechanisms.
2000	COP-6 is held in The Hague, Netherlands.

Source: UNFCCC (see Suggested Reading).

greenhouse warming several times; its 1979 report concluded that CO_2 emissions would lead to warming in the next century. Also in 1979, the World Meteorological Organization (WMO) and other U.N. agencies held the First World Climate Conference, where CO_2 was a major topic. Later that year, the WMO established the World Climate Program to coordinate research. That in turn led to a 1985 conference in Villach, Austria, that demonstrated a grow-

ing consensus among scientists on the probability of a warmer climate caused by greenhouse gases (GHGs)—CO_2 and several other gases generated by industry and agriculture.

This prospect was deeply troubling to anyone with an interest in the environmental ethic. From its beginnings, the science of ecology has been based on the concept of an intricate system of balances. Research increasingly showed a high risk that industrial development and rising standards of living were tipping, perhaps irrevocably, one crucial balance in a way that could affect all life on the planet.

But until the mid-1980s, the initiative remained almost entirely with the professionals—scientists, public officials, and international bureaucrats running agencies devoted to weather and climate. The advocacy organizations that had transformed environmental policy over the previous two decades, carrying it to the highest levels of politics in North America and Europe, were largely absent from the issues of climate change.

One reason was that few of the advocacy organizations considered themselves equipped for international politics. They had grown strong in the struggles over domestic issues, and their techniques were adapted to the political and legal processes of their own countries. But climate change would have to be addressed internationally, and re-equipping the environmental movement to deal with the very different institutions of international cooperation seemed a daunting challenge.

In 1987, the scene changed suddenly with a great triumph for environmental protection in a closely related matter, the worldwide campaign to preserve the stratospheric ozone layer. The ozone treaty suggested enormous new possibilities for worldwide environmental action. Because it established a model strategy that advocacy groups and governments followed closely in their approach to climate change, it is useful to recall how and why that was accomplished.

Stratospheric Ozone Accord

In the early 1970s, scientific evidence had begun to suggest that a family of synthetic chemical compounds, chlorofluorocarbons (CFCs), was eroding a layer of ozone that lay in the upper atmosphere. The ozone layer provided a vital protection against harmful ultraviolet radiation from the sun. Unscreened, this radiation could cause skin cancers and other damage in people exposed to it.

The CFCs had many uses but were most commonly used as propellants of aerosols in products such as hair sprays. The suggestion of a connection

to cancer immediately set off boycotts by consumers and a race for substitutes among manufacturers. The U.N. Environment Program called for worldwide action. In 1978, the United States and several other countries banned most aerosol sprays. A succession of meetings and scientific reports followed.

These meetings and reports culminated in 1985 at the Vienna Convention for the Protection of the Ozone Layer, an agreement that was deliberately written loosely to attract the widest possible support with the fewest possible arguments over details. It committed governments to take only unspecified actions and to do so only if the chemicals involved should be found to have adverse effects. The idea was not to bind governments to a precise program but to start a process of research and consultation that could be adapted as further findings and necessities became clear. That strategy was brilliantly successful.

Two months after the convention was signed, a British research team reported the first actual observations of thinning of the ozone layer from the Antarctic. These findings revived public anxiety about the health effects that was fueled by the discovery that the scale of damage was considerably greater than the models had predicted. This development, incidentally, was a warning that environmental change is not always gradual or predictable, even in a process that is under careful study.

By late 1986, the United States was calling for international controls. It began pushing for a freeze and an eventual phase-out of production of the gases that attack the ozone layer. The American chemical industry gave powerful support to the idea of a world agreement. For the American producers, CFCs were not a major product, and they were unwilling to jeopardize their public reputations to defend them. As a practical matter, they understood that the choice was between a ban worldwide or one only in the United States, and they very much preferred rules that would also apply to their European competitors. As for competition, the American companies believed that they were ahead of the Europeans in the search for substitutes for the CFCs, and world limits on CFCs would increase the competitive value of their lead. This difference between the

American and European industries' positions is the chief explanation for the difference between the governments' positions. Throughout this period, the United States forcefully pushed for action while most of the European countries dragged their feet.

Despite its backing from business, the movement toward a binding agreement generated fierce infighting in the conservative Republican administration in Washington. Some officials saw it as a precedent for international regulation. In the spring of 1987, the dispute went public when Secretary of the Interior Donald Hodel suggested that the dangers of ultraviolet radiation were greatly exaggerated and that people could easily protect themselves by wearing hats and sunglasses. His comment brought a deluge of ridicule down on the administration. The question was finally put to President Ronald Reagan at a White House meeting in June. To the great surprise of the antiregulators, the President sided with Secretary of State George Shultz and U.S. Environmental Protection Agency (EPA) Administrator Lee Thomas in supporting binding limits on CFCs. Perhaps Hodel had forgotten that two years earlier, the President had undergone surgery to remove skin cancer from his nose.

Three months after that White House meeting, negotiators in Montreal, Quebec, Canada, completed a second treaty on CFCs—technically, a protocol to the first one—and most of the industrial countries quickly signed it. Unlike the Vienna Convention, the Montreal Protocol contained firm and legally binding limits on CFC consumption. There was to be an immediate freeze, followed by a 20% decrease (from the 1986 level) in 1993, then a 50% decrease in 1998. The Montreal Protocol was a powerful signal that effective international action to protect the environment was possible.

Toward a Climate Treaty

In 1988, climate change and global warming became a widespread public concern and a political issue, no longer confined to meetings of scientists and specialists. In the background was the rising average global temperature. By the late 1980s, it was clear that the decade would be the warmest in the century or so

for which reliable measurements were available. Several events brought this fact and its implications to public attention.

In the United States, Senator Tim Wirth of Colorado, who had been interested in climate change for some time, had grown increasingly exasperated by the country's refusal to notice it. In late June of 1988, amidst growing concern about a severe and widespread drought in the South and the Midwest, he called a hearing on a day forecast to be spectacularly hot. As the temperature reached 98 °F that afternoon, Wirth called a series of experts to testify. One of them, James E. Hansen, director of the National Aeronautics and Space Administration's Institute for Space Studies, told the committee that NASA was 99% certain that the cause of the warming trend was synthetic gases, not natural variation. "It is time to stop waffling so much and say the evidence is pretty strong that the greenhouse effect is here," Hansen said to a reporter for the *New York Times*, which put the story on its front page ("Global Warming Has Begun, Expert Tells Senate," June 24, 1988). Hansen's testimony had an immediate impact because he was the first scientist of his stature to declare that the rising temperatures were very probably related to burning fossil fuel.

Four days later, a conference called by the Canadian government opened in Toronto, attended by not only scientists but also politicians from dozens of countries. Although it grew out of the succession of meetings and reports sponsored by the U.N. agencies, it marked the stage in this international process where the meetings expanded beyond the U.N. agencies and the usual specialists. It also was at this point that the discussion turned to specific preventive action. In its final statement, the Toronto conference called for a 20% reduction in global emissions of CO_2 by 2005.

This statement turned out to be highly influential, setting its key goal in terms of a reduction of the volume of emissions in the near term. That target was not based on any economic analysis, because at that time, economists were only beginning to study the subject. The conference knew that a 20% reduction would not be sufficient to stabilize the concentration of CO_2 in the atmosphere, but it did not want to set a figure so high that governments would dismiss it as implausible. On the other hand, it wanted to choose a figure that would show a serious intent to make fundamental changes.

It is never a simple matter to trace the origins of people's ideas about what shows serious intent. But 20% was also the figure that the Montreal Protocol—negotiated a year earlier, to great applause, in the same country—set as the first cut in world CFC consumption.

In mid-1988, at their annual summit meeting, the heads of the seven big industrial democracies' governments called for a framework treaty to limit the world's production of CO_2, and negotiations were soon under way. In December 1988, the U.N. General Assembly approved the establishment of an expert Intergovernmental Panel on Climate Change (IPCC) to provide an authoritative review of the science of global warming to inform the talks.

But this time, reversing their positions in the ozone negotiations, Europe pushed for rapid action and the United States resisted. One reason was that, as a matter of principle, hostility to regulation and interference with markets remained strong in the Bush administration. Another reason was that, this time, the United States was being forcefully pressed by its industries to delay. In Europe, most of the indigenous energy industries are relatively small and are often tied closely to their governments. In the United States, the coal, oil, gas, and power producers are huge, and their interests are defended by not only corporations but also influential labor unions.

The first IPCC reports appeared in late 1990, showing broad agreement among scientists in the field that the possibility of global warming at least had to be taken seriously. The Europeans, urging action, cited the reports' warnings of the possible consequences of higher temperatures, whereas the Bush administration replied by pointing to their emphasis on the scientific uncertainties. In the negotiations over the U.N. Framework Convention on Climate Change (UNFCCC), the United States flatly opposed any firm targets for reductions of CO_2 emissions. The Europeans were able to gain an acknowledgement that, at least in principle, reductions were desirable. The final language set a voluntary goal of cutting emissions back to the 1990 level by 2000, but like the Vienna Convention on ozone, the UNFCCC contained no enforceable commitments. Even that purely aspirational goal represented substantial movement from the United States' original inclination to do no more than study the climate. The text of the UNFCCC was completed in time to be signed with great ceremony by nearly every country on Earth in June 1992 at the United Nations' huge and colorful Conference on Environment and Development in Rio de Janeiro, Brazil.

Development of an Action Plan

At this point, if the politics of global warming had followed the ozone model, increasingly ominous scientific reports would have pushed the diplomats and politicians rapidly from the voluntary commitments of the UNFCCC to tight and obligatory cuts in emissions of CO_2 and the other GHGs. But that didn't happen, for several reasons.

First, the campaign to protect the ozone layer was driven by a fear of cancer. No similarly compelling motive pushed a climate agreement forward. Vice President Al Gore and many others spoke of the possibilities of terrible storms and epidemics, but scientists said that the evidence was not conclusive. Rather than becoming more precise and urgent, as the ozone findings had done, the science of global warming remained unclear on many important points. Most scientists concluded that it would not be possible for years, perhaps decades, to predict with any assurance how fast the world would warm, or what consequences might come of it. The scientific uncertainties deflated the impulse toward action.

It also was true that in the ozone case, action had meant reducing the consumption of certain products for which technology was rapidly providing satisfactory substitutes. But most of the world's energy comes from burning fossil fuels, and at present, there is no way to burn fossil fuels without releasing CO_2 to the atmosphere. Treating exhaust gases to remove CO_2 and storing or recycling it are not yet economically practical or even technically feasible on a large scale. How rapidly the world could turn to other sources of energy, and at what cost, were subjects for intense controversy throughout the 1990s. The answers remained sufficiently unclear that many political leaders hesitated to impose policies that attempted to change people's long-established habits in the use of energy (such as driving) and that seemed likely to cause great disruption in economic life.

President Bill Clinton, who took office in early 1993, had sharply criticized President George Bush's reluctance to take action on greenhouse emissions. In February, he proposed a broad tax on all energy consumption, and in April, to celebrate Earth Day, he announced that his administration would adopt measures to stabilize emissions at 1990 levels as the Rio treaty had urged. But the President was soon distracted by the great struggle over his budget. Congress, hostile to the idea of an energy tax from the beginning, pared it down to a mere increase of 4.3 cents a gallon in the gasoline tax—too little to have any significant effect on consumption. When the specific details of the President's Climate Change Action Plan were presented in fall 1993, they turned out to be entirely voluntary. The Clinton administration did not intend to take on the massive and politically costly campaign that would have been necessary to change Americans' accustomed practices in using energy.

But neither were the European governments willing at that time to undertake difficult and drastic energy programs. The European Union tentatively proposed a tax on carbon emissions, but it got no farther than President Clinton's energy tax. By the mid-1990s, it was clear that of the world's major industrial powers, only three would meet the Rio goal of getting their emissions back to 1990 levels by 2000. Russia would meet it because the old Soviet economy, profligate in its use of energy, was collapsing. Germany would meet it because it had shut down most of the grossly inefficient power plants in its formerly Communist eastern region. And Britain, to save money, was cutting down its subsidies to coal and swinging its power production toward less carbon-intensive natural gas. Nowhere in any of the large economies was there any sign of a serious and purposeful effort to reduce CO_2 emissions for environmental reasons.

In early 1995, under these unpromising circumstances, the United Nations held the first Conference of Parties to the Rio treaty (COP-1) in Berlin, Germany. The purpose was to assess progress toward the grand promises made at Rio. Noting that progress was exceedingly modest, the 120 countries represented at COP-1 agreed to begin work on a further agreement to strengthen their commitments by setting specific targets for emissions limits and reductions by certain years such as 2005, 2010, and 2020. These targets and timetables were to apply to the industrial countries, but not the developing countries.

A few months later, the IPCC brought out its second survey of the science of global warming; the tone was much more conclusive than five years earlier. In a widely quoted line, it declared, "The balance of evidence suggests that there is a discernible human influence on global climate." However, this sentence was preceded by a warning about the present state of knowledge: "Our ability to quantify the human influence on global climate is currently limited because the expected signal is still emerging from the noise of natural variability, and because there are uncertainties in key factors" (see Suggested Reading). The reference to "a discernible human influence" was enough to encourage the politicians and diplomats who were working for a stronger treaty. But it wasn't enough to change many minds among the opposition.

At the next negotiating session, the second Conference of Parties (COP-2), in the summer of 1996, the United States announced a clear and important change of policy. Tim Wirth, the former senator, was now an assistant secretary of state and head of the American delegation. The United States would support legally binding limits on emissions, he said, if other countries did.

In Berlin, the Conference of Parties had set a deadline for the negotiations, agreeing to draft a treaty at the third Conference of Parties (COP-3), which would be held in 1997 in Kyoto, Japan. The Europeans kept pressing for substantial action. In Britain, where an election campaign was under way, the Labour Party pledged in its manifesto to take the lead in the world environmental movement by supporting a 20% reduction by 2010. Labour's huge victory in May gave additional force to its demands.

Meanwhile, a rapidly growing economic literature, especially in the United States, was making the discussion of alternative control strategies very much more complicated and was multiplying the trade-offs with which policymakers were confronted. A comparison with the ozone case is, once again, instructive. In late 1987, shortly after the Montreal Protocol on the ozone layer was signed, EPA published a cost–benefit analysis of an 80% reduction in CFC use. The costs would amount to $31 billion, the EPA found, but the benefits would be $6.4 trillion— nearly all of it representing the value of cancer deaths averted, at $3 million/death (see Hammitt in Suggested Reading). These figures only added momentum to the drive for further sweeping reductions in CFCs. But similar attempts at cost–benefit analysis of CO_2 reductions were ambiguous and much less compelling. They depended on assumptions that were open to challenge, and the long-term nature of climate change exposed the calculations to further doubts.

One widely cited paper argued persuasively that the cost of controlling CO_2 levels could be much lower without compromising long-term environmental protection if emissions reductions were delayed and commenced gradually before accelerating (see Wigley and others in Suggested Readings). Another paper suggested that society would be better served by policies that began by fixing the cost of making reductions, rather than fixing the volume to be reduced and leaving the cost unknown. By 1997, a substantial economic literature had emerged in the United States that challenged the emerging Kyoto strategy and its emphasis on relatively near-term goals expressed as reductions in volumes of emissions.

For politicians, environmental advocacy groups, and a growing public audience, the basic idea of goals and timetables had been laid down at the 1988 Toronto conference and seemed self-evident: The goals were to be expressed in reductions of GHG by volume, and the timetables were to be in the near term, with deadlines only a decade or so away, because anything farther in the future sounded like mere procrastination.

Other papers emphasized that flexibility of location could cut costs enormously. Flexibility meant, for example, allowing trading in emission rights. The sulfur dioxide trading program in the United States had been a great success in holding down the cost of combating acid raid, and the same principle would apply to CO_2. It made good sense to the White House, but traditionally, Europeans are suspicious of the market as an instrument of public policy. Even among Americans, debate continued between economists seeking efficiency and regulators who suspected that flexibility was simply a synonym for a loophole. Most of the environmental movement agreed with the regulators, creating an uncomfortable disagreement between the administration and one of its most conspicuous constituencies.

By the spring of 1997, President Clinton was aware that although his administration was committed to legally binding limits on CO_2 emissions, he was going to have difficulty getting legislation through Congress. In 1993, when he was at the height of his authority, a Congress controlled by Democrats had refused to enact his proposed energy tax; in 1997, Congress was in the hands of conservative Republicans who were not only hostile to taxes but also skeptical about climate change in general. The President knew that relatively few Americans were even aware of the issue of global warming and that they were sharply divided over whether to do anything about it. But his administration had promised to impose legally enforceable limits on itself and the rest of the industrial world at the coming Kyoto conference. The President's dilemma tightened over the spring, as the conference approached. A wide range of lobbies were engaged with the environmentalists on one side and industrial organizations and conservative political groups on the other. By this time, the news media gave regular coverage to climate change.

European criticism of the United States' energy policy—or lack of it—was no longer confined to small meetings of professional negotiators. President Clinton was the host of the 1997 meeting of the heads of the seven big industrial democracies. Control of GHGs was a prominent topic there, and French President Jacques Chirac publicly pushed

Clinton for a specific American commitment. A week earlier, the European Union had agreed on a reduction of 15% from 1990 levels by 2010. At a news conference, Chirac pointedly observed that, per capita, the United States' GHG emissions were three times as high as France's.

From that meeting, the seven leaders took themselves to New York, where the U.N. General Assembly was assessing the meager progress toward the goals set by the Rio conference five years earlier. There, led by the new prime minister of Britain, Tony Blair, the Europeans—including Chirac and Chancellor Helmut Kohl of Germany—again demanded, publicly, more aggressive action by the United States.

President Clinton's speech that week to the General Assembly was the next step in his campaign to educate the American public and create a constituency for action. He offered a dire view of the future and called for a new era in technology (Presidential Document 973, June 26, 1997):

> The science is clear and compelling. We humans are changing the global climate.... Climate changes will disrupt agriculture, cause severe droughts and floods and the spread of infectious diseases, which will be a big enough problem for us under the best of circumstances in the twenty-first century. There could be 50 million or more cases of malaria a year....
>
> We must create new technologies and develop new strategies like emissions trading that will both curtail pollution and support continued economic growth.... We will work with our people, and we will bring to the Kyoto conference a strong American commitment to realistic and binding limits that will significantly reduce our emissions of greenhouse gases.

But the President still gave no hint of the level at which he wanted to place those limits. While some environmental organizations chided him for delay, public reaction to the speech was otherwise modest. The previous day, the President had acted to tighten

the standards on smog and soot in the air that Americans breathe, a decision for which he received strong public support. In contrast to smog, the threat of global warming still seemed remote and abstract to most people.

The next step in the President's education campaign was a White House conference that he and Vice President Gore held in late July, a month after the U.N. speech. "We see the train coming," the President told the conference, "but most Americans in their daily lives can't hear the whistle blowing" (Presidential Document 1116, July 24, 1997).

In response, the next day, the Senate passed a resolution (95 to 0) admonishing the President not to sign any treaty limiting American emissions unless the treaty also committed the developing countries to similar restrictions. The resolution reflected widespread fears that tightening environmental restrictions in this country would send manufacturing plants and jobs overseas to countries with less rigorous emissions restrictions or none at all. The two chief sponsors of this resolution were Robert C. Byrd of West Virginia, a very senior Democrat vigilant in his protection of his state's coal mining industry, and Chuck Hagel of Nebraska, a newly elected Republican.

With the Byrd–Hagel resolution, the role of the developing countries became central to the politics of a treaty. Although the industrial countries, including the former Soviet bloc, produced three-fifths of the world's CO_2 emissions, emissions were rising much faster in the developing countries. Some Americans cited projections suggesting that China would overtake the United States as the largest emitter around 2020 and argued that any agreement omitting China would be useless. The developing countries replied that the threat of global warming had been created over two centuries by the countries that were now rich, and it was up to them to address it without adding burdens to the poor. Furthermore, they added, by any reasonable definition of equity, it was not emissions per country that counted but rather emissions per capita. Even under the most expansive scenarios for Chinese growth, it would be a very long time indeed before the Chinese produced as much CO_2 per capita as Americans did.

In the fall, the President held another White House conference. Vice President Gore was there again, along with much of the Cabinet. President Clinton declared that he was convinced that the science was accurate. Once again, he said that the United States must be prepared to commit itself to binding targets. However, he added, all the world's nations must participate. The United States wanted fair but significant contributions by all countries. This caveat was new, a concession to the Senate. But it raised severe difficulties for the negotiators working on the text of the Kyoto Protocol. Most of the developing countries were adamantly opposed to any restrictions on their emissions, at the very least until they had seen a serious and substantial effort by the rich economies. The Europeans complained that President Clinton was changing the terms agreed in the Berlin Mandate two years earlier. This new requirement was being introduced by the United States in the final weeks of a long and cumbersome process that involved more than 140 governments.

Whereas President Clinton's speeches and conferences had succeeded in broadening public interest in the subject of climate change, they had also attracted powerful and articulate opponents. According to one study of public opinion, large majorities of Americans believed that global warming was occurring and that government had a responsibility to take action. However, at the same time, the intense debate of the subject by the White House and its adversaries had sharpened the differences between the two parties. Most Republicans continued to feel that the science was too uncertain to justify a large, expensive program to change the national economy. The industries and labor unions that would be touched by emissions restrictions initiated their own counter-campaign against any world agreement.

Finally, in late October 1997, less than six weeks before the Kyoto conference was to open, President Clinton laid out the American position. The United States would support a legally binding target of returning its emissions of GHGs to the 1990 level by the five-year period 2008–12. He also spoke of additional reductions in the years beyond 2012 but did not offer specific numbers. President Clinton again made it clear that he did not propose to challenge a unanimous Senate on the inclusion of the developing countries. The negotiators were left to write around a major disagreement among the governments.

The Kyoto Protocol

The Kyoto conference was a huge affair, bringing together more than 10,000 people—officials who represented nearly 170 countries, press, and lobbyists of every political persuasion from the greenest of green to coal black. The main focus was on targets and timetables, that is, how much the annual emissions of GHGs were to be reduced by industrialized countries and by when. The conference appeared to be deadlocked between the Americans, who were ready to return only to 1990 emissions levels, and the Europeans, who wanted deeper cuts. At the last moment, when the conference seemed about to collapse, President Clinton sent Vice President Gore on a rescue mission to make a one-day appearance and tell the American delegation that it could give a little more. By stopping the clock and working through the last night of the scheduled conference into the following day, the negotiators were able to draft a text that everyone could accept. The United States agreed to a target, for average emissions over the period 2008–12, that was 7% below its 1990 emissions. The European Union was to reduce emissions 8%. Emissions from all of the industrial countries together would go down 5.2%.

The conference proceeded by consensus, not by decisions taken on formal votes, and it achieved consensus only by leaving many issues out. One conspicuous example was the limitation of developing countries' emissions. On another point of intense interest to the Clinton administration, the provisions for trading emissions permits among the industrialized countries that agreed to cap their total emissions—known as the Annex I countries—were incomplete. (The reference is to Annex I of the 1992 Framework Convention. To be legally precise, the relevant list for emissions trading is actually the similar but slightly different Annex B of the Kyoto text.) The protocol also provided for a Clean Development

Mechanism (CDM) through which industrial countries could earn emissions credits by financing projects that reduced emissions in developing countries, but again, details were scarce. The United States was counting on permit trading to hold down the costs of compliance. President Clinton told Congress that he would not send the Kyoto Protocol to the Senate for the necessary vote on ratification until American concerns about both developing countries' participation and trading were satisfactorily addressed in subsequent negotiations.

Other omissions in the draft represented additional issues too difficult to be dealt with in the conference. For example, there was no mention of who was to measure and verify countries' emissions, nor was there any suggestion of what might happen if a country failed to meet its commitments or of sanctions to enforce those commitments. These lapses showed the distance between the bold pioneering concept that the Kyoto Protocol represented and a tight legal agreement that could withstand the criticism of its enemies.

The text also contained vague references to financial aid to poor countries, without any of the specifics on which the finance ministers of the donor countries would insist before it went into effect. Although the developing countries were not required to cut their emissions, they were committed to keeping emissions inventories and setting up programs to mitigate their emissions. The Kyoto text obligated the industrial countries to provide funds to cover the costs of that work, as well as to finance the transfer of environmentally sound technology to the countries that lacked it. Regarding the CDM, the text promised that a share—unspecified—of the costs of these projects was to be taken to cover administrative expenses, as well as to help the most vulnerable developing countries meet the costs of adapting to climate change. All of the hard questions regarding amounts of money and control of the spending were left unanswered in the rush to get a quick agreement and end the conference on a note of harmony.

Beyond these omissions in the text lay a broader issue. The Kyoto Protocol called for substantially lower emissions of GHGs by most of the industrial countries, and few of them had any clear strategy to get there. The most conspicuous example was the United States. With no change in policy and with expected economic growth, according to the administration's own projection, the country's emissions would be more than 30% higher in 2010 than in 1990. But the treaty would require the emissions to be 7% lower than in 1990, a reduction of nearly one-third the business-as-usual volume in less than 15 years—a massive change in a very short time.

In Washington, the debate was largely over the costs and burdens of compliance. Janet Yellen, the chair of the President's Council of Economic Advisers, testified before a succession of congressional committees in March 1998 that the costs would be low. The administration's figures were about $14–23/ton of carbon or its equivalent in the other GHGs, assuming all the potential flexibilities in the Kyoto Protocol were fully and frictionlessly implemented. By the years 2008–12, it would amount to an increase of 3–5% in retail energy prices or, for the average household, $70–110 a year. In fact, Yellen noted, these increases might not be noticeable at all because they would be offset by the drop in electric power rates that the administration expected to result from the deregulation of the electric utilities.

Although her basic estimates were low, they lay within the range of plausibility. But they depended on certain assumptions which, while not spelled out, could be inferred from the mathematical models that the administration was using. The most important of these assumptions, it appeared, was that the United States would in fact make few reductions in its own emissions but would depend on an international trading system to buy about 85% of the emissions permits it needed to meet its Kyoto target.

The administration's low estimates assumed not only that there would be worldwide trading of emissions permits but also that the trading would be as efficient as, say, trading on the New York Stock Exchange. In the absence of international trading, the cost of meeting the Kyoto target in the United States would be four times as high, other economists soon showed with calculations from the same model that the administration was using (see Edmonds and others in Suggested Reading). Trading, as well as the limits and conditions that might be imposed on it, would make a huge difference in the costs of compliance. But, as already noted, the protocol contained only brief and general language about trading, even among Annex I countries. Regarding the rest of the world, it envisioned only project-based emission reduction credits through the CDM.

Relying heavily on undefined mechanisms was both open to criticism on economic grounds and inflammatory as foreign policy. American industry was more efficient in using fuel than most of the developing countries. It was obviously cheaper to make emissions reductions in the developing world. But with so much uncertainty about the practical operation of the Kyoto Protocol's flexibility mechanisms, the administration figures could hardly be regarded as firm estimates. Meanwhile, officials in both Europe and the developing countries reacted with hostility to the implication that the United States intended to use its financial power to enable itself to stay close to a business-as-usual track of steadily rising emissions. In May 1999, the European Union adopted the position that no more than one-half of any country's permits should come from abroad.

This strategy of buying permits abroad also proved to be controversial in the United States. Environmentalists immediately observed that, because of the severe economic decline in the former Soviet states, Russia and the other countries would have enormous numbers of emissions permits that they could sell without reducing their actual emissions. They derided the concept as "hot air." Meanwhile, economists noted that buying permits on the scale envisioned by the Clinton plan would not be inexpensive. The previous year, two analysts—using figures less optimistic than Dr. Yellen's but still well within the range of the possible—had pointed out that this outflow of dollars would be sufficient to affect currencies' exchange rates and to distort international trade (see McKibbin and Wilcoxen in Suggested Reading). Congressional representatives soon began to express concern that the trading provisions of the Kyoto Protocol would result in large payments to foreign countries.

Trading was not the only point of contention. In making its cost estimates, the Clinton administration had also used highly favorable assumptions re-

garding the speed and ease with which the economy would adjust to emissions limits. Industrial lobbies quickly commissioned studies that, using different assumptions on both trading and economic adjustment, showed that the costs of compliance might well be ten times as high as the administration's figures. Nor was the attack solely from Republicans. Rep. John Dingell (D-Michigan), ranking minority member of the Commerce Committee and a power in the House of Representatives, called for a sweeping renegotiation of the Kyoto Protocol, declaring that in its present form, it would do less to protect the environment than to promote commercial advantages for other countries at American expense.

Since the Kyoto Conference

While this debate roared along in Washington, the negotiators who had written the Kyoto text were thinking about the omissions in it. Soon it became conventional to defend the treaty by describing it as "a work in progress," suggesting that it could not be fairly judged until it had been completed. The fourth Conference of Parties (COP-4) was held in Buenos Aires, Argentina, in November 1998 to consider the process of bringing the outstanding issues to resolution.

It ended without much progress on the political differences among the major countries, but it set a plan of action for addressing them over the next two years. Two developing countries created a flurry by announcing that they intended to place themselves under enforceable emissions limits. One was Argentina, the host of the meeting, which was anxious to show movement on the most difficult of these disputes. The other was Kazakhstan, which—like the rest of the former Soviet Union—had suffered a sharp industrial depression and would have permits to sell. Both were denounced by the other developing countries for breaking ranks, and no others followed suit.

By this time, the United Nations had provided the negotiating process with a substantial professional staff and headquarters in a handsome house on the bank of the Rhine River in Bonn, Germany. This staff oversaw a heavy schedule of meetings on the Kyoto issues. Two or three times a year, the Conference of Parties' subsidiary bodies met, and they in turn commissioned specialized workshops, all leading up to the annual conference. This procedure generated a series of sophisticated and useful discussions of the technical points, even while the major political questions remained unresolved.

At the fifth Conference of Parties (COP-5), held in Bonn in November 1999, the negotiators agreed that the following year, they would try to prepare the Kyoto Protocol to go into force. The rules necessary to implement the protocol were under vigorous negotiation at the sixth Conference of Parties (COP-6), held in November 2000 in The Hague, the Netherlands. Many governments, especially among the Europeans, hoped for a final agreement that would lead to the protocol's entry into force by 2002. But these talks ended in continued disagreement and are to resume in 2001.

To put the Kyoto Protocol into force would require ratification by 55 countries, which together represent 55% of the 1990 emissions of the industrial countries. Most countries have signed the treaty, including the United States, but a signature in this case is a mere gesture with no legal significance. By mid-2000, only 22 countries had actually ratified the protocol—none of them industrial powers. Ratification procedures vary from one country to another. In the United States, two-thirds of the Senate must vote approval. Because the United States contributed slightly less than 40% of the industrial countries' 1990 emissions, it is possible that the protocol could take effect without American participation. In early 2000, a vigorous movement got under way among the protocol's supporters to put it into force in 2002—the 10th anniversary of the Rio conference—most notably in western Europe.

A Final Comment

The Kyoto Protocol is by far the most complex environmental treaty that governments have ever attempted. It does not affect merely one restricted class of products, like the Montreal Protocol on stratospheric ozone. It does not affect merely one industry, like a fishing treaty. In a world that runs on fossil fuel, the Kyoto Protocol reaches nearly every industry, nearly all forms of transportation, and most households. The lack of progress in the negotiations since the Kyoto conference reflects chiefly a need among governments and citizens, not only in the United States, to reread and reconsider a document that would deeply affect every industrial economy. This reconsideration proceeds while emissions continue to grow and the technical debate goes on regarding the risks of climate change, the costs of different response options, and the prospects for sustained international cooperation in the face of divergent national interests.

Although President Clinton's 1997 campaign to generate support for an agreement on climate change was perhaps less successful than he had hoped, he certainly succeeded in making Americans more conscious of the prospect of global warming. Although governments around the world have promised more than they know how to achieve, a broader public has now joined the discussion. Governments of rich countries are spending very large amounts of money on climate research. The leader in this enterprise, the United States, has for several years been putting nearly $2 billion a year into the science of the climate. The flow of new findings is impressive.

A rising worldwide temperature over the past century is a reality. More and faster warming over the next century is a high probability. The world has not yet decided how to deal with this phenomenon. But it has begun—slowly, uncertainly, and contentiously—to think about it.

Suggested Reading

Benedick, Richard Elliot. 1998. *Ozone Diplomacy: New Directions in Safeguarding the Planet* (Second edition). Cambridge, MA: Harvard University Press.

Bernstein, Paul M., and W. David Montgomery. 1998. *How Much Could Kyoto Really Cost? A Reconstruction and Reconciliation of Administration Estimates*. Washington, DC: Charles River Associates.

Edmonds, J.A., and others. 1997. *Return to 1990: The Cost of Mitigating U.S. Carbon Emissions in the Post-2000 Period*. PNNL-11819. October. Washington, DC: Pacific Northwest National Laboratory.

Elliott, Lorraine. 1998. *The Global Politics of the Environment*. New York: New York University Press.

Grubb, Michael (with Christian Vrolijk and Duncan Brack). 1999. *The Kyoto Protocol: A Guide and Assessment*. London, U.K.: Royal Institute of International Affairs.

Grubb, Michael. 1997. Technology, Energy Systems and the Timing of CO_2 Emissions Abatement: An Overview of Economic Issues. *Energy Policy* 25(2): 159–72. (This article is a European response defending the approach subsequently adopted at Kyoto.)

Hammitt, James K. 1997. Stratospheric-Ozone Depletion. In *Economic Analyses at EPA: Assessing Regulatory Impact,* edited by Richard D. Morgenstern. Washington, DC: Resources for the Future.

IPCC (Intergovernmental Panel on Climate Change). 1996. *Climate Change 1995: The Science of Climate Change*. Contribution of Working Group I to the Second Assessment Report of the Intergovernmental Panel on Climate Change. New York: Cambridge University Press.

Krosnick, Jon A., and Penny S. Visser. 1998. *The Impact of the Fall 1997 Debate about Global Warming on American Public Opinion*. Washington, DC: Resources for the Future.

Manne, Alan, and Richard Richels. 1996. The Berlin Mandate: The Costs of Meeting Post-2000 Targets and Timetables. *Energy Policy* 24(3): 205–10.

McKibben, Warwick J., and Peter J. Wilcoxen. 1997. A Better Way to Slow Global Climate Change. Brookings Policy Briefing 17. June. Washington, DC: Brookings Institution.

Mintzer, Irving M., and J. Amber Leonard (eds.). 1994. *Negotiating Climate Change: The Inside Story of the Rio Convention*. Cambridge, U.K.: Cambridge University Press for the Stockholm Environment Institute.

OECD (Organisation for Economic Co-operation and Development). 1999. *Action Against Climate Change: The Kyoto Protocol and Beyond*. Paris, France: OECD.

Paterson, Matthew. 1996. *Global Warming and Global Politics*. London, U.K.: Routledge.

Pizer, William A. 1997. Prices vs. Quantities Revisited: The Case of Climate Change. Discussion Paper 98-02. Washington, DC: Resources for the Future.

Presidential Documents. Various issues. 33 Weekly Compilations. Washington, DC: U.S. Government Printing Office.

Toman, Michael A., Richard D. Morgenstern, and John Anderson. 1999. The Economics of "When" Flexibility in the Design of Greenhouse Gas Abatement Policies. *Annual Review of Energy and the Environment* 24: 431–60.

UNFCCC (Framework Convention on Climate Change). 1999. *Guide to the Climate Change Negotiation Process*. http://www.unfcc.de/resource.

Weyant, John P., and Jennifer N. Hill. 1999. Introduction and Overview. *The Energy Journal*, Special Issue (The Costs of the Kyoto Protocol: A Multi-Model Evaluation): vi–xiiv.

Wigley, T.M.L., R. Richels, and J.A. Edmonds. 1996. Economic and Environmental Choices in the Stabilization of Atmospheric CO_2 Concentrations. *Nature* 379 (6562): 240–43.

The Energy–CO$_2$ Connection
A Review of Trends and Challenges

Joel Darmstadter

A major—and undisputed—challenge that faces every country is access to energy on terms that facilitate economic growth while respecting environmental integrity. This challenge affects different countries to different degrees; even countries with highly developed economies such as the United States cannot avoid making energy choices that involve trade-offs between these sometimes conflicting goals.

The United States and other advanced industrialized countries have in varying degrees met this challenge under circumstances where the environmental problems were *localized* (for example, automotive pollution) or posed a *regional* threat (for example, acidification). Not surprisingly, ways of dealing with *global* problems have proven to be far less tractable.

In this chapter, I do not address the conundrum of designing and implementing workable multinational policies to restrain global greenhouse gas (GHG) emissions. Rather, my more limited purpose is to spotlight the intimate connection between carbon dioxide (CO$_2$)—the GHG of primary focus in the climate change debate—and energy consumption. The crux of the message is that any attempts to curb CO$_2$ emissions that fail to consider the need for changes in energy production and consumption (rates of growth, mix of fuels, and composition of demand) are likely to be futile. I emphasize the United States in my discussion; however, given the global nature of the GHG issue, I also direct some

attention to growth trends and prospects of less industrialized and developing economies.

GHG Emissions, CO$_2$, and Energy: Basic Relationships

Serious discussion about climate change invariably involves serious discussion about energy consumption. The coupling is well grounded: Even though some precursors believed to be responsible for global warming are unrelated to energy use (for example, the application of nitrogen fertilizer, which releases nitrous oxide) and certain energy activities have no direct impact on greenhouse warming (such as nuclear power), a major focus of the global warming debate is the role of fossil energy. Tables 1 and 2 show why. The energy sector accounts for some 86% of total U.S. GHG emissions (measured in carbon equivalents—see the box on page 26) and energy-related CO$_2$ alone for 81%. Within the energy sector, coal and petroleum (and the sectors in which the two fuels predominate, electric power generation and transport) give rise to a major portion of the country's CO$_2$ emissions. Roughly the same broad pattern prevails in all industrialized countries.

However, the policy emphasis on energy's contribution to GHG emissions reflects not only the quantitative importance of energy but also the possibilities for abatement through incentives and

Table 1. Sources of U.S. Greenhouse Gas Emissions, 1997.

Emissions (million metric tons carbon equivalents)

Sector	CO_2	Methane	Other	Total
Energy	1,466	58	29	1,553
Others	22	122	117	261
Total	1,488	180	146	1,814

Note: See Box 1 for additional information about how emissions are measured.

Source: U.S. EPA (see Sources for Data and Projections).

Table 2. U.S. Energy Consumption and CO_2 Emissions, 1997.

Source and consuming sector	Energy consumption (quads)	CO_2 emissions (million tons of carbon)[a]
By energy source		
Fossil		
Coal	20.9	533
Natural gas	22.6	319
Petroleum	37.0	613
Total fossil	80.5	1,466
Nonfossil		
Nuclear	6.7	NA
Hydro	3.9	NA
Other	3.1	NA
Total nonfossil	13.7	NA
Total[b]	**94.2**	**1,466**
By sector		
Fossil		
Electric power	22.3	532
Industry	21.9	307
Transportation	24.7	446
Residential/ commercial	10.9	168
Total fossil[b]	80.5	1,466
Nonfossil		
Electric power	11.2	NA
Other sectors	2.5	NA
Total nonfossil[b]	13.7	NA
Total[b]	**94.2**	**1,466**

Notes: The electric power sector is treated as a consumer of energy sources and an emitter of CO_2. An alternative treatment would bypass the power sector and ascribe its energy use to ultimate consumers of electricity. 1 quad = 1 quadrillion (10^{15}) Btu; NA = not applicable.

[a] Includes small unallocable amounts emitted in U.S. territories, not shown separately. Excludes small amounts attributable to nonfossil (biogenic) resources.

[b] Figures for totals may not exactly correspond to the sum of the constituent parts because of rounding.

Source: U.S. DOE/EIA *Annual Energy Review 1997* (see Sources for Data and Projections).

regulations that reshape the energy system. The electricity sector in particular can avail itself of different generation technologies, siting options (centralized or distributed), and energy sources for producing power. Especially telling is the fact that coal, which in recent years has fueled close to 60% of U.S. electricity production, contains about 80% more carbon per unit of heat energy than natural gas and 30% more than oil. Of course, technical and cost factors are critical in defining the comparative options, as are other considerations, such as the nature of public policies for local environmental improvement.

Determinants of CO_2 Emission Trends

Demographic, economic, and technological factors underlie trends in CO_2 emissions. Table 3 and Figure 1 point out the role of these key elements. They indicate that reduced levels or growth rates for CO_2 emissions from the energy sector can come about in various ways, which can be consolidated into four key elements: population, GDP/person, energy/GDP, and CO_2/energy (where GDP is gross domestic product).

From this framework, a straightforward arithmetic identity can be constructed:

$$CO_2 \text{ emissions} = \text{Population} \times \frac{\text{GDP}}{\text{Person}} \times$$

$$\frac{\text{Energy consumption}}{\text{Unit GDP}} \times \frac{CO_2 \text{ emissions}}{\text{Unit energy consumption}}$$

Measuring Greenhouse Gases
Radiative Forcing, Residence Time, and Global Warming Potential

In Table 1, methane (CH_4) and other greenhouse gas (GHG) emissions (principally nitrous oxide [NO_2] and certain fluorocarbon compounds) are expressed in carbon equivalents. Using this metric as a common denominator unit makes it possible to aggregate the different kinds of emissions and thereby get a sense of the relative importance of each contributor to greenhouse warming.

The aggregation procedure is based on the fact that each of the principal GHGs has two important attributes. The first is radiative forcing—that is, the extent to which the gas magnifies the greenhouse effect. The second is residence time in the atmosphere—that is, how long the gas exercises its climatic influence.

An example will help clarify how these attributes relate to climate change. The radiative forcing of CH_4 is 30 times greater than that of CO_2. However, the estimated residence time of CO_2 is around 250 years, whereas CH_4 (and the climatic effects it causes) will dissipate in only 10–15 years. To capture the combined effect of both factors, climate analysts have developed the notion of global warming potential (GWP). Its purpose is to "provide a simple measure of the relative radiative effects of the emissions of various greenhouse gases." The IPCC has chosen 100 years as its basic yardstick for calculating GWP; however, shorter or longer periods can be assumed as well. Under the 100-year assumption, methane's GWP is valued at 21 relative to 1 for CO_2. Other gases have a still much higher GWP—for example, NO_2 has a value of 310. Although GWP is an important tool for quantifying the relative effects of GHGs, the volume of emissions for each particular gas must also be taken into account. The volume of CO_2 emissions is so high that non-CO_2 emissions are far less consequential.

GWPs also do not—and are not meant to—reflect the different economic impacts of different gases over time. Economists argue that transforming the GWP into some index of economic damage would provide a more discriminating basis for decisions regarding the priority attention to controlling one or another greenhouse gas.

To clarify this point, consider a scenario in which future damages were expected to closely coincide with a peak in global temperature. Normally, the incremental emissions of CO_4 (or another gas with a short atmospheric residency time) would have a negligible effect compared with that of a long-lived greenhouse gas such as CO_2; however, their relative importance would invert as the peak temperature approached. In such a situation, the effects of CH_4 emissions would be more damaging than those of CO_2.

Developing an economic damage index that reflects both the peak-temperature phenomenon and the application of an appropriate discount rate is important for refining assessments of the benefits and costs of different greenhouse gas control initiatives. Whereas knowing the damages associated with, say, CO_2 versus CH_4 would provide a useful clue in assessing the benefits of damage reduction, it would obviously also be important to compare the costs of those control initiatives. Thus, even if CH_4 was likely to cause serious damages, it still might prove more economically efficient to concentrate on controlling CO_2 emissions if they were judged more tractable than, say, altering rice cultivation practices to limit CH_4 release.

Throughout this paper, quantitative measures of CO_2 emissions are expressed in terms of the carbon content of the fuel being burned. The mass of CO_2 gas equals 3.667 the mass of elemental carbon.

Very small amounts of CO_2 are released from nonenergy production. The principal example is cement manufacturing, which accounts for around 2% of global emissions. Notwithstanding that fact, for the purposes of this paper, statistical convenience dictated relating energy use to total CO_2 emissions, irrespective of origin. The broad findings I present are negligibly affected by this inconsistency.

Table 3. Average Annual Rates of Change in the United States and Major World Regions, 1973–90 and 1990–97.

Determinant of CO$_2$ emissions	Years	United States	Other OECD countries	Total OECD countries	Former Soviet Union	Other	World
Population	1973–90	0.98	0.81	0.85	0.86	2.07	1.74
	1990–97	1.00	0.69	0.76	0.12	1.73	1.46
GDP/population	1973–90	1.68	2.40	2.16	2.57	0.60	1.17
	1990–97	1.44	1.79	1.70	−7.81	2.44	0.93
Energy consumption/GDP	1973–90	−1.89	−2.13	−2.06	−0.76	2.13	−0.90
	1990–97	−0.80	−0.93	−0.88	1.64	0.04	−1.14
CO$_2$ emissions/energy consumption	1973–90	−0.27	−0.02	−0.13	−0.55	−1.10	−0.33
	1990–97	−0.24	−0.63	−0.46	−0.12	−1.36	−0.69
CO$_2$ emissions	1973–90	0.46	1.01	0.77	2.10	3.72	1.67
	1990–97	1.39	0.90	1.11	−6.30	2.83	0.54

Notes: To the extent possible, the gross domestic product (GDP) figures used in these calculations represent U.S. dollar estimates of other countries' GDPs based on so-called purchasing power parity (PPP), rather than market exchange rate means of conversion. Although the PPP estimates are often very approximate, they track more faithfully what people's income actually commands in terms of goods and services consumed. For example, China's PPP-based GDP is estimated at three or more times the level calculated using market exchange rates. OECD = Organisation for Economic Co-operation and Development.

Sources: The data in this table were constructed using several sources as the primary building blocks. In some cases, separate and not totally consistent data sets over different time intervals had to be spliced together. Additionally, in several instances, proxy indicators served as the basis for calculating the item of interest. For example, estimated CO$_2$ emissions in 1973 for certain world regions required using consumption data for different energy resources multiplied by their unit carbon content. (See British Petroleum Company, ORNL, U.S. DOE/EIA *Annual Energy Review 1997,* U.S EPA, World Bank, and World Resources Institute in Sources for Data and Projections.)

Thus, other things being equal, slower population growth means less growth in CO$_2$ release, whereas higher GDP per capita signifies a greater volume of CO$_2$ emitted. The energy-to-GDP ratio (also referred to as energy intensity) is a measure of an economy's aggregate energy intensity that reflects the structural, technological, and energy use characteristics of the society. All else unchanged, a falling energy intensity means less CO$_2$ emitted. The forces that can contribute to a decline in energy intensity—such as efficiency improvements through use of combined-cycle technology in generating electricity—are critical elements in the analysis of CO$_2$ abatement strategies. Finally, the CO$_2$-to-energy consumption ratio (sometimes called carbon intensity) spotlights the effect of a changing mix of energy types consumed in terms of carbon characteristics. Clearly, an important issue in the determination of CO$_2$ mitiga-

tion possibilities and costs is the ease or difficulty of altering that mix away from carbon-intensive components such as coal toward low-carbon or carbon-free resources such as natural gas, solar, or nuclear power. In one way or another, the four determinants of CO$_2$ emissions listed in Table 3 and shown in Figure 1 must enter into any analyst's attempt to determine the feasibility of CO$_2$ mitigation.

CO$_2$ Emission Trends and Implications

In looking at the weight exerted by these four factors in the growth of U.S. and global CO$_2$ emissions during the past quarter of a century, we see that the United States recorded "desirable" trends in terms of a declining energy intensity and a diminishing degree of carbon intensity in the use of energy resources (Table 3). However, neither improved energy

Figure 1. Average Annual Rates of Change in the United States and Major World Regions, 1973–90 and 1990–97.

a) Population

b) Ratio of GDP to Population

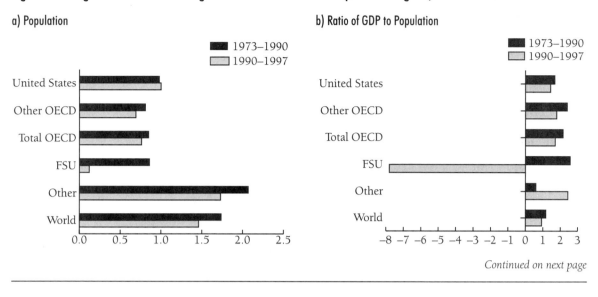

Continued on next page

efficiency nor the increased degree of "decarbonization" was sufficient to counter the combined effect of economic and demographic growth. Consequently, overall CO_2 emissions have continued to rise steadily in the United States.

It is worth recalling that the earlier of the two periods shown in Table 3 (1973–90) was marked by significant energy price increases that, reinforced by several conservation policy measures, helped curb the growth of fuel consumption, most notably in automotive transportation, but also in residential and business activities. For example, the fuel efficiency of passenger cars improved by more than 50% between 1973 and 1990. More conscious conservation practices contributed to reduced use of energy in the average household by more than 25% between 1978 and 1987; and energy consumption per square foot of commercial building space declined by 30% between 1979 and 1992. The industrial sector, no doubt sensitive to the effect of rising costs on its competitive standing, reduced its ratio of energy use to gross output by one-third during the first 15 years after the 1973–74 oil price shock.

In contrast to these earlier trends, the situation worsened for most of the 1990s, during which the annual rate of increase in U.S. CO_2 emissions more than doubled—largely because of a striking decline in both aggregate and sectoral energy efficiency advances (Table 4). In numerous sectors, efficiency gains ceased entirely, as buoyant income growth reinforced attractive energy prices to spur energy-related expenditures. (Although Tables 3 and 4 are divided at 1990, I recognize that the mid-1980s, marking a sharp break in real energy prices, would have provided a somewhat more meaningful benchmark from which to map the subsequent disincentive in energy conservation. But many of the non-U.S. data needed to construct the table were benchmarked at 1990.)

The dramatically increased popularity of sport utility vehicles and minivans (which weigh in at the low end of the energy efficiency scale) is a telling example of the forces at work. Rising incomes made the capital outlay for the vehicles affordable, and the cheap gasoline prices prevailing in the mid- to late 1990s—as low in real terms as before the oil market upheavals in the early 1970s—minimized their operating costs. In the absence of policies that constrain travel and/or tighten fuel economy, rising gasoline use is inevitable.

Figure 1. (continued)

c) Ratio of Energy to GDP

d) Ratio of CO₂ to Energy

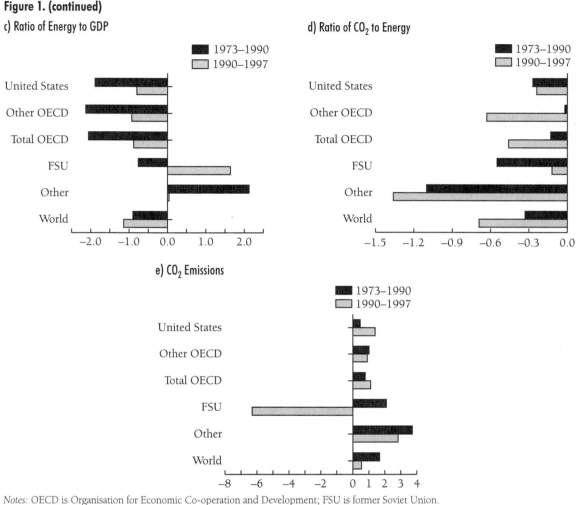

e) CO₂ Emissions

Notes: OECD is Organisation for Economic Co-operation and Development; FSU is former Soviet Union.
Source of all graphs: Same as Table 3.

These energy efficiency trends were mirrored in other parts of the economy, as Table 4 shows. Still, even before the oil-price spikes of 1999–2000 added a new factor—and one of uncertain duration—to the picture, there was some evidence that, low energy prices notwithstanding, the U.S. industrial sector was in the process of generating some significant energy savings. My conjecture about future U.S. energy and CO₂ emission trends touches on this development.

As these energy intensity trends proceeded, modest reductions in overall carbon intensity continued

(Table 3). One important factor driving this trend has been the rising share of electricity generation fired by natural gas, which was 12.5% in 1990 and had reached 14.1% by 1997. This shift contributed to making the electric power sector the only example of those shown in Table 4 whose CO₂ emissions did not worsen in recent years. But such changes could moderate to only a limited extent the accelerated rate of increase in CO₂ emissions.

Table 3 shows that what happened in the United States (and to a considerable extent in other OECD

Table 4. Selected Sectoral Trends in U.S. Energy Consumption and CO$_2$ Emissions, 1980–96.

	Average annual rate of change (%)	
Sector and indicator	1980–90	1990–96
Passenger cars		
Energy consumption	0.1	0.3
Miles/gallon	2.4	0.8
CO$_2$ emissions	0.1	0.3
Motor vehicles[a]		
Energy consumption	0.9	1.5
Miles/gallon	2.1	0.5
CO$_2$ emissions	0.9	1.5
Industry		
Energy consumption	0.5	1.7
Energy/gross output	−1.7	−1.5
CO$_2$ emissions	−0.7	0.8
Residential sector		
Energy consumption	−0.1	1.7
Energy/household	−1.5	2.1
CO$_2$ emissions	0.2	2.1
Commercial buildings[b]		
Energy consumption	1.5	1.4
Energy/square footage	−2.2	−0.2
CO$_2$ emissions	1.5	2.1
Electric utilities		
Fossil energy consumption	0.9	0.9
Kilowatt-hours generated (fossil)	1.0	1.1
Kilowatt-hours generated (total)	2.1	1.5
Fossil energy/fossil kilowatt-hours	−0.1	−0.2
Fossil energy/total kilowatt-hours	−1.2	−0.6
CO$_2$ emissions	1.3	1.3

[a] Includes passenger cars as well as vans, sport utility vehicles, and light trucks.

[b] Values are for 1979–89 and 1989–95.

Sources: Council of Economic Advisers, U.S. DOE/EIA *Annual Energy Review 1997* (see Sources for Data and Projections).

countries) occurred much more pointedly elsewhere in the world. This non-OECD group comprises countries of the former Soviet Union and some 100 "Other" countries, the majority of which are more or less in a state of underdevelopment—a situation borne out by the well-known disparities in per capita income (about 18% of the average OECD level), energy use (13%), and CO$_2$ emissions (13%). (The reason for separating the former Soviet Union from other non-OECD countries is that exceedingly poor former Soviet Union performance during the past decade distorts any totals in which they are included.) The combination of developing countries' energy-to-GDP and CO$_2$-to-energy ratios would have produced an annual growth rate in CO$_2$ emissions of just 1% during 1973–90 and a decline of 1.3% during 1990–97. Actual growth rates in carbon emissions of 3.7 and 2.8% during the two respective periods attest to the weight that demographic and economic forces can exert on trends in emissions.

As evidence that we are dealing with phenomena and trends that are not straightforward or predictable, Table 3 also reveals a striking, almost anomalous statistic. In contrast to its increase during 1973–90, energy intensity in the Other countries remained almost unchanged during the 1990s, even though a continuing momentum toward industrialization (not to mention more affordable energy prices) would have suggested intensification.

The reason for that striking statistic, it turns out, is that in China (which exerts a strong weight in the Other group), energy consumption grew at substantially below half the more than 11% GDP growth rate. (By contrast, India's energy consumption grew 21% faster than its GDP.) If true—and there is some doubt about the reliability of Chinese energy statistics—the country has evidently been able to increase energy efficiency significantly in its manufacturing and other sectors to achieve that overall outcome. Without such energy efficiency improvement in China, Table 3 would have shown marked increase in the energy-to-GDP ratio (rather than 0.04%) and a faster rise in CO$_2$ emissions, with proportionately less of that rise attributable to demographic and economic forces. Even if population growth in the developing world does slow, the imperative pursuit of increases in per capita income in these countries will likely make the search for a serious dent in global CO$_2$ emissions a major challenge for the pattern of energy supply and use in richer and poorer countries—both in energy's relationship to growth in the overall economy and in the carbon content of the energy consumed.

Conjectures about Future Developments

The historical record I have sketched thus far provides an appropriate backdrop for speculation about future developments in the interplay among population, GDP, energy consumption, and CO$_2$ emissions. One could spin out a diverse set of outcomes predicated on a diverse set of assumptions. However, my purpose here is more limited. Reverting to the framework of Table 3, I consider the circumstances under which one could envisage a significant slowdown in the growth of CO$_2$ emissions in the United States and elsewhere. Somewhat arbitrarily, I conduct this exercise within a 20-year time frame, looking out to 2020. A much shorter period would have exposed the unrealistic nature of contemplating significant short-term changes in factors characterized by a substantial degree of inertia in both their physical and policy dimensions; for example, coal-fueled power plants and numerous other kinds of energy infrastructure may require a 30-year amortization period. To contemplate a much longer time frame would require assumptions about such things as technological change and behavioral dynamics, with uncertainties that are simply too great to account for in the present broad-brush discussion.

The U.S. Scene

Consider specifically the circumstances surrounding a U.S. resolve to stabilize its CO$_2$ emissions at 1997 levels over the next two decades. To explore this possibility, suppose first that the U.S. population continues to grow at about 0.8% annually, a rate projected by the U.S. Census Bureau (see U.S. Department of Commerce in Sources for Data and Projections). Suppose also that per capita income grows at the modest pace of 1.5% yearly. (GDP or in-

come cannot actually be treated as independent of what is assumed about energy, because unusually costly energy use disincentives can penalize overall economic performance. I sidestep this "feedback loop" problem here; doing so does not affect the thrust of the discussion.) According to the earlier equation describing the components of emissions, it follows that the combined result of these two assumptions would be a 2.3% annual growth in emissions, *if,* in combination, the energy intensity of GDP and the carbon intensity of energy use remained unchanged. Conversely, to hold emissions constant with these population and GDP growth figures would require decreases in either energy intensity of GDP or carbon intensity of energy use averaging 2.3% per year. However, achieving such results without noticeable increases in energy prices seems questionable.

Energy Intensity Outlook. An important part of the energy intensity issue involves recognizing the potential contribution of "autonomous" technological advances—improvements in the energy-to-GDP ratio that occur over time and in a profit-seeking environment, even with a flat energy price trajectory and without any particular encouragement via public policy initiatives. Such technological progress allows manufacturers and other energy users to make do with less energy, even without the prospect of having to pay more for energy or being forced to economize by government dictate. The size of this autonomous component is subject to dispute; some analysts assert that all change is induced by specific and identifiable factors, not a reflection of an inherent technological phenomenon. Most mainstream economists would place the figure for autonomous technical change between 0.5% and 1.0% per year.

One intriguing prospect, which could augment energy-saving technological advances, revolves around the role of the Internet in both commerce and industry. Growing anecdotal evidence indicates that the Internet can enhance both economic and energy efficiency. Telecommuting is one obvious and commonly cited example of reduced motor fuel consumption. Another example relates to online purchase of goods, which can improve the economic and energy efficiency of inventory management and distribution processes relative to a more decentralized distribution network. (Whether a society bereft of bookstores in which to browse improves one's quality of life is a more subjective issue.) Space-saving electronic media could significantly cut down on square-footage requirements of new commercial structures, attenuating the input of construction-related energy. Manufacturers' supply chains—already benefiting from just-in-time inventory controls introduced in recent decades—could be further streamlined, with transactions coupled to electronic channels of communication. On the other hand, some Internet-based transactions can stimulate energy use. Internet surfing to pinpoint lower airline fares or electricity costs will give rise to increased, rather than decreased, energy demand.

Romm suggests that electronic commerce explains a significant part of the country's energy intensity drop during 1997–98, despite low energy prices (see Suggested Reading). However, the extent to which this is the case, and the extent to which it will be a sustained influence on energy conservation in the years ahead, remain open questions.

Carbon Intensity Outlook. Numerous options for lowering carbon intensity may also be possible. For example, the hydrogen fuel cell releases far less carbon than the gasoline-powered internal combustion engine but still faces a substantial period of development before it will qualify as a technologically reliable, safe, and economically competitive contender in the automotive market. Although unrealistic from today's vantage point, the prospects of a revived nuclear industry based on vastly improved safety characteristics should not be ruled out. Also, several renewable energy systems have succeeded in lowering their electricity generation costs to a remarkable degree. Indeed, by the 1990s, these technologies— wind power in particular—had managed to reduce their generation costs to levels even lower than had been projected by "green" advocates (credited, at the time, more for enthusiasm than realism) 25 years earlier. The fact that, in contrast to falling costs, *market share* continued to elude renewable technologies was largely due to the falling real price of fossil

energy resources, making them the fuels of choice in electricity production.

Role for Energy Prices. Let us now suppose that improvement in energy efficiency were to occur at 1% per year, the top end of the range generally embraced by economists. To stabilize CO_2 emissions given our assumptions for population and income growth then would require a "decarbonization" rate—a decline in the carbon intensity of energy use—of 1.3% per year. This is well above the historical experience over the past quarter century of less than 0.3% (Table 3). Conversely, emissions stabilization with a decline in carbon intensity of 0.3% would require a decline in energy intensity of GDP of 2.0%, more than twice what seems plausible for autonomous efficiency improvement.

Thus, unless some "new economy" phenomenon were to drastically alter energy–economy relationships relative to the past, it seems highly unlikely that autonomous trends could lead to emissions stabilization. One must turn instead to the possibility of energy price increases or regulations that stimulate energy efficiency and decarbonization, both directly and through the development and diffusion of new technologies. In particular, technologies such as the hydrogen fuel cell or wind turbines probably would mature faster if fossil fuel prices were higher. But a narrowing of the wedge between fossil and renewable energy costs seems less likely to happen because of rising real costs of conventional energy—increasing scarcity is a prospect that seems continually to recede with the horizon—than as the result of policy-driven economic disincentives to emit CO_2.

Past experience illustrates the powerful effect of energy prices on energy use. For example, as the annual rate of decline in electricity prices during 1960–73 (3.4%) gave way to a rise of 2.4% during 1973–85, electricity sales growth dropped from 7.2% to 2.4% per year. In the case of gasoline, a real price decline of 1.4% per year during 1960–73 was followed by a rise of 2.4% per year during 1973–85, and a turnaround in sales from an increase of nearly 4% annually during 1960–73 to almost flat consumption over the subsequent 12 years. (Although 1973–85 was a period of significant macroeconomic dislocation, GHG policies implemented gradually, with foresight and complementary macroeconomic buffering, need not produce comparable dislocation.)

Apart from the effect of structural factors in the economy that signify lower energy intensity (for example, growth of the service sector, or as noted earlier, the increase in Internet transactions), debate continues over the potential for cost-effective energy conservation even without the spur of higher prices. Undoubtedly, such opportunities exist in various sectors of the economy. The reasons for their lagging application are variously given as imperfect information, distorted credit and capital market arrangements, unnecessarily high transaction costs, and a failure of public R&D support where the target of such support can be justified as serving the public good rather than simply displacing or duplicating private R&D activity. However, the magnitude of the missed—and attainable—energy savings remains contentious, with most economists taking a fairly conservative view of the potential.

The Global Scene

Although the burden of adjustment for the United States (and the rest of the OECD member countries) in stabilizing emissions is challenging but (with some fortuitous mix of policy and technological prospects) arguably manageable, other parts of the world face a more formidable task. Some perspective on this point is provided in Table 5, which presents the "business as usual" (or reference case) set of pertinent indicators to 2020 from the U.S. Department of Energy's Energy Information Administration.

As already pointed out in connection with the U.S. scenario discussion, demographic and economic forces are major determinants of the outlook for CO$_2$ emissions. Notwithstanding an expected slowing of population growth, the assumption of a sharply accelerated rise in per capita income puts significant pressure on CO$_2$ emissions. The projections in Table 5 also show a reduction in energy intensity of somewhat more than 1% annually, a respectable rate by historic norms. The carbon content of total energy also continues to diminish, though more slowly than in the past. For example, the pro-

jections presume that coal will remain the energy source of choice in the developing regions of Asia.

Taking all the various factors into account, the projections in Table 5 show CO$_2$ growing at 3.5% per year for the developing world, somewhat under 3% per year if eastern European countries are included. Suppose that a 1% reduction in the rate of growth in emissions were deemed a desirable target to pursue. Given assumed population and income growth, achieving this target would put a substantial burden of adjustment on the already-declining energy-to-GDP ratio and the CO$_2$ intensity of energy use. Moreover, the "High Economic Growth" scenario in the Energy Information Administration's projections raises the rate of growth in CO$_2$ emissions by about 1% above the values shown in Table 5. These figures indicate the limits to what can be expected from autonomous energy sector adjustments in terms of slower emissions growth.

Nonetheless, economically exploitable energy-saving opportunities beyond those indicated in Table 5 should not be casually dismissed. Market-determined rather than subsidized prices for fuels and power promote greater energy conservation and confer real overall economic benefits to society. Tax-

---

Table 5. Projected Average Annual Rates of Change for "Business-As-Usual" Scenario in Non-OECD Countries, 1997–2020.

Indicator	Annual rate of change (%)	
	Including eastern Europe	Excluding eastern Europe
Population	1.2	1.2
GDP/population	3.6	3.5
Energy/GDP	−1.3	−1.1
CO$_2$/energy	−0.7	−0.1
CO$_2$	2.7	3.5

Notes: Countries of the former Soviet Union are excluded. Because many eastern European countries are frequently excluded from statistics on developing countries of the world, the table shows data with and without eastern Europe. See text for a detailed discussion.

Source: U.S. DOE/EIA *Annual Energy Review 1997* (see Sources for Data and Projections).

ing the worst forms of pollution could contribute to more rational energy use. Sound rules and institutions encouraging private foreign investment could enhance the international diffusion of energy-saving technology. Some developing and transitional countries already are making progress along these lines, and there is room for additional "win–win" improvements.

But after all is said and done, a serious commitment to CO_2 mitigation in the United States and abroad requires a willingness to consider some major changes in the use and mix of energy resources. These far-reaching changes—which will extend beyond the next couple of decades considered here—will not be easy and probably won't be cheap. They will require considerable political will as well as technological acumen.

Suggested Reading

Ausubel, Jesse H. 1995. Technical Progress and Climatic Change. *Energy Policy* 23(4/5): 411–16.

Hammitt, J.K., A.K. Jain, J.L. Adams, and D.J. Wuebbles. 1996. A Welfare-Based Index for Assessing Environmental Effects of Greenhouse-Gas Emissions. *Nature* 381(6580): 301–3.

IPCC (Intergovernmental Panel on Climate Change). 1996. *Climate Change 1995: The Science of Climate Change*. Contribution of Working Group I to the Second Assessment Report of the Intergovernmental Panel on Climate Change. New York: Cambridge University Press, 21–22.

———. 1996. *Climate Change 1995: Economic and Social Dimensions of Climate Change*. Contribution of Working Group III to the Second Assessment Report of the Intergovernmental Panel on Climate Change. New York: Cambridge University Press.

Jaffe, Adam B., Richard G. Newell, and Robert N. Stavins. 1999. Energy-Efficient Technologies and Climate Change Policies: Issues and Evidence. Climate Issue Brief 19. Washington, DC: Resources for the Future.

McVeigh, James, Dallas Burtraw, Joel Darmstadter, and Karen Palmer. 2000. Winner, Loser, or Innocent Victim? Has Renewable Energy Performed as Expected? *Solar Energy* 68(3): 237–55.

Nakićenović, N. 1996. Freeing Energy from Carbon. *Daedalus* 125(3): 95–112.

Romm, Joseph (with Arthur Rosenfeld and Susan Herrmann). 1999. The Internet Economy and Global Warming. Washington, DC. Center for Energy and Climate Solutions/Global Environment and Technology Foundation. http://www.cool-companies.org.

Sources for Data and Projections

British Petroleum Co. Various issues. *BP Statistical Review of World Energy* (annual). London, U.K.: British Petroleum Co.

Council of Economic Advisers. 1999. *Economic Report of the President*. February. Washington, DC: U.S. Government Printing Office.

ORNL (Oak Ridge National Laboratory). 1994. *Trends '93: A Compendium of Data on Global Change*. Oak Ridge, TN: Carbon Dioxide Information Analysis Center.

U.S. Department of Commerce. 1999. *World Population Profile: 1998*. February. Washington, DC: U.S. Department of Commerce, Census Bureau.

U.S. DOE/EIA (Department of Energy, Energy Information Administration). 1999. *International Energy Annual 1997*. April. Washington, DC: U.S. Government Printing Office.

———. 1999. *International Energy Outlook 1999*. March. Washington, DC: U.S. Government Printing Office.

———. 1998. *Annual Energy Outlook 1999*. December. Washington, DC: U.S. Government Printing Office.

———. 1998. *Annual Energy Review 1997*. July. Washington, DC: U.S. Government Printing Office.

U.S. EPA (Environmental Protection Agency). 1999. *Inventory of U.S. Greenhouse Gas Emissions and Sinks 1990–1997*. March. Washington, DC: U.S. EPA.

World Bank. 1999. *1999 World Development Indicators*. Washington, DC: World Bank.

World Resources Institute (with World Bank and United Nations). 1998. *World Resources 1998–99: A Guide to the Global Environment*. Oxford, U.K.: Oxford University Press.

How Much Climate Change Is Too Much?
An Economics Perspective

Jason F. Shogren and Michael A. Toman

Having risen from relative obscurity as few as 10 years ago, climate change now looms large among environmental policy issues. Its scope is global; the potential environmental and economic impacts are ubiquitous; the potential restrictions on human choices touch the most basic goals of people in all nations; and the sheer scope of the potential response—a significant shift away from using fossil fuels as the primary energy source in the modern economy—is daunting. The magnitude of these changes has motivated experts the world over to study the natural and socioeconomic effects of climate change as well as policy options for slowing climate change and reducing its risks. The various options serve as fodder for often testy negotiations within and among nations about how and when to mitigate climate change, who should take action, and who should bear the costs.

Lurking behind these policy activities is a deceptively simple question: How much climate change is acceptable, and how much is "too much"? (The other key question is, Who is going to pay for mitigating the risks?) The lack of consensus on this issue reflects the uncertainties that surround it and differences in value judgments regarding the risks and costs.

In this chapter, we review the economic approach to the question of how much climate change is too much. The economic perspective emphasizes the evaluation of benefits and costs broadly defined while addressing uncertainties and important considerations such as equity. We also consider some important criticisms of the benefit–cost approach. Then, we discuss the key factors that influence the benefits and costs of mitigating climate change risks. This discussion leads to a review of findings from the many quantitative "integrated assessment" models of climate change risks and response costs. This review does not lead to a simple answer to our overarching question about how much climate change is too much. But we do identify several good reasons for taking a deliberate but gradual approach to the mitigation of climate change risks.

The issues we cover are both diverse—ranging from the economics and philosophy of long-term cost-benefit analysis, to modeling strategies for representing climate change risks and greenhouse gas abatement costs—and, at times, somewhat complex. We have tried to be fairly comprehensive while seeking to make the discussion as accessible as possible.

This chapter is adapted from Chapter 5 of *Public Policies for Environmental Protection* (Second Edition), edited by Paul R. Portney and Robert N. Stavins, published by Resources for the Future in 2000.

Overview of the Risks and Response Costs

Life on Earth is possible partly because some gases such as carbon dioxide (CO_2) and water vapor, which naturally occur in Earth's atmosphere, trap heat—like a greenhouse. CO_2 released from use of fossil fuels (coal, oil, and natural gas) is the most plentiful human-created greenhouse gas (GHG). Other gases—which include methane (CH_4), chlorofluorocarbons (CFCs; now banned) and their substitutes currently in use, and nitrous oxides associated with fertilizer use—are emitted in lower volumes than CO_2 but trap more heat. Human-made GHGs work against us when they trap too much sunlight and block outward radiation. Scientists worry that the accumulation of these gases in the atmosphere has changed and will continue to change the climate.

The risk of climate change depends on the physical and socioeconomic implications of a changing climate. Climate change might have several effects:

- Reduced productivity of natural resources that humans use or extract from the natural environment (for example, lower agricultural yields, smaller timber harvests, and scarcer water resources).
- Damage to human-built environments (for example, coastal flooding from rising sea levels, incursion of saltwater into drinking water systems, and damage from increased storms and floods).
- Risks to life and limb (for example, more deaths from heat waves, storms, and contaminated water, and increased incidence of tropical diseases).
- Damage to less managed resources such as the natural conditions conducive to different landscapes, wilderness areas, natural habitats for scarce species, and biodiversity (for example, rising sea levels could inundate coastal wetlands, and increased inland aridity could destroy prairie wetlands).

All of these kinds of damage are posited to result from changes in long-term GHG concentrations in the atmosphere. Very rapid rates of climate change could exacerbate the damage. The adverse effects of climate change most likely will take decades or longer to materialize, however. Moreover, the odds that these events will come to pass are uncertain and not well understood. Numerical estimates of physical impacts are few, and confidence intervals are even harder to come by. The rise in sea level as a result of polar ice melting, for instance, is perhaps the best understood, and the current predicted range of change is still broad. For example, scenarios presented by the Intergovernmental Panel on Climate Change (IPCC) in *Climate Change 1995: The Science of Climate Change* (see Suggested Reading) indicate possible increases in sea level of less than 20 cm to almost 100 cm by 2100 as a result of a doubling of Earth's atmospheric GHG concentrations. The uncertainty in these estimates stems from not knowing how temperature will respond to increased GHG concentrations and how oceans and ice caps will respond to temperature change. The risks of catastrophic effects such as shifts in the Gulf Stream and the sudden collapse of polar ice caps are even harder to gauge.

Unknown physical risks are compounded by uncertain socioeconomic consequences. Cost estimates of potential impacts on market goods and services such as agricultural outputs can be made with some confidence, at least in developed countries. But cost estimates for nonmarket goods such as human and ecosystem health give rise to serious debate.

Moreover, existing estimates apply almost exclusively to industrial countries such as the United States. Less is known about the adverse socioeconomic consequences for poorer societies, even though these societies arguably are more vulnerable to climate change. Economic growth in developing countries presumably will lessen some of their vulnerability—for example, threats related to agricultural yields and basic sanitation services would decline. But economic growth in the long term could be imperiled in those regions whose economies depend on natural and ecological resources that would be adversely affected by climate change. Aggregate statistics mask considerable regional variation: Some areas probably will benefit from climate change while others lose.

In weighing the consequences of climate change, it is important to remember that humans adapt to risk to lower their losses. In general, the ability to adapt contributes to lowering the net risk of climate change more in situations where the human control over relevant natural systems and infrastructure is greater. Humans have more capacity to adapt in agricultural activities than in wilderness preservation, for example. The potential to adapt also depends on a society's wealth and the presence of various kinds of social infrastructure, such as educational and public health systems. As a result, richer countries probably will face less of a threat to human health from climate change than poorer societies that have less infrastructure. Beyond this general point, the potential for adaptation to reduce climate change risks continues to be debated.

GHGs remain in the atmosphere for tens or hundreds of years. GHG concentrations reflect long-term emissions; changes in any one year's emissions have a trivial effect on current overall concentrations. Even significant reductions in emissions made today will not be evident in atmospheric concentrations for decades or more. This point is important to keep in mind in deciding when to act—we do not have the luxury of waiting to see the full implications of climate change before taking ameliorative action. Many observers characterize responding to the risks of climate change as taking out insurance; nations try to reduce the odds of adverse events occurring through mitigation, and to reduce the severity of negative consequences by increasing the capacity for adaptation once climate change occurs. The insurance analogy underscores both the uncertainty that permeates how society and policymakers evaluate the issue and the need to respond to the risks in a timely way.

In constructing a viable and effective risk-reducing climate policy, policymakers must address hazy estimates of the risks, the benefits from taking action, and the potential for adaptation against the uncertain but also consequential cost of reducing GHGs. Costs of mitigation matter, as do costs of climate change itself. One must consider the consequences of committing resources to reducing climate change risks that could otherwise be used to meet other human interests, just as one must weigh the consequences of different climatic changes.

Why Consider the Costs and Benefits of Climate Policy?

Responding effectively to climate change risks requires society to consider the potential costs and benefits of various actions as well as inaction. By costs we mean the opportunity costs of GHG mitigation or adaptation—what society must forgo to pursue climate policy. Benefits are the gains from reducing climate change risks by lowering emissions or by enhancing the capacity for adaptation. An assessment of benefits and costs gives policymakers information they need to make educated decisions in setting the stringency of a mitigation policy (for example, how much GHG abatement to undertake, and when to do it) and deciding how much adaptation infrastructure to create.

It is important to consider the costs and the benefits of climate change policies because all resources—human, physical, and natural—are scarce. Policymakers must consider the benefits not obtained when resources are devoted to reducing climate change risks, just as they must consider the climate change risks incurred or avoided from different kinds and degrees of policy response. Marginal benefits and costs reveal the gain from an incremental investment of time, talent, and other resources into mitigating climate risks, and the other opportunities forgone by using these resources for climate change risk mitigation. It is not a question of *whether* to address climate change but *how much* to address it.

Critics object to a benefit–cost approach to climate change policy assessment on several grounds. Their arguments include the following:

- The damages due to climate change, and thus the benefits of climate policies to mitigate these damages, are uncertain and thus inherently difficult to quantify given the current state of knowledge. Climate change also could cause large-scale irreversible effects that are hard to address in a simple benefit–cost framework. Therefore, the estimated benefits of action are biased downward.

- Climate mitigation costs are uncertain and could escalate rapidly from too-aggressive emission control policies. Proponents of this view are indicating a concern about the risk of underestimating mitigation costs.

- Climate change involves substantial equity issues—among current societies and between current and future generations—that are questions of morality, not economic efficiency. Policymakers should be concerned with more than benefit–cost analysis in judging the merits of climate policies.

As these arguments indicate, some critics worry that economic benefit–cost analysis gives short shrift to the need for climate protection, whereas others are concerned that the results of the analysis will call for unwarranted expensive mitigation.

Both groups of critics have proposed alternative criteria for evaluating climate policies, which can be seen as different methods of weighing the benefits and costs of policies given uncertainties, risks of irreversibility, the desire to avoid risk, and distributional concerns. For example, under the *precautionary principle*, which seeks to avoid "undue" harm to the climate system, cost considerations are absent or secondary. Typically, the idea is that climate change beyond a certain level simply involves too much risk, if one considers the distribution of benefits and costs over generations.

Knee-of-the-cost-curve analysis, in contrast, seeks to limit emission reductions to a point at which marginal costs increase rapidly. Benefit estimation is set aside in this approach because of uncertainty. The approach implicitly assumes that the marginal damages from climate change (which are the flip side of marginal benefits from climate change mitigation) do not increase much as climate change proceeds and that costs could escalate rapidly from a poor choice of emissions target.

The benefit–cost approach can address both uncertainty and irreversibility. We do not mean to imply that estimates in practice are always the best or that how one evaluates and acts on highly uncertain assessments will not be open to philosophical debate. For example, as people become more informed about climate change, it is safe to presume that the importance they attach to the issue will change. Critics of the economic methodology argue that this process reflects in part a change in preferences through various social processes, not only a change in information. Moreover, under conditions of great uncertainty, the legitimacy of a policy decision may depend even more than usual on whether the processes used to determine it are deemed inclusive and fair, as well as on the substantive evidence for the decision.

But it is fundamentally inaccurate to see analysis of economic benefits and costs from climate change policies as inherently biased because of uncertainty and irreversibility. Nor should benefit–cost analysis be seen as concerned only with market values accruing to developed countries. One of the great achievements in environmental economics over the past 40 years has been a clear demonstration of the importance of nonmarket benefits, which include benefits related to the development aspirations of poorer countries. These values can be given importance equal to that of market values in policy debates.

Our advocacy that benefits and costs be considered when judging climate change policies does not mean we advocate a simple, one-dimensional benefit–cost test for climate change policies. In practice, decisionmakers can, will, and should bring to the fore important considerations about the equity and fairness of climate change policies across space and time. Decisionmakers also will bring their own judgments about the relevance, credibility, and robustness of benefit and cost information and about the appropriate degree of climate change and other risks that society should bear. Our argument in favor of considering both benefits and costs is that policy deliberations will be better informed if good economic analysis is provided.

The alternative decision criteria advanced by critics also are problematic in practice. The definition of "undue" is usually heuristic or vague. The approach is equivalent to assuming a sharp spike, or peak, in damages caused by climate change beyond the proposed threshold. It may be the case, but not enough evidence yet exists to assume this property (let alone to indicate at what level of climate change such a spike would occur). On the other hand, with knee-of-the-curve analysis, benefits are ignored so there is no assurance of a sound decision either.

Benefits and costs are unavoidable. How their impacts are assessed is what differentiates one approach from another. We maintain throughout this discussion that the assessment and weighing of costs and benefits is an inherent part of any policy decision.

Equity and Fairness Issues

The fairness of climate change policies to today's societies and to future generations continues to be at the core of policy debates. These issues go beyond what economic benefit–cost analysis can resolve, though such analysis can help illustrate the possible distributional impacts of different climate policies. In this section, we focus first on intergenerational equity issues. Then, contemporaneous equity issues are addressed.

Advocates of more aggressive GHG abatement point to the potential adverse consequences of less aggressive abatement policies for the well-being of future generations as a moral rationale for their stance. They assert that conventional discounting—even at relatively low rates—may be inequitable to future generations by leaving them with unacceptable climate damages or high costs from the need to abate future emissions very quickly. Critics also have argued that conventional discounting underestimates costs in the face of persistent income differences between rich and poor countries. Essentially, the argument is that because developing countries probably will not close the income gap over the next several decades, and because people in those countries attach higher incremental value to additional well-being than people in rich countries, the effective discount rate used to evaluate reductions in future damages from climate change should be lower than that applied to richer countries.

Supporters of the conventional approach to discounting on grounds of economic efficiency argue just as vehemently that any evaluation of costs and benefits over time that understates the opportunity cost of forgone investment is a bad bargain for future generations because it distorts the distribution of investment resources over time. These supporters of standard discounting also argue that future generations are likely to be better off than the present generation, casting doubt on the basic premise of the critics' concerns.

Experts attempting to address this complex mixture of issues increasingly recognize the need to distinguish principles of equity and efficiency, even though there is as yet no consensus on the practical implications for climate policy. We can start with the observation that anything society's decisionmakers do today—abating GHGs, investing in new seed varieties, expanding health and education facilities, and so on—should be evaluated in a way that reflects the real opportunity cost, that is, the options forgone both today and in the long term. This answer responds to the critics who fear a misallocation of investment resources if climate policies are not treated similarly to other uses of society's scarce resources.

Long-term uncertainty about the future growth of the economy provides a rationale for low discount rates on grounds of economic efficiency. The basic argument is that if everything goes well in the future, then the economy will be productive, the rate of return on investment will remain high, and the opportunity cost of displacing investment with policy today likewise also will be high. However, if things do not go so well and the rate of return on capital is low because of climate change or some other phenomenon, then the opportunity cost of our current investment in climate change mitigation versus other activities also will be low.

But economic efficiency only means a lack of waste given some initial distribution of resources. Specifically how much climate change mitigation to undertake is a different question, one that refers to the distribution of resources across generations. The answer depends on how concerned members of the current generation are about the future in general, how much they think climate change might imperil the well-being of their descendants, and the options at their disposal to mitigate unwelcome impacts on future generations. For example, one could be very concerned about the well-being of the future but also believe that other investments—such as health and education—would do more to enhance the well-being of future generations. Not surprisingly, experts and policymakers do not agree on these points.

We turn next to a brief discussion of international equity issues associated with climate change. The most immediate aspect of this debate involves the international distribution of responsibility for reducing GHGs and the associated costs. Developing nations have many pressing needs, such as potable water and stable food supplies, and less financial and technical capacity than rich countries have for mitigating GHGs. These nations have less incentive to agree to a policy that they see as imposing unacceptable costs.

Beyond this question are even more vexing issues surrounding the distribution of climate change risks. As already noted, it is likely that developing countries are both relatively more vulnerable to climate change than advanced industrialized countries and have less adaptive capacity; however, these disadvantages likely will be reduced in the future with further economic development. These differences are only beginning to be accounted for in climate change risks assessments. Analyses that consider only aggregate benefits and costs of climate change mitigation, without addressing the distribution of these benefits and costs, miss an important dimension of the policy problem. For example, the absolute magnitude of avoided costs from slowing climate change may be smaller in developing countries simply because per capita incomes are lower. But the implication that climate change mitigation should be given short shrift just because it mainly affects poorer people is ethically troubling.

Differences in perceptions about what constitutes equitable distributions of effort complicate any agreement. No standard exists for establishing the equity of any particular allocation of GHG control responsibility. Simple rules of thumb, such as allocating responsibility based on equal per capita rights to emit GHGs (advantageous to developing countries) and allocations that are positively correlated to past and current emissions (advantageous to developed

countries) are unlikely to command broad political support internationally.

What Do Existing Economic Analyses Say?

Analyzing the benefits and costs of climate change mitigation requires understanding biophysical and economic systems as well as the interactions between them. Integrated assessment (IA) modeling combines the key elements of biophysical and economic systems into one integrated system (Figure 1). IA models strip down the laws of nature and human behavior to their essentials to depict how more GHGs in the atmosphere raise temperature and how temperature increase induces economic losses. The models also contain enough detail about the drivers of energy use and interactions between energy and economy that the economic costs of different constraints on CO_2 emissions can be determined.

Researchers often use IA models to simulate a path of carbon reductions over time that would maximize the present value of avoided damages (that is, the benefits of a particular climate policy) less mitigation costs. As noted earlier, considerable controversy surrounds this criterion for evaluation.

A striking finding of many IA models is the apparent desirability of imposing only limited GHG controls over the next 20 or 30 years. According to the estimates in most IA models, the costs of sharply reducing GHG concentrations today are too high relative to the modest benefits the reductions are projected to bring.

The benefit of reducing GHG concentrations in the near term is estimated in many studies to be on the order of $5–25 per ton of carbon (see for example the papers by Nordhaus and Tol in Suggested Reading). Only after GHG concentrations have increased considerably do the impacts warrant more effort to taper off emissions, according to the models.

Even more striking is the finding of many IA models that emissions should rise well into this century. In comparison, the models indicate that policies pushing for substantial near-term control, such as the Kyoto Protocol, involve too much cost, too soon, relative to their projected benefits. Critics

Figure 1. Climate Change and Its Interaction with Natural, Economic, and Social Processes.

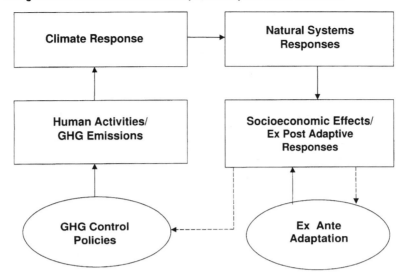

Note: The key components of an integrated assessment model are illustrated. Solid lines represent physical changes; dotted lines represent policy changes.

Source: Darmstadter and Toman (see Suggested Reading).

argue that IA models inadequately address several important elements of climate change risks: uncertainty, irreversibility, and risk of catastrophe. Assessing the weight of these criticisms requires us to explore the influences on the economic benefits and costs of climate protection.

Influences on the Benefits

The IPCC Second Assessment Report concluded that climate change could pose some serious risks. The IPCC presented results of studies showing that the damaging effects of a doubling of GHG concentrations in the atmosphere could cost on the order of 1.0–1.5% of gross domestic product (GDP) for developed countries and 2.0–9.0% of GDP for developing countries (see also Frankhauser and others in Suggested Reading). Reducing such losses is the benefit of protecting against the negative effects of climate change.

Several factors affect the potential magnitude of the benefits. One is the potential scale and timing of

damages avoided. Although IA models differ greatly in detail, most have economic damage representations calibrated to produce damages resulting from a doubling of atmospheric GHG concentrations roughly of the same order as the IPCC Second Assessment Report. This point is worth keeping in mind when evaluating the results. The models increasingly contain separate damage functions for different regions. Generally, the effects in developing countries are presumed to be worse than those in the developed world, as in the IPCC Second Assessment Report. For the most part, these costs would be incurred decades into the future. Consequently, the present value of the costs would be relatively low today.

Assumptions about adaptation also affect estimates of potential benefits. Some critics of the earlier IPCC estimates argue that damages likely will be lower than predicted because expected temperature increases from a doubling of atmospheric GHG concentrations probably will be less than projected, ecosystems seem to be more resilient over the long term than the estimates suggest, human beings can adapt more than was supposed, and damages are not likely to increase proportionally with GDP. The implication is that the optimal path for GHG control (in a present value sense) should be even less aggressive than the IA results indicate. These new assessments remain controversial. One ongoing question concerns the cost of adjusting to a changing climate versus the long-term cost of a changed climate. Another is whether the effects of climate change (for example, in encouraging the spread of human illness through a greater incidence of tropical diseases, reducing river flows that concentrate pollutants, and increasing the incidence of heat stress) are being underestimated.

A third factor affects benefits: Damage costs not only are uncertain but also involve a chance of a catastrophe. However, a general finding from IA models is that GHG reductions should be gradual, even if damages are larger than conventionally assumed. A risk of catastrophe provides a rationale for more aggressive early actions to reduce GHG concentrations, but the risk has to be very large to rationalize near-term actions as aggressive as those envisioned in the Kyoto Protocol in a present-value IA framework. Part of the reason for this finding is that the outcome with the lowest cost also is the most likely to occur. IA models also do not incorporate risk-averse attitudes, which would provide a stronger rationale for avoiding large costs. Moreover, discounting in the models reduces the effective impact of all but the most catastrophic costs after a few decades.

Irreversibility of GHG emissions is yet another factor influencing the benefits of GHG abatement. Because GHG emissions persist in the atmosphere for decades, even centuries, the resulting long-term damages strengthen the rationale for early and aggressive GHG control. Moreover, given that some damage costs from adjusting to a changed climate depend on the *rate* of climate change, immediate action also might be valuable. To date, however, the importance of this factor has not been conclusively demonstrated; the gradual abatement policies implied by the IA models do not seem likely to greatly increase the speed of further climate change.

Finally, policies that reduce CO_2 also can yield ancillary benefits in terms of local environmental qual-

ity improvement, such as fewer threats to human health and reduced damage to water bodies from nitrogen deposition. The magnitudes of these ancillary effects remain fairly uncertain. They are lower to the extent that more environmental improvement would occur anyway, in the absence of GHG policy. They also depend on how GHG policies are implemented (for example, a new boiler performance mandate that encouraged extending the lives of old, dirty boilers would detract from the environment).

Influences on the Costs

Estimates of the cost of mitigating GHG emissions vary widely. Some studies suggest that the United States could meet its Kyoto Protocol target at negligible cost; other studies claim that the United States would lose at least 1–2% of its GDP each year. A study by the Energy Modeling Forum helped explain the range of results in assessing the costs to meet the Kyoto Protocol policy targets (see Weyant and Hill in Suggested Reading). For example, the carbon price (carbon tax or emissions permit price) needed to achieve the Kyoto Protocol emissions target in the United States with domestic policies alone ranges from about $70 per metric ton of carbon to more than $400 per ton (in 1990 dollars) across the models. The corresponding GDP losses in 2010 range from less than 0.2% to 2.0% relative to baseline. (The percentages of GDP are not reported in Weyant and Hill but implied from graphs presented there.) Carbon prices are put in perspective by relating them to prices for common forms of energy, as listed in Table 1.

The results reported by Weyant and Hill and previous assessments of GHG control costs reflect different views about three key assumptions that drive the estimated costs of climate policy: stringency of the abatement policy, flexibility of policy instruments, and possibilities for development and diffusion of new technology. First, as one would expect, the greater the degree of CO_2 reduction required (because the target is ambitious, baseline emissions are high, or both), the greater the cost.

Costs of GHG control depend on the speed of control as well as its scale. Wigley and others (see

Table 1. Implications of a Carbon Tax for U.S. Gasoline and Coal Prices.

	Price ($)		
Commodity	1997 average	With $100/ton carbon tax	With $400/ton carbon tax
Bituminous coal	26.16	87.94	273.28
Motor gasoline	1.29	1.53	2.26

Note: Coal price is national average annual delivered price per ton to electric utilities; gasoline price is national average annual retail price per gallon.

Sources: U.S. DOE (see Suggested Reading).

Suggested Reading) showed that most long-term target GHG concentrations could be achieved at substantially lower present value costs if abatement were increased gradually over time, rather than rapidly, as envisaged under the Kyoto Protocol. Subsequent elaboration of this idea has shown that, in principle, cost savings well in excess of 50% could be achieved by using a cost-effective strategy for meeting a long-term concentration target versus an alternative path that mandates more aggressive early reductions (see the 1997 paper by Manne and Richels in Suggested Reading). These cost savings come about not only because costs that come later are discounted more but also because less existing capital becomes obsolete prematurely. There is an irreversibility problem associated with premature commitment to a form and scale of low-emissions capital, just as irreversibility is associated with climate change. The former irreversibility implies lower costs with a slower approach to mitigation.

Another important factor in assessing the costs of CO_2 control is the capacity and willingness of consumers and firms to substitute alternatives for existing high-carbon technologies. Substitution undertaken depends partly on the technological ease of substituting capital and technological inputs for energy inputs and partly on the cost of lower-carbon alternatives. Some engineering studies suggest that 20–25% of existing carbon emissions could be eliminated at low or negligible cost if people switched to new technologies such as compact fluorescent light bulbs, improved thermal insulation, efficient heating and cooling systems, and energy-efficient appliances. Economists counter that the choice of energy technology offers no free lunch (for further discussion, see Chapter 17, Energy-Efficient Technologies and Climate Change Policies). Even if new technologies are available, many people are unwilling to experiment with new devices at current prices. Factors other than energy efficiency also matter to consumers, such as quality, features, and the time and effort required to learn about a new technology and how it works. People behave as if their time horizons are short, perhaps reflecting their uncertainty about future energy prices and the reliability of the technology.

In addition, the unit cost of GHG control in the future may be lower than in the present, as a consequence of presumed continuation in trends toward greater energy efficiency in developed and developing countries (as well as some increased scarcity of fossil fuels). These trends will be enhanced by policies that provide economic incentives for GHG-reducing innovation. It is possible that the cost associated with premature commitment to irreversible long-lived investments in low-emissions technologies is more important in practice than climatic irreversibility, at least over the medium term. The reason is that sunk investments cannot be undone if climate change turns out to be less serious than might be expected, whereas society can accelerate GHG control if it learns that the danger is greater than estimated. The strength of this point depends in part on how irreversible low-GHG investment is and on the costs of irreversible climate change. In addition, critics of this view argue that without early action to reduce GHG emissions, markets for low-emission technologies would not develop and societies would lock in to continued use of fossil fuel–intensive energy systems.

Still another important factor is the flexibility and cost-effectiveness of the policy instruments imposed, both domestically and internationally. For example, Weyant and Hill's review showed that the flexibility to pursue CO_2 reductions anywhere in the Annex I countries (the industrialized countries that would cap their total emissions under the Kyoto Protocol) through some form of international emissions trading system could lower U.S. costs to meet the Kyoto Protocol target by roughly 30–50%. Less quantitative analysis has been done of alternative domestic policies. Nevertheless, it can be presumed from studies of the costs of abating other pollutants that cost-effective policies will lower the cost of GHG abatement, perhaps significantly. In contrast, constraints on the use of cost-effective policies—for example, the imposition of rigid technology mandates in lieu of more flexible performance standards—will raise costs, perhaps considerably. This factor often is neglected in analyses of domestic abatement activity that consider only the use of cost-effective policies such as emissions permit trading, although use of such policies is hardly foreordained. Ignoring this factor means that the costs reported in the economic models probably understate the costs societies will actually incur in GHG control. By the same token, studies of international policies that assume ideal conditions of implementation and compliance are overoptimistic.

A subtle but important influence on the cost of GHG control is whether emission-reducing policies also raise revenues (such as a carbon tax) and what is done with those revenues. When revenue generated by a carbon tax or other policy is used to reduce other taxes (a process commonly referred to as revenue recycling), some of the negative effect on incomes and labor force participation of the increased cost of energy is offset. However, it may be more effective at stimulating employment and economic activity in countries with chronically high unemployment than in the United States. The issue of revenue recycling applies also to policies that would reduce CO_2 through carbon permits or "caps." If CO_2 permits are auctioned, then the revenues can be recycled through cuts in existing taxes; freely offered CO_2 permits do not allow the possibility of revenue recycling. The difference in net social costs of GHG control in the two cases can be dramatic. Reducing CO_2 emissions with auctioned permits and revenue recycling can have net costs less than the benefits of GHG control indicated by the IA models (for further discussion, see Chapter 11, Revenue Recycling and the Costs of Reducing Carbon Emissions). In contrast, with a system of freely provided CO_2 permits,

any level of emissions reduction yields environmental benefits (according to the IA models) that fall short of society's costs of abatement.

Most cost analyses presume that the relevant energy and technology markets work reasonably efficiently (other than the commonly recognized failure of private markets to provide for all the basic R&D that society wants, because this is a kind of public good). This assumption is more or less reasonable for most developed industrial economies. Even in these countries, one can identify problems such as direct and indirect energy subsidies that encourage excessive GHG emissions. Problems of market inefficiency are far more commonplace in the developing countries and in countries in transition toward market systems; accordingly, one expects incremental CO_2 control costs to be lower (even negative) in those countries. However, the institutional barriers to accomplishing GHG control in these economic systems may negate the potential efficiency gains.

Thus far, our discussion had focused on CO_2 control. Because CO_2 is only one of several GHGs, and because CO_2 emissions can be sequestered or even eliminated by using certain technologies, emissions targets related to climate change can be met in several ways. Some recent analyses suggest that the costs of other options compare very favorably with the costs of CO_2 reduction. For example, counting the results of forest-based sequestration and the reduction of non-CO_2 gases toward total GHG reduction goals could lower the cost to the United States of meeting its Kyoto Protocol emissions target by roughly 60% (see Reilly and others in Suggested Reading). But care is needed in interpreting some of the cost estimates. In particular, low estimates for the cost of carbon sequestration may not adequately capture all the opportunity cost of different land uses (see Chapter 13, Carbon Sinks in the Post-Kyoto World).

Uncertainty, Learning, and the Value of New Information

Another key factor in choosing the timing and intensity of climate change mitigation is the opportunity to learn more about both the risks of climate change and the costs of mitigation. Several studies show that the value of more and better information about climate risks is substantial. This value arises because one would like to avoid putting lots of resources into mitigation in the short term, only to find out later that the problems related to climate change are not serious. However, one also would like to minimize the risk of doing too little mitigation in the short term, only to find out later that very serious consequences of climate change will cost much more to avert because of the delay.

Manne and Richels, as well as Kolstad, showed that it generally pays to do a little bit of abatement in the short run under these conditions—to hedge against the downside without making too rapid a commitment. One virtue of some delay in emissions control is that it allows us to learn more about the severity of the risk of climate change and the options for responding to it. If the risk turns out to be worse than expected, mitigation can be accelerated to make up for lost time. To be sure, the strength of this argument depends on how costly it is to accelerate mitigation and on the degree of irreversibility of climate change. Analysts will continue to debate these points for some time to come.

Concluding Remarks

In this chapter, we have explained that benefits and costs matter, for reasons of both efficiency and equity, and that benefits and costs must and can be considered in the context of the uncertainties that surround climate change. Economic analyses provide several rationales for pursuing only gradual abatement of GHG emissions. Because damages accrue gradually, catastrophes are uncertain and far off in the future, and unit mitigation costs are likely to fall over time (especially with well-designed climate policies), it makes sense to proceed slowly. To the extent that innovation is slower than desired with this approach, government programs targeted at basic R&D can help. The IA models indicate that rapid abatement does not maximize the present value of all society's resources.

We have not argued that current benefit–cost analyses are the last word on the subject. Opportu-

nities certainly exist to improve the measurement of benefits and costs and to track the incidence of costs and risks across groups and over time. In practice, policy decisions will turn on a broader set of considerations than a single expected benefit–cost ratio. However, the arguments in favor of purposeful but gradual reduction in GHGs seem strong.

Economic analysis also could be used to justify not only a slower approach to GHG mitigation but also a less stringent long-term target. Here is where the potential conflict can arise between individuals' narrower economic self-interests and their concern for the well-being of future generations. Determining the right long-term policy goals ultimately requires us to address our attitudes toward intergenerational equity as well as to better understand the scale of environmental and economic risks that different climate policies imply for future generations. A more gradual GHG policy over the next 10–20 years does not preclude any but the most environmentally stringent targets, while potentially increasing the political acceptability of increasingly demanding mitigation measures. These considerations warrant renewed attention as the international community continues to grapple with the problem of finding a climate policy it can really live with.

Suggested Reading

Azar, Christian, and Thomas Sterner. 1996. Discounting and Distributional Considerations in the Context of Global Warming. *Ecological Economics* 19: 169–84.

Burtraw, Dallas, Alan Krupnick, Karen Palmer, Anthony Paul, Michael Toman, and Cary Bloyd. 1999. Ancillary Benefits of Reduced Air Pollution in the U.S. from Moderate Greenhouse Gas Mitigation Policies in the Electricity Sector. RFF Discussion Paper 99-51. September. Washington, DC: Resources for the Future.

Darmstadter, Joel, and Michael A. Toman (eds.). 1993. *Assessing Surprises and Nonlinearities in Greenhouse Warming: Proceedings of an Interdisciplinary Workshop.* Washington, DC: Resources for the Future.

Frankhauser, Samuel, Richard S. J. Tol, and David W. Pearce. 1998. Extensions and Alternatives to Climate Change Impact Valuation: On the Critique of IPCC Working Group III's Impact Estimates. *Environment and Development Economics* 3(Part 1): 59–81.

Ha-Duong, M., Michael J. Grubb, and Jean-Charles Hourcade. 1997. Influence of Socioeconomic Inertia and Uncertainty on Optimal CO_2 Emission Abatement. *Nature* 390: 270–73.

Hahn, R. W., and R. N. Stavins. 1999. What Has Kyoto Wrought? The Real Architecture of International Tradeable Permit Markets. RFF Discussion Paper 99-30. March. Washington, DC: Resources for the Future.

Howarth, Richard H. 1996. Climate Change and Overlapping Generations. *Contemporary Economic Policy* 14: 100–111.

———. 1998. An Overlapping Generation Model of Climate–Economy Interactions. *Scandinavian Journal of Economics* 100(3): 575–91.

IPCC (Intergovernmental Panel on Climate Change). 1996. *Climate Change 1995: The Science of Climate Change.* Contribution of Working Group I to the Second Assessment Report of the Intergovernmental Panel on Climate Change. New York: Cambridge University Press.

———. 1996. *Climate Change 1995: Impacts, Adaptations, and Mitigation of Climate Change: Scientific-Technical Analysis.* Contribution of Working Group II to the Second Assessment Report of the Intergovernmental Panel on Climate Change. New York: Cambridge University Press.

———. 1996. *Climate Change 1995: Economic and Social Dimensions of Climate Change.* Contribution of Working Group III to the Second Assessment Report of the Intergovernmental Panel on Climate Change. New York: Cambridge University Press.

———. 1998. *The Regional Impacts of Climate Change: An Assessment of Vulnerability.* New York: Cambridge University Press.

Interlaboratory Working Group (IWG). 1997. *Scenarios of U.S. Carbon Reductions: Potential Impacts of Energy Technologies by 2010 and Beyond.* Report LBNL-40533 and ORNL-444. September. Berkeley, CA, and Oak Ridge, TN: Lawrence Berkeley National Laboratory and Oak Ridge National Laboratory.

Kolstad, Charles D. 1996. Learning and Stock Effects in Environmental Regulation: The Case of Greenhouse Gas Emissions. *Journal of Environmental Economics and Management* 31: 1–18.

Manne, Alan S. 1996. Hedging Strategies for Global Carbon Dioxide Abatement: A Summary of the Poll Results EMF 14 Subgroup—Analysis for Decisions under Uncertainty. In *Climate Change: Integrating Science, Economics, and Policy,* edited by Nebojsa Nakićenović and others. Laxenburg, Austria: International Institute for Applied Systems Analysis.

Manne, Alan S., and Richard Richels. 1992. *Buying Greenhouse Insurance: The Economic Costs of CO_2 Emission Limits.* Cambridge, MA: The MIT Press.

———. 1997. On Stabilizing CO_2 Concentrations—Cost-Effective Emission Reduction Strategies. *Environmental Modeling & Assessment* 2(4): 251–65.

Mendelsohn, Robert, and James E. Neumann (eds.). 1999. *The Impact of Climate Change on the United States Economy.* Cambridge, U.K.: Cambridge University Press.

Narain, Urvashi, and Anthony Fisher. 1999. Irreversibility, Uncertainty, and Catastrophic Global Warming. Gianni Foundation Working Paper 843. Berkeley, CA: University of California, Department of Agricultural and Resource Economics.

Nordhaus, William D. 1993. Rolling the "DICE": An Optimal Transition Path for Controlling Greenhouse Gases. *Resource and Energy Economics* 15(1): 27–50.

———. 1998. *Roll the DICE Again: The Economics of Global Warming.* December 18. New Haven, CT: Yale University.

Nordhaus, William D., and Zili Yang. 1996. A Regional Dynamic General-Equilibrium Model of Alternative Climate-Change Strategies. *American Economic Review* 86(4): 741–65.

Parry, Ian W. H., Roberton C. Williams III, and Lawrence H. Goulder. 1999. When Can Carbon Abatement Policies Increase Welfare? The Fundamental Role of Distorted Factor Markets. *Journal of Environmental Economics and Management* 37: 52–84.

Peck, Stephen C., and Thomas J. Teisberg. 1993. Global Warming Uncertainties and the Value of Information: An Analysis Using CETA. *Resource and Energy Economics* 15(1): 71–97.

———. 1999. The Optimal Choice of Climate Change Policy in the Presence of Uncertainty. *Resource and Energy Economics* 21(3–4): 255–87.

Portney, Paul R., and John P. Weyant (eds.). 1999. *Discounting and Intergenerational Equity.* Washington, DC: Resources for the Future.

Reilly, John M., Ronald Prinn, J. Harrisch, J. Fitzmaurice, H. Jacoby, D. Kicklighter, J. Melillo, P. Stone, A. Sokolov, and C. Wang. 1999. Multi-Gas Assessment of the Kyoto Protocol. *Nature* 401: 549–55.

Roughgarden, Tim, and Stephen H. Schneider. 1999. Climate Change Policy: Quantifying Uncertainties for Damages and Optimal Carbon Taxes. *Energy Policy* 27: 415–29.

Schelling, Thomas C. 1995. Intergenerational Discounting. *Energy Policy* 23: 395–401.

Shogren, Jason, and Michael Toman. 2000. Climate Change Policy. In *Public Policies for Environmental Protection*, 2nd ed., edited by Paul R. Portney and Robert N. Stavins. Washington, DC: Resources for the Future.

Smith, J. B., N. Bhatti, G. V. Menzhulin, R. Benioff, M. I. Budyko, M. Campos, B. Jallow, and F. Rijsberman (eds.). 1996. *Adapting to Climate Change: Assessments and Issues*. New York: Springer-Verlag.

Tol, Richard S. J. 1995. The Damage Costs of Climate Change toward More Comprehensive Calculations. *Environment and Resource Economics* 5: 353–74.

———. 1999. The Marginal Costs of Greenhouse Gas Emissions. *Energy Journal* 20(1): 61–81.

Toman, Michael A. 1998. Sustainable Decision-Making: The State of the Art from an Economics Perspective. In *Valuation and the Environment: Theory, Method and Practice*, edited by Martin O'Connor and Clive Spash. Northampton, MA: Edward Elgar, 59–72.

U.S. DOE/EIA (Department of Energy, Energy Information Administration). 1999. *Annual Energy Review 1998*. July. Washington, DC: U.S. Government Printing Office.

———. 1999. *Annual Energy Outlook 2000*. December. Washington, DC: U.S. Government Printing Office.

———. 1998. *Annual Energy Review 1997*. July. Washington, DC: U.S. Government Printing Office.

Weitzman, Martin L. 1999. Just Keep Discounting, but .... In *Discounting and Intergenerational Equity*, edited by Paul R. Portney and John P. Weyant. Washington, DC: Resources for the Future.

Weyant, John P., and Jennifer N. Hill. 1999. Introduction and Overview. *The Energy Journal*, Special Issue (The Costs of the Kyoto Protocol: A Multi-Model Evaluation): vi–xiiv.

Wigley, Thomas M. L., Richard Richels, and James A. Edmonds. 1996. Economic and Environmental Choices in the Stabilization of Atmospheric CO_2 Concentrations. *Nature* 379(6562): 240–43.

Appendix
The Costs of
the Kyoto Protocol

Sarah A. Cline

The commitments called for in the Kyoto Protocol would result in the reduction of greenhouse gas (GHG) emissions from Annex I countries to approximately 5% below their 1990 levels during 2008–12. The concept of emissions trading was incorporated into the protocol to help reduce the costs to individual countries, but the specific conditions of trading have yet to be established. To examine the possible results from various emissions trading scenarios, many researchers have developed models to assess the possible outcomes.

The wide range of results from these studies are attributable to the different model structures, baselines, and assumptions incorporated into the models. Traditionally, climate change models have been categorized as two types. Top-down models take a macroeconomic view to address the economic impact of a climate policy and analyze aggregate data, whereas bottom-up models focus on technology costs and performance data. Recently, many researchers have developed hybrid models that merge the characteristics of both models. A 1999 issue of *The Energy Journal* presents the results from several analyses conducted as part of the Stanford Energy Modeling Forum (see Suggested Reading). These studies present the economic and energy impacts of the Kyoto Protocol under different trading scenarios.

First, I discuss differences in model structure and assumptions. Next, I focus specifically on the cost studies presented in *The Energy Journal* and their re-

sults. Finally, I present an overview of the cost estimates obtained from the models and the factors that explain the differences among them.

Model Structure

Traditionally, climate change models have been classified as either bottom-up models or top-down, economy-wide models. Bottom-up models take a more disaggregated approach, specifically with respect to the energy sector; they contain a great deal of information about the engineering performance and engineering costs of using different energy fuels and technologies. Top-down, economy-wide models may be based on macroeconomic forecasting or numerical general equilibrium frameworks. The general equilibrium models often are less detailed than forecasting frameworks but seek to provide a characterization of the behavior of economic actors (firms and households) that is consistent with the basic principles of economic theory. Because of the recent development of hybrid models that merge both types, it is useful to determine other distinguishing model characteristics.

One distinction among models is the extent to which the energy sector is described. The detail of bottom-up models facilitates analysis of technological potential for energy efficiency improvement and interfuel substitution. Top-down models have begun to include more energy and technological detail as well. Some models contain fossil fuel consumption and production information as well as details from the electricity industry. Other models focus on carbon as an input into the economy with a more aggregated industry structure.

Another distinction between models is the extent to which nonenergy sectors of the economy are described. Top-down models generally have been more detailed in this area, allowing for a broader analysis of policy impacts on the economy. They include a range of sectors and industries and focus on the interactions of these firms within the economy.

Models may also differ according to the level that behavior is endogenized in the equations or given as exogenous assumptions. Endogenization of behavior has been common in top-down models, whereas bottom-up models have been more likely to use exogenous assumptions to describe behavior.

Some climate change models address the issue from a macroeconomic perspective. They may consider aspects of the economy such as international capital flows, unemployment, and financial effects that are not included in many other models.

Climate Change Models

The analyses presented in *The Energy Journal,* Kyoto Special Issue, were conducted through the Stanford Energy Modeling Forum. I present 13 models that vary in specifications, assumptions, and baseline projections. All the models discussed here are multinational in coverage. The four core scenarios for emissions trading analyzed by most of the models included no international trading, full Annex I trading, "double bubble" trading (two trading groups of countries trading only within their group: one consisting of Japan, the United States, Canada, New Zealand, and the Russian Federation and the other consisting of western and eastern Europe), and full global trading. Descriptions of the models and their results follow.

Multisector General Equilibrium Models with Limited Energy Detail

Multisector general equilibrium models as a class tend to focus on the interactions between firms in different industries. These models generally allow for

interaction between sectors and trade in nonenergy goods. They generally do not include a large amount of energy sector detail or macroeconomic factors such as unemployment and financial market effects.

MIT Emissions Prediction and Policy Assessment (MIT-EPPA) is a dynamic, recursive general equilibrium model that divides the world into twelve regions, each of which contains eight production sectors and four consumption sectors. Trade is allowed in energy and nonenergy producer goods between regions. The model is calibrated to 1985 benchmark data and then solved on a five-year time step. Population growth, autonomous energy improvement, increases in labor productivity, and the availability of natural resources are exogenous assumptions. Each of the regional economies is modeled as having two kinds of capital: capital in fixed proportions, and malleable capital, which indicates that its input proportions may change as a result of price changes. It is assumed that the emissions targets do not change throughout the analysis period. The model considers only fossil fuel CO_2.

The resulting U.S. carbon prices from the EPPA model are very susceptible to changes in capital malleability when capital proportions are more rigid. Carbon prices increase sharply in the initial years and drop later in the commitment period. Malleable capital shows a different trend, however, with a lower carbon price in the initial years and a more gradual increase throughout the analysis period to 2030. Emissions trading could lower commitment costs, and Europe and Japan would experience a larger gain than the United States because of a higher initial carbon price absent international carbon trading. An analysis was conducted using three different times of ratification of the protocol. The difficulty of reaching the Kyoto Protocol targets increased substantially as time passed. A later ratification date for the Kyoto Protocol led to higher carbon prices in all of the Annex I regions, because less time remained to adjust to the Kyoto Protocol constraints.

WorldScan is a general equilibrium model that divides the world into 12 regions and includes macroeconomic feedbacks through prices. The structure of the model incorporates investment in capital stocks every year, with a one-year planning horizon. Trade is allowed among sectoral commodities and is differentiated by region. This model also differentiates between existing capital and new capital. Only limited adjustments to reduce the costs of inputs can be made to existing capital, whereas new capital can be changed more readily.

The results of the WorldScan Annex I trading scenario show that all Organisation for Economic Cooperation and Development (OECD) countries would import permits, and the United States would import a smaller percentage of its target than the rest of the OECD countries. The carbon price in the Annex I trading scenario is lower than in the no trading case for the United States and the European Union. The club (or double bubble) trading scenario results in the European Union paying a higher permit price. The former Soviet Union loses in this scenario because it no longer is able to sell permits to the European Union, and the rest of the countries in the second club will gain because of a lower permit price than in the Annex I trading case. When restrictions are placed on the amount of permits that may be exported, demand for permits will decrease because of an increase in permit price. This scenario leads to losses for the European Union, the former Soviet Union, and eastern Europe. However, the United States will generally benefit from ceilings.

WorldScan results show that the Clean Development Mechanism (CDM) would increase global emissions compared with the no trading case. Because the countries hosting CDM projects do not have national emissions targets, there is a possibility of carbon leakage from Annex I countries to the host countries. There is also the possibility that because of the operation of regional energy markets, the host countries still will experience lower energy prices because of reduced Annex I energy demand. With these lower prices, energy demand in the host countries may increase, thus increasing overall emissions. Carbon prices under this scenario would be lower for all Annex I regions because of the flexibility in domestic targets.

Global General Equilibrium Growth (G-Cubed) is similar in structure to the other multisector general equilibrium models but also includes some short-term macroeconomic effects. It consists of eight re-gional models, each of which contains a household, government, and financial sector. The regions are linked by flows of goods and assets between regions. Production is represented by twelve industries and two producing sectors. Labor is modeled to be mobile between sectors within a region but not between regions. Involuntary unemployment is also accounted for in this model.

Annex I trading in G-Cubed results in lower marginal costs of mitigation than in the no trading case for all of OECD countries. The United States purchases permits to account for 29% of its abatement costs, which results in a 30% decrease in marginal costs of abatement. Permit prices increase after 2010 as a result of economic growth and a decrease in the supply of permits from the former Soviet Union. The double bubble trading scenario results in costs for the rest of the OECD countries that are similar to those in the no trading case, and the rest of the Annex I countries experience significantly lower marginal abatement costs. The United States benefits from this scenario as a result of decreased permit prices and capital flows from the rest of the OECD countries. The global trading scenario results in a decrease in carbon permit prices for all of the OECD regions. The effects on the Annex I economies are relatively small compared with the results of the other trading scenarios. The less developed countries suffer a loss in gross domestic product (GDP) in the global trading scenario compared with the other cases in which they did not participate in emissions reduction. This loss results from the lack of benefits, such as capital inflows, lower oil prices, and exchange rate appreciation that these countries received in the other trading scenarios.

Macroeconomic Models

Oxford's Global Macroeconomic and Energy is the only true macroeconomic forecasting model of the 13 models in *The Energy Journal* Special Issue. It includes macroeconomic variables such as inflation, GDP, employment, unemployment, exchange rates, and the trade balance and considers near-term reactions to economic shocks. The Oxford model describes energy demand in detail for 22 regions and incorporates only major macroeconomic variables for another 50 regions. The energy sector is represented by six fuel types, and energy demand is disaggregated into four sectors for the major industrialized economies. Some additional parameters considered in the model that may affect abatement costs include the annual emissions growth projected between 1990 and 2010, population of the region, coal demand as a percentage of the primary energy use, and initial emissions levels.

Results of the Oxford model indicate that with no international trading, Canada, France, and Italy could experience the highest carbon taxes or permit prices to achieve reductions outlined in the Kyoto Protocol by 2010. In the no trading scenario, the United States experiences a relatively low carbon price compared with most of the other countries because of the high possibility of carbon substitution for the present dependence on coal. In the Annex I trading scenario, the permit price is lower than in the no trading case for all countries except Russia. This case results in Russia being the only permit seller. The United States purchases the largest amount of permits, whereas Canada purchases the largest amount compared with the size of its GDP. The projected output is higher under the permit trading scenario than in the no trading case for all countries except Russia and China. This higher total world output indicates an overall world welfare gain because of permit trading. The double bubble scenario results in a lower permit price for non-European Union countries than the Annex I trading case does. The carbon prices for the European Union in this case are similar to those in the no trading scenario. The decrease in the U.S. GDP is less in the double bubble case than in the Annex I trading scenario.

Detailed Energy Sector Models

Some models describe consumption and supply in the energy sector in substantial detail. These models may consider the electricity industry, energy prices, and future technology as well as both supply and demand of fossil fuel and renewable energy use. Generally, nonenergy industries are aggregated into a fairly simple description of the nonenergy part of the economy, and industry interaction and unemployment are excluded.

MERGE3 is an intertemporal market equilibrium model that focuses on the energy sector. The model considers nine geographic regions and uses 10-year time intervals through 2050 and 25-year time intervals through 2100. International and intertemporal trade in carbon emission rights and international trade in oil, gas, and energy-intensive basic materials are included. Technical diffusion and price and nonprice-induced conservation are also incorporated into the model.

MERGE3 presents results from three scenarios: no trading, Annex I trading plus CDM, and global trading. Of the three scenarios, no trading results in the highest cost of emissions rights for the United States. The cost of emissions rights for the United States was lower under the scenario of Annex I trading plus CDM than with the no trading case and lowest with global trading. The permit price is shown to increase in 2020 because the economies of eastern Europe and the former Soviet Union are expected to have grown so much by that time that they will no longer have excess permits to sell. Limits in the amount of permits that a region is allowed to buy or sell increase the price of emissions rights as well as GDP losses for the United States.

Carbon Emissions Trajectory Assessment (CETA) considers different ways to allocate permits between the Annex I countries and non-Annex I countries. CETA includes economic growth, global warming, energy consumption, energy technology, and global warming costs. Damages due to global warming are disaggregated into market damages and nonmarket damages. International trade is allowed in carbon emission permits, aggregate output, oil, gas, and synthetic fuels. Exogenous labor input, endogenous capital stock, and energy use are considered in regional output. CO_2 emissions are determined by the type and amount of energy inputs used in each region.

Through the global warming damage function, the CETA model determines a "bargaining range" for which both of the regions would prefer an international trading agreement. The bargaining range in the model is described as the percentage of the optimal emissions path for each region over time. The bargaining range is determined by allocating permits to non-Annex I countries equal to between 70 and

115% of non-Annex I countries' optimal emissions. The bargaining range was recalculated for 2040 and 2090, and both years resulted in a broader bargaining range. An international welfare analysis of the Kyoto Protocol found that the protocol results in an inefficient welfare distribution between the two regions. In non-Annex I countries, welfare is improved with the Kyoto Protocol over no emissions control, and the largest improvement results when they are allowed to sell permits to Annex I countries. Annex I countries experience a welfare loss in the Kyoto case compared with no control. Permit trading between Annex I and non-Annex I countries results in welfare effects very similar to those of no permit trading for Annex I countries, probably because of oil price effects. Annex I/non-Annex I trading results in an increase of global oil prices compared with the no trading scenario. Because Annex I countries import a large amount of oil, the gains they experience from emissions trading are offset to a large extent by increases in oil expenditures.

Global Relationship Assessment to Protect the Environment (GRAPE) is a nonlinear dynamic intertemporal optimization model. It consists of five modules: energy, climate, land use, macroeconomics, and environmental impacts. The base year is 1990, and the analysis runs through 2100.

Results of GRAPE show that with business as usual, the percentage of coal used in the energy sector is expected to increase. Oil use is expected to decrease in the future, whereas natural gas, biomass, and methanol from coal production are expected to increase to some extent. In the Annex I trading case, the percentage of nuclear, natural gas, wind, and photovoltaics is larger than with no trading. Under global trading, nuclear power expands in non-Annex I countries but decreases in Annex I regions. In the Annex I trading scenario, all regions import permits except eastern Europe and the former Soviet Union. The global trading case shows that all Annex I regions are importers after 2030. Global trading results in a lower permit price than in the Annex I trading case.

Multisector Models with a Detailed Energy Sector

Some models include multiple sectors and regions as well as a relatively detailed description of the energy sector. These models generally model trade in nonenergy goods.

Asian Pacific Integrated Model (AIM) is a recursive dynamic equilibrium model that divides the world into 21 regions. It includes seven energy goods and four nonenergy goods. AIM models a production, household, and government sector in each of the geographic regions. Trade in intermediate goods is included. Output is modeled using primary factors, intermediate goods, and energy.

Six post-Kyoto scenarios were addressed using AIM. The results with no trading show that marginal costs in 2010 were the greatest for New Zealand followed by Japan, the European Union, and the United States. Even though it experiences a low carbon price, the United States suffers the greatest loss in GDP because of its carbon-intensive economy. The global trading scenario reduces marginal costs compared with the other models. Double bubble trading results in a decrease in GDP for the European Union compared with no trading because of its inability to purchase low-cost permits from eastern Europe and the former Soviet Union under this scenario.

The *Global Trade and Environment Model (GTEM)* is a dynamic general equilibrium model that divides the world into 18 regions. Production is divided into sixteen goods, three of which—coal, gas, and petroleum and coal products—are responsible for producing CO_2. The three primary factors of production included in the model are capital, land, and labor. GTEM also models trade flows between regions. Emissions reductions are assumed to occur gradually until reaching the region's Kyoto target in 2010. In the no international trading case, a per-unit CO_2 tax is imposed when emissions restrictions are in place. Any revenue from this tax is returned to the economy in a lump-sum payment, resulting in no effect on the economy. In the international trading scenario, a homogeneous tax or carbon emissions penalty is assumed for each region to meet the emissions target. Each country's emissions reduction commitment then can be interpreted as emissions permits that can be traded among other regions. The model assumes that any income received from the sale of carbon permits is added to the country's GDP.

Under the no international trading scenario, GTEM results show that Canada has the highest carbon emissions penalty because of its high emissions reduction commitment. In the Annex I trading case, the penalties are reduced compared with no trading for all regions except the former Soviet Union and eastern Europe (both of which had low penalties under the no trading scenario). The former Soviet Union and eastern Europe experience increased emissions and an income transfer from other countries because of the sale of emissions permits. In the double bubble scenario, the European Union experiences a higher penalty than with Annex I trading as a result of the loss of access to former Soviet Union emission permits. All countries experience gains from trading that are absent with no international trading. The economic impacts of emissions reduction are only partly correlated with the magnitude of the carbon emission penalty. The carbon intensity of the country's economy also is a major factor. Non-Annex I regions benefit from emissions reduction due to carbon leakage. Developing countries can produce carbon-intensive goods at lower costs than Annex I countries, thus leading to carbon leakage. A comparison of economic costs (measured as change in GNP) with and without carbon leakage relative to the reference case indicates that the impacts of eliminating carbon leakage on the GNP of Annex I countries would be negligible.

The *Second Generation Model* (SGM) is a computable general equilibrium model that projects economic activity, energy consumption, and carbon emissions. It divides the world into twelve geographic regions and models production using nine producing sectors and twelve production inputs. In addition, it describes the electricity generation sector in great detail using at least five subsectors, each with a separate production function. To represent the basket of six gases included in the Kyoto Protocol, the SGM uses both non-CO_2 GHG trajectories and CO_2 emissions trajectories.

In addition to the trading approaches studied in most models, the SGM examines two other scenarios: competitive permit supply, in which there are multiple permit holders and no one seller can affect the permit price by withholding permits from the market; and monopolistic permit supply, in which only a few permit sellers are able to affect permit prices. With a competitive permit supply, Annex I trading results in lower permit prices compared with no trading for all regions except the former Soviet Union and eastern Europe. In addition, the change in energy consumption of permit purchasers from 1990 to 2010 is larger with Annex I trading compared with the no trading scenario. A monopolistic permit supply results in an increased permit price compared with the competitive permit price. However, the Annex I trading permit price is still lower under this scenario than in the no trading case. Growth in energy consumption is restricted under the monopolistic supply case compared with the competitive supply case. Under the double bubble trading case, the permit price is about the same as in the no trading case for the European Union but lower than with Annex I competitive trading for other regions. A full global trading scenario results in lower permit prices for all buyers of permits compared with no trading and Annex I trading under both monopolistic and competitive supply scenarios. The bulk of the additional permits available under global trading are supplied by China.

The model also is used to analyze situations in which the cost of mitigating non-CO_2 GHGs is not proportional to CO_2 mitigation costs. In the limiting case of infinite cost for non-CO_2 gases, the Annex I competitive trading permit price increases over the proportional cost scenario. The application of land-use change emissions credits increases the base emissions credits for the former Soviet Union and eastern Europe and decreases the permit price significantly for Canada under no trading. The Annex I trading permit price is also projected to decrease when land-use change emissions credits are allowed. It is expected that this program would have limited long-term effects as forests reach steady-state carbon-to-land ratios.

Multisector Multiregion Trade (MS-MRT) is a dynamic general equilibrium model that disaggregates industry to consider different energy intensities. It divides the world into ten regions and includes six industries, four energy forms, and two nonenergy forms. The model uses an intertemporal budget constraint to limit the long-term trade deficit.

Results from MS-MRT indicate that limits on emissions without international trading would have a negative impact on the welfare of industrialized and oil-producing countries. Economic well-being improves under Annex I trading for all regions except China and India. Compared with the no trading scenario, all regions were better off under global trading. All regions except eastern Europe and the former Soviet Union also were better off with the global trading case than the Annex I trading scenario. Permit price decreases under global trading for Annex I countries compared with no trading and Annex I trading, whereas permit and energy prices increase for non-Annex I countries. The negative effect of emissions trading on developing countries results from increased costs in developed countries leading to increased export prices. In addition, the income and import demand of developed countries decrease, thus causing import prices to fall. Generally, the results show that limits on emissions trading reduce global economic welfare; however, individual countries may gain or lose depending on the restrictions. The CDM has little impact on the economic welfare of most countries compared with Annex I trading because of permit trading restrictions. Under the CDM scenario, most countries experience economic welfare levels close to those in the Annex I trading case.

Economic Growth/Carbon Input Models

Some models have a simple representation of economic activity that includes "carbon" as a main input. These models usually have a single aggregate abatement cost function for each region based on aggregating all industries. They generally omit interactions between industries and include trade only in the fossil energy composite (carbon) and carbon emission rights.

RICE-98 is an integrated assessment model that divides the world into 13 regions. It contains two components: one that models the economy, and a separate climate component. Energy is included in the model as a production input called carbon

energy. Emissions included in the model consist of only industrial CO_2.

Carbon price estimates obtained by RICE-98 are lowest for the optimal scenario (a case that sets emission levels to minimize the net benefits of emissions reductions) and the global trading scenario. Restrictions on trading result in increased carbon prices. Annex I trading results in a higher price compared with global trading. The projected price with trading allowed only among OECD countries is lower than the Annex I trading case but higher than the no trading scenario. The results also show a large loss in national income in Annex I countries for all of the Kyoto scenarios relative to no action. Russia, eastern Europe, and non-Annex I countries receive the greatest benefits from the protocol. The United States is predicted to experience the greatest increase in production costs. Trading is expected to reduce overall costs of abatement, but the United States again bears much of the cost because of the purchase of emissions permits from Russia, China, and other less developed countries.

The *Climate Framework for Uncertainty, Negotiation, and Distribution (FUND)* is an integrated assessment model that divides the world into nine regions and runs from 1950 to 2200 in time steps of one year. Endogenous factors included in the model are methane and nitrous oxide, global mean temperature, atmospheric CO_2 concentrations, and the impacts of CO_2 reductions and climate change on the economy and population.

Although FUND does not model emissions trading per se, it does show that flexibility in the location of emissions reductions through the cooperation of certain countries to reach emissions reduction targets greatly decreases the costs of GHG reductions. The case in which each country works independently to reach its emissions reduction commitment exhibits the highest cost, followed by the double bubble scenario, Annex I only, and then Annex I and Asia; world cooperation entails the lowest cost. Restrictions in the level of cooperation also were examined. Cooperation among all Annex I parties has the lowest cost, whereas the cases that limit the amount of emissions reductions that can be "bought" or "sold" produce higher costs.

Cost Estimates

A comparison of cost estimate results and reference emissions for each of the models is presented in an overview by Weyant and Hill (see Suggested Reading). The model results across each of the four trading scenarios (no trading, Annex I trading, double bubble trading, and global trading) are presented for four OECD regions: the United States; the European Union; Japan; and Canada, Australia, and New Zealand.

One issue that arises when comparing results across different models is the ambiguity surrounding how "cost" is defined. Some measures commonly used to measure costs in these models include GDP, discounted present value of consumption, equivalent variation, and direct cost. Direct cost includes only net compliance outlays and thus understates total economic cost by not considering other negative impacts. Equivalent variation and discounted present value of consumption are better measures of overall economic cost. However, to account for the finite time horizons used in the models, one must make an adjustment to these measures that represents the change in the value of the terminal capital stock. The assumptions included in the valuation of the terminal capital stock result in an approximate calculation of welfare into the infinite future, beyond the model's terminal date. GDP is a problematic cost measure because it includes both measures of consumption and investment in the calculation. In comparing these different cost measures, one measure may project a gain for a country under a specific scenario, whereas another measure may project a loss for the same situation. The rankings of different policy scenarios also differ depending on the cost measure used.

The models presented in *The Energy Journal* Special Issue tend to show the same trend in carbon prices across the four trading scenarios. Although the magnitude of the price varies greatly from model to model, the results for the United States; Japan; and Canada, Australia, and New Zealand generally show the highest price for no trading and much lower prices for the other three scenarios. The price for Annex I trading has the largest tax of the remaining three scenarios, followed by double bubble trading and global trading, which has the lowest tax. The case for the European Union differs from the other countries in that in most cases, the price for the double bubble trading scenario is approximately equal to the price under no trading because of the absence of permits for purchase from the former Soviet Union in the E.U. bubble. Without the ability to purchase permits from the former Soviet Union, the European Union will need to meet a larger portion of its emissions target through domestic actions.

The resulting loss in GDP under the different trading scenarios also is similar across models. Again for the United States; Japan; and Canada, Australia, and New Zealand, no trading results in the greatest GDP loss, followed by Annex I trading, double bubble trading, and global trading, in that order. For the double bubble trading scenario, the European Union exhibits a GDP loss that is generally equal to or greater than the loss experienced under no trading. Annex I trading reduces the GDP loss for the United States by 21–65% relative to no trading, and global trading reduces the GDP loss for the United States by 53–90% relative to no trading. Half of the Annex I trading results for cost savings are in the 50–60% range. The pattern is similar for other non-European Union countries, whereas in the European Union, the results tend to show a greater benefit from Annex I trading; most results reduce GDP loss compared with no trading by more than 50%. The range of GDP loss reduction under the global trading scenario for the European Union is similar to that for the United States. In most cases, for the European Union, the double bubble trading scenario tends to increase or be approximately equal to the GDP loss compared with no trading.

The range of carbon leakage estimates varies significantly among the different models, most likely due to differences in import substitution elasticities. The amount of carbon leakage is projected to decrease under the Annex I trading scenario in many models, but the magnitude of the decrease differs among models. WorldScan also projects carbon leakage to occur under the CDM because of lower energy prices in local markets even when global energy use increases.

The slopes of the marginal cost curves for CO_2 reductions in the different models indicate the implications of more stringent emissions targets. These slopes depend on the reference case emissions level, the various substitution and demand elasticities used in the model, and the rate of energy demand adjustments. The Oxford macroeconomic model tends to have one of the steeper marginal cost curves for most of the regions. Other models that tend to have relatively steep marginal cost curves include GTEM and MS-MRT, two multisector models with a detailed energy sector; MERGE3, which is modeled with a detailed energy sector; and MIT-EPPA, a multisector general equilibrium model.

As one would expect, higher carbon emissions in the reference case increase the estimated costs of implementing the Kyoto Protocol or other emissions targets. The reference case emissions are influenced by assumptions about external factors such as population, economic output, energy used, and carbon emitted per unit of energy. Including macroeconomic costs of unemployment and financial markets in the model tends to induce higher costs as well; Weyant and Hill (see Suggested Reading) show the Oxford model as exhibiting the highest carbon price in most scenarios. Additional cost influences internal to the models include the rates at which the models allow for changes in energy demand and energy inputs, and the substitution and demand elasticities. Lower costs generally result from models with high elasticities and rapid adjustment dynamics, whereas high long-run elasticities and slow adjustment or low long-run elasticities and rapid adjustment dynamics generally lead to high costs. All of these factors interact to affect the value of the costs estimated in a given model. As a consequence, there is no simple relationship between the general modeling approach used and the cost estimates that result from that model. Any particular modeling approach can generate higher or lower costs depending on other assumptions going into the analysis.

Suggested Reading

Bernstein, Paul, W.D. Montgomery, Thomas Rutherford, and G. Yang. 1999. Effects of Restrictions on International Permit Trading: The MS-MRT Model. *The Energy Journal*, Special Issue (The Costs of the Kyoto Protocol: A Multi-Model Evaluation): 221–56.

Bollen, Johannes, Arjen Gielen, and Hans Timmer. 1999. Clubs, Ceilings, and CDM. *The Energy Journal*, Special Issue (The Costs of the Kyoto Protocol: A Multi-Model Evaluation): 177–206.

Cooper, Adrian, S. Livermore, V. Rossi, J. Walker, and A. Wilson. 1999. Economic Implications of Reducing Carbon Emissions: The Oxford Model. *The Energy Journal*, Special Issue (The Costs of the Kyoto Protocol: A Multi-Model Evaluation): 335–66.

Jacoby, Henry, and Ian S. Wing. 1999. Adjustment Time, Capital Malleability and Policy Cost. *The Energy Journal*, Special Issue (The Costs of the Kyoto Protocol: A Multi-Model Evaluation): 73–92.

Kainuma, Mikiko, Y. Matsuoka, and T. Morita. 1999. Analysis of Post-Kyoto Scenarios: the Asian Pacific Integrated Model. *The Energy Journal*, Special Issue (The Costs of the Kyoto Protocol: A Multi-Model Evaluation): 207–20.

Kurosawa, A., H. Yagita, Z. Weisheng, K. Tokimatsu, and Y. Yanagisawa. 1999. *The Energy Journal*, Special Issue (The Costs of the Kyoto Protocol: A Multi-Model Evaluation): 157–76.

MacCracken, Chris, J. Edmonds, S. Kim, and R. Sands. 1999. Economics of the Kyoto Protocol. *The Energy Journal*, Special Issue (The Costs of the Kyoto Protocol: A Multi-Model Evaluation): 25–72.

Manne, Alan, and Richard Richels. 1999. The Kyoto Protocol: A Cost-Effective Strategy for Meeting Environmental Objectives? *The Energy Journal*, Special Issue (The Costs of the Kyoto Protocol: A Multi-Model Evaluation): 1–23.

McKibbin, Warwick, M. Ross, R. Shackleton, and P. Wilcoxen. 1999. Emissions Trading, Capital Flows and the Kyoto Protocol. *The Energy Journal*, Special Issue (The Costs of the Kyoto Protocol: A Multi-Model Evaluation): 287–334.

Nordhaus, William, and Joseph Boyer. 1999. Requiem for Kyoto: An Economic Analysis. *The Energy Journal*, Special Issue (The Costs of the Kyoto Protocol: A Multi-Model Evaluation): 93–130.

Peck, Stephen, and Thomas Teisberg. 1999. CO_2 Emissions Control Agreements: Incentives for Regional Participation. *The Energy Journal*, Special Issue (The Costs of the Kyoto Protocol: A Multi-Model Evaluation): 367–90.

Tol, Richard S. J. 1999. Kyoto, Efficiency, and Cost-Effectiveness: Applications of FUND. *The Energy Journal*,

Special Issue (The Costs of the Kyoto Protocol: A Multi-Model Evaluation): 131–56.

Tulpulé, Vivek, S. Brown, J. Lim, C. Polidano, H. Pant, and B. Fisher. 1999. The Kyoto Protocol: An Economic Analysis Using GTEM. *The Energy Journal*, Special Issue (The Costs of the Kyoto Protocol: A Multi-Model Evaluation): 257–86.

Weyant, John P., and Jennifer Hill. 1999. Introduction and Overview. *The Energy Journal*, Special Issue (The Costs of the Kyoto Protocol: A Multi-Model Evaluation): vii–xiiv.

Part 2

Impacts of Greenhouse Gas Emissions

5 Agriculture and Climate Change

Pierre Crosson

Many scientists now agree that continued accumulation of heat-trapping greenhouse gases in the atmosphere will eventually lead to changes in regional climates as well as the global climate. The agreement is expressed in the 1996 report of the Intergovernmental Panel on Climate Change (IPCC; see Suggested Reading), an international body of leading natural and social scientists sponsored by the U.N. Environment Programme and the World Meteorological Organization. According to the panel's report, an increase in atmospheric concentrations of greenhouse gases equivalent to a doubling of atmospheric carbon dioxide ($2\times CO_2$) will force a rise in global average surface temperature of 1.0–3.5 °C by 2100. Average precipitation also will rise as much 10–15% because a warmer atmosphere holds more water.

The general circulation models (GCMs) that the IPCC used to analyze climate change are in reasonably good agreement that with $2\times CO_2$ the global average temperature will rise within the range of 1.0–3.5 °C, as indicated above. The models also agree reasonably well that the northern latitudes will warm more than the tropics. With respect to all other regional changes, however, agreement among the models is poor. Because human activities and ecological systems are highly variable among regions, this lack of agreement greatly complicates the task of estimating the impacts of the changes on activities of interest to humans.

Despite this limitation, much useful work has been done on estimating the potential impacts of different climate change scenarios. In this chapter, I examine the potential impacts of climate change on agriculture on a global scale in general and in the United States in particular. Even if the reader's interest lies only in the impact on the United States, the global scale must be considered. U.S. agriculture is inextricably entwined with agriculture worldwide. What might happen nationally cannot be understood without taking into account impacts elsewhere in the world.

Global Impacts

The IPCC report estimates climate change impacts on grain production at the global level and then focuses on the estimated effect on the developed countries (DCs) of North America and Europe as well as on the less-developed countries (LDCs) of Asia, Africa, and Latin America. (Grain is often used as a proxy for all food because it accounts for more than one-half of all food calories consumed in the world.) The sources of the IPCC estimates are the three different GCMs, which reflect four different scenarios for estimating the impact of climate change on grain production:

- **Scenario 1** disregards any adjustments that farmers might make to offset the impacts of climate change on grain production and disregards the effects on production of an atmosphere richer in CO_2. (CO_2 is essential to plant growth, and much experimental work shows that higher concentrations of it in the atmosphere in fact stimulate such growth).

- **Scenario 2** incorporates the CO_2 enriching effect on growth.
- **Scenario 3** includes both the CO_2-enriching effect and the effect of modest adjustments that farmers could make using currently known practices (for example, shifting to a different variety of the same crop and changing the planting date by less than one month in response to a change in the length of the growing season).
- **Scenario 4** includes the CO_2 effect on growth, the modest adjustments to farming just mentioned, as well as more ambitious adjustments, such as shifting to an entirely different crop, changing the planting date by more than one month, and using more irrigation.

Note: The farming adjustments considered in the IPCC scenarios apparently did not include developing entirely new crop varieties designed to be more productive under changed climate conditions. However, research done on the impacts of climate change in the U.S. Midwest indicates that such new technologies could potentially offset much of the negative effects of climate change on crop production. And because the climate change contemplated is not expected to be fully realized until sometime in the

second half of the next century, plenty of time remains for researchers to develop the new technologies needed to make this most advanced type of adjustment.

The IPCC analyses of these four scenarios are summarized in Table 1. The range in each entry reflects differences in the results obtained with the various climate models. Notably, the CO_2 fertilization effect substantially reduces yield losses and may even lead to net increases in grain output in DCs as a whole. Smaller but significant offsets are obtained by allowing for adaptive behavior by farmers. However, notwithstanding these adjustments and offsets, climate change is indicated by the IPCC report to reduce grain yields in developing nations, underscoring the greater vulnerability of these countries.

The sharp difference in impact that climate change is expected to have on grain production in developed as opposed to LDCs has two main causes. The first one might be called the *physical* factor. As mentioned earlier, the GCMs estimate that the high latitudes will warm more than the tropics. Most of the DCs are in the northern latitudes, and their agriculture would benefit from the longer growing seasons that a warmer climate would bring. Most LDCs, on the other hand, include much terrain in the tropics where the negative effects of a warmer climate would not be offset by other favorable trends.

The second reason might be called the *ecostructural* factor. The IPCC notes that the DCs have more economic resources that can be devoted to helping farmers adjust to climate change than LDCs do. In

Table 1. Estimated Changes in Grain Production as a Result of Climate Change.

Scenario	Change (%)		
	World	Developed countries	Developing countries[a]
1. No offsetting effects considered	−11 to −20	−4 to −24	−14 to −16
2. CO_2 fertilization effect	−1 to −8	−4 to +11	−9 to −11
3. CO_2 fertilization and modest farmer adaptation	0 to −5	+2 to +11	−9 to −13
4. CO_2 fertilization and ambitious farmer adaptation	−2 to +1	+4 to +14	−6 to −7

[a] Asia, Africa, and Latin America.

Source: IPCC (see Suggested Reading).

addition, the institutional structures of the DCs appear to be more efficient in mobilizing the resources needed to pursue specific social objectives—whether they be adjustments to climate change or anything else—than those in the LDCs.

If the GCMs are right in predicting generally beneficial climate change in the northern latitudes, then the physical factor that accounts for the difference in impacts on the DCs and the LDCs would seem to be pretty much fixed. But the effect of the ecostructural factor may be more malleable. In eastern and Southeast Asia, and to a lesser extent in southern Asia, agricultural performance over the past 10–15 years has been impressive. Farmers have adopted new, more productive technologies as they have become available, and production (both per person and per hectare) has increased. This strong agricultural performance has been part of a generally impressive economic performance in the countries of those regions.

It is not clear why some Asian countries have been so much more successful than countries in Latin America, and especially in Africa. Their success does suggest, however, that the ecostructural weaknesses so common now among the LDCs are not fixed for all time. The Asian experience offers some promise that, given time and incentive to improve their material standard, farmers in other LDCs can and will seize the opportunities presented. This prospect provides some reason to hope that by the time that climate change begins to impinge negatively on LDCs, these countries will have developed a capacity to adjust to it well beyond what they could accomplish under present conditions. If so, the differences between the DCs and LDCs in terms of the effects of climate change on grain production could be much lower than the IPCC report suggests.

Several studies generally support the findings of the 1996 IPCC report about the global impacts of climate change on agriculture. A 1995 study from the U.S. Department of Agriculture (USDA) (see Darwin and others in Suggested Reading), for example, indicates that the overall impacts would be small, taking into account adjustments in agriculture and other sectors of the economy made possible by wide trading opportunities among countries. Specifi-

cally, the study showed that, given these trading opportunities, gross world economic product in the face of climate change would be 0.2% less—or 0.1% more—than it would be in the absence of climate change (see Darwin and others in Suggested Reading). Allowing for trading opportunities and farm-level adjustments, including the ability of farmers to move land into and out of production depending on the economic effects of climate change, the study found that world cereal production would increase 0.2–1.2%. *These results did not include the positive production effects of CO_2 enrichment.*

Rosenzweig and Parry (see Suggested Reading) also estimated changes in cereal prices resulting from climate-induced changes in production. The direction of change is consistent with well-established knowledge about price–production relationships in agriculture: Prices are what economists call *inelastic*; that is, a given percent change in production is associated with a significantly greater "opposite direction" percent change in price. A given percent decline in production because of climate change would result in a greater percent increase in prices, and conversely, a given percent increase in production because of climate change would result in a greater percent decrease in prices.

U.S. Impacts

Darwin and others (1995) found that in the United States, both crop and animal output would fall under the type of long-term climate change likely to occur in response to 2×CO_2. The model used in the study took into account climate change impacts elsewhere in the world and the consequent changes in U.S. trading opportunities in agricultural commodities. The crop production declines in the United States would be small, from 0.8% to 3.4%. Livestock production would fall between 0.5% and 1.3%. *These estimates do not reflect the positive crop production effects of CO_2 enrichment.*

These estimates reflect farm-level adjustments that farmers would find economical among the practices available in 1990, including shifting land into and out of production. However, they do not include all adjustments likely to be available by the time climate change impacts might become significant, say, around 2050. Additional adjustment possibilities almost surely will include those that agricultural research establishments would develop in response to emerging evidence that the climate is changing. With these new adjustment possibilities available to farmers, the impacts of climate change on U.S. crop and animal output would likely be less harmful or even positive. Such an outcome was indicated by Mendelsohn and others (see Suggested Reading) and by Resources for the Future (RFF) researchers who focused on four states: Missouri, Iowa, Nebraska, and Kansas (known as the MINK study). The RFF study (see Easterling and others in Suggested Reading) is of particular interest in that the effects of climate change were calibrated to actual experience during the drought years of the 1930s, rather than to the results of GCMs.

Policy Issues and Caveats

In thinking about policy issues related to prospects for global agricultural development, especially in the LDCs (where, studies show, more than 90% of the increase in global demands for food will occur during the next 30 or 40 years), the prospective impact of climate change is at most of secondary importance. Studies done to date show that the impact on the already struggling LDCs is likely to be negative, but not disastrous (the IPCC reports that production would be down only 6–8% after accounting for on-farm adjustments). Moreover, by the time climate change impacts become significant in the middle of the next century, LDCs should be in much better shape to deal with the impacts than they are now. This will be especially the case if the world trading system in agricultural commodities remains as robust as it presently is and if the global impacts of climate change on agriculture are small (or positive). I return to these caveats later.

In addition, the amount of time before climate change impacts occur is expected to be long relative to the time needed to develop technological and managerial responses. Many of the farm-level responses incorporated into the impact models described earlier are already known to farmers and suppliers of farm inputs. They could be adopted in one or two years. To develop entirely new technologies and practices better adapted to the changed climate, the elapsed time from beginning of research to the availability of results to farmers would be some 10–20 years. Thus, if significant impacts on agriculture are not likely to be felt for another 30 or 40 years, there is time to develop technological responses, *if investments in agricultural research do not lag*. Only the development of large surface irrigation projects involves a time span comparable to that expected before the impacts of climate change on agriculture are felt. And most irrigation systems developed over the past decade or so operate by pumping groundwater. These systems require much less time to develop than surface systems do.

Quite apart from the relatively long-term issue of climate change impact, many LDCs—especially in Africa but also to some extent in Latin America and parts of Asia—face immediate problems that are severe. These problems inhibit achievement of sustainable agricultural systems that can meet rising demands for food and other agricultural commodities at socially acceptable economic and environmental costs into the indefinite future. Natural resource degradation is serious in some parts of those countries, but recent studies indicate that, in general, and contrary to a widely held view, degradation of land and water resources is not a major threat to agricultural sustainability in the LDCs. The critical issue, rather, is whether in the immediate future and over the next several decades these countries can develop the capacity to increasingly expand the knowledge base needed to achieve sustainable agricultural systems.

The needed knowledge is embodied in people, technology, and institutions. During the past 30 years, food output per person has increased 15–20% in the LDCs as a whole (but not in Africa). Farmers are better educated and trained; new technologies—those embodied in the Green Revolution are the outstanding examples—have been developed and widely adopted by farmers, and institutional performance has improved as people have become more aware of the importance of markets and secure property rights in providing farmers the incentives they need to adopt new technology.

Now, however, evidence suggests that the systems that generated the powerful increases in the three kinds of knowledge are in jeopardy. In Africa, for example, where supplies of the three kinds of knowledge are in particularly short supply, investments in rural education, after advancing smartly in the 1960s and 1970s, declined sharply in the 1980s and have not yet shown much evidence of a turnaround. Spending on agricultural research in that region also has declined in recent years, precisely when it should have been increasing robustly if Africa is to achieve sustainable agricultural systems. According to studies done at the International Food Policy Research Institute (see Alston and others in Suggested Reading), agricultural research spending elsewhere in the developing world also is either declining in absolute amounts or the rates of increase in such spending are down sharply.

These threats to the knowledge base needed for continued progress in LDC agriculture are immediate and of major importance. If the threats can be overcome, LDC agriculture will prosper, and by the time the climate may have changed significantly, those countries will be in a reasonably strong position to deal with the consequences. If the threats are not overcome, the resulting economic, social, and political consequences over the next few decades will make the ramifications of climate change seem insignificant.

Implicit Assumptions

The assertion that, from a policy standpoint, the agricultural consequences of climate change on LDCs are relatively less important than other problems of agricultural development in those countries hinges on four assumptions that are, so far, implicit. One is that the GCMs used in climate change research give a reasonably accurate account of the changes that might occur, at least on global and continental scales. However, enormous uncertainty still surrounds most aspects of climate change, particularly its characteristics on subcontinental and smaller regional scales. It is on these scales that "the rubber hits the road," that is, we need to know in some detail how the climate might change. We do not.

A second implicit assumption is that the climate will change in what climate researchers call a linear fashion. That is, it will evolve without major ups and downs from what it is today to whatever it will be at equilibrium with $2{\times}CO_2$ warming sometime in the second half of the twenty-first century. The assumed gradual progression of the process underlies the thought that society will have time to adjust to whatever climate change may bring. However, the assumption of linear change may prove to be unfounded. Some evidence from the ancient climate record suggests that, occasionally, for unknown reasons, the world's climate has changed in a rather short and chaotic fashion. If global warming produced such a response, then the consequences for agriculture could be more severe. Even linear climate change could increase the frequency of extreme weather events, with more pronounced periods of drought and flooding. These possibilities are not picked up in the relatively benign scenarios of the IPCC and other researchers.

The conclusion that the impact of climate change on global and LDC agriculture will prove less important than other issues also assumes that the impacts will be limited to those resulting from $2{\times}CO_2$ warming. But the focus on $2{\times}CO_2$ is simply an analytical convenience adopted by climate researchers. Nowhere is it written that the atmospheric accumulation of CO_2 and other greenhouse gases must stop at an equivalent of $2{\times}CO_2$. Unless measures are taken to eventually bring the emissions of these gases to a level where they can be absorbed by the oceans and the terrestrial biosphere, they will continue to accumulate in the atmosphere and continue to warm the Earth beyond what might occur with $2{\times}CO_2$. In this case, all the studies of climate change consequence reviewed in this chapter probably would prove to be irrelevant.

Finally, the conclusion that the climate change impact on LDC agriculture is of relatively small importance assumes that LDCs will continue to make good economic progress and that the world trading system in agricultural products will be no less robust than it is now. Both of these assumptions underlie the argument that, by the time climate change begins to pose a threat to their agriculture, LDCs will be in much better shape to deal with the threat than they are now. If either of the assumptions fails, then the conclusion probably would no longer be warranted.

These caveats must be kept in mind when thinking about climate change and its consequences for global and LDC agriculture. Nonetheless, we must go with what we think we presently know about these consequences. What we think we know supports the conclusion that DC agriculture may in fact benefit from the kind of climate change likely to result from $2\times CO_2$ warming. And, although climate change is likely to damage LDC agriculture, the LDCs face other immediate threats that far outweigh the distant ones that climate change may bring.

Suggested Reading

Alston, J., P. Bailey, and J. Roseboom. 1998. Financing Agricultural Research: International Investment Patterns and Policy Perspectives. *World Development* 26(6): 1057–71.

Crosson, Pierre. 1989. Climate Change and Mid-Latitudes Agriculture: Perspectives on Consequences and Policy Responses. *Climate Change* 15(1): 51–73.

Darwin, R., M. Tsigas, J. Lewandrowski, and A. Raneses. 1995. *World Agriculture and Climate Change: Economic Adaptations.* Agricultural Economic Report 703. Washington, DC: U.S. Department of Agriculture.

Easterling, W., III, P. Crosson, N. Rosenberg, M. McKinney, L. Katz, and K. Lemon. 1992. Agricultural Impacts of and Response to Climate Change in the Missouri-Iowa-Nebraska-Kansas (MINK) Region. *Climatic Change* 24: 23–61.

IPCC (Intergovernmental Panel on Climate Change). 1996. *Climate Change 1995: Impacts, Adaptations, and Mitigation of Climate Change: Scientific-Technical Analysis.* Contribution of Working Group II to the Second Assessment Report of the Intergovernmental Panel on Climate Change. New York: Cambridge University Press.

Mendelsohn, Robert. 1999. *The Greening of Global Warming.* Washington, DC: American Enterprise Institute Press, American Enterprise Institute for Public Policy Research.

Mendelsohn, Robert, and others. 1994. The Impact of Climate Change on Agriculture: A Ricardian Analysis. *American Economic Review* 84(4): 753–71.

Rosenberg, Norman J. 1993. *Towards an Integrated Assessment of Climate Change: The MINK Study.* Boston, MA: Kluwer Academic Publishers.

Rosenzweig, Cynthia, and Martin Parry. 1994. Potential Impact of Climate Change on World Food Supply. *Nature* 367: 133–38.

Water Resources and Climate Change

Kenneth D. Frederick

Human efforts to alter the hydrological cycle date back to ancient times. Prayer, dance, human and animal sacrifice, and other rituals have been believed to bring rain. Cloud seeding is a more scientific but still uncertain attempt to induce precipitation. Although it is questionable whether any of these *intentional* efforts have significantly altered precipitation patterns, the balance of evidence now suggests that humans are *inadvertently* influencing the global climate and thereby altering the hydrological cycle.

The most recent scientific assessment by the Intergovernmental Panel on Climate Change (IPCC; see Contribution of Working Group I in Suggested Reading) concludes that, since the late nineteenth century, anthropogenic emissions of greenhouse gases (GHGs) such as carbon dioxide (CO_2), which trap heat in the atmosphere, have contributed to an increase in global mean surface air temperatures of about 0.3–0.6 °C (0.5–1.1 °F). Moreover, based on the IPCC's midrange scenario of future GHG emissions and aerosols and the best estimate of climate sensitivity, a further increase of 2 °C (3.6 °F) is expected by 2100. In this chapter, I examine the likely impacts of greenhouse warming on the supply and demand for water and the resulting socioeconomic implications.

Climate Impacts on Water Supplies

IPCC Results

Hydrological changes associated with greenhouse warming—whether it will rain more or less, for instance—are more speculative than temperature projections, especially at the regional and local geographic levels that are of interest to water planners. IPCC analysis (see Suggested Reading) suggests that greenhouse warming will have the following effects on water supplies:

- The timing and regional patterns of precipitation will change, and the number of days of intense precipitation per year probably will increase.
- General circulation models (GCMs) used to predict climate change suggest that an increase of 1.5–4.5 °C (2.7–8.1 °F) in global mean temperature would increase global mean precipitation about 3–15%.
- Although the regional distribution is uncertain, precipitation is expected to increase in higher latitudes, particularly in winter. This conclusion extends to the middle latitudes in most GCM results.
- Potential evapotranspiration (ET)—water evaporated from the surface and transpired from plants—increases with air temperature. Conse-

quently, even in areas with increased precipitation, higher ET rates may lead to reduced runoff, implying a possible reduction in renewable water supplies.

- More annual runoff caused by increased precipitation is likely in the high latitudes. In contrast, some basins in lower latitudes may experience large reductions in runoff and increased water shortages as a result of a combination of increased evaporation and decreased precipitation.
- Flood frequencies are likely to increase in many areas, although the amount of increase for any given climate scenario is uncertain, and impacts will vary among basins. Floods may become less frequent in some areas.
- The frequency and severity of droughts could increase in some areas as a result of decreased total rainfall, more frequent dry spells, and higher ET rates.
- The hydrology of arid and semiarid areas is particularly sensitive to climate variations. Relatively small changes in temperature and precipitation in these areas could result in large percent changes in runoff, increasing the likelihood and severity of droughts and/or floods.
- Seasonal disruptions might occur in the water supplies of mountainous areas if more precipitation falls as rain than snow and if the length of the snow storage season is reduced.
- Water quality problems may increase where there is less flow to dilute contaminants introduced from natural and human sources.

Regional Uncertainties

Even the direction of regional changes in precipitation and runoff are uncertain. The American Association for the Advancement of Science Panel on Climate Variability, Climate Change, and the Planning and Management of U.S. Water Resources (see Waggoner in Suggested Reading) estimated the range of regional equilibrium values (neglecting transient delays and adjustments) for an equivalent doubling of CO_2 from preindustrial levels as -3 to $+10$ °C (5.4–18 °F) for temperature, -20 to $+20\%$ for precipitation, -10 to $+10\%$ for evapotranspiration, and -50 to $+50\%$ for runoff. Subsequent advances in

global climate modeling have done little to reduce the uncertainty regarding the impacts of increasing atmospheric GHGs on regional water supplies.

Changes in runoff, the source of a region's renewable water supply, are the direct result of changes in precipitation and evaporation (which is strongly influenced by temperature). Studies simulating the effects of climate changes on hydrologic processes have been performed for several river basins and subbasins. Although these studies estimate the impacts on water resources, they offer no guidance regarding the likelihood that the assumptions underlying the modeling will be realized. Nevertheless, these studies are instructive about the possible magnitude of and the uncertainty surrounding the hydrological implications of greenhouse warming.

Estimates of the impacts of alternative temperature and precipitation changes on annual runoff in several U.S. river basins are presented in Table 1. Simulation studies suggest that relatively small changes in temperature and precipitation can have large effects on runoff. With no change in precipitation and a 2 °C (3.6 °F) increase in temperature, estimated runoff declines by 2–12%; with a 4 °C (7.2 °F) increase, runoff declines by 4–21% simply because of increased evaporation. Although runoff usually is more sensitive to changes in precipitation than changes in temperature, a 10% increase in precipitation does not fully offset the negative impacts on runoff attributable to a 4 °C (7.2 °F) increase in temperature in three of the five rivers for which this climate scenario was studied.

The IPCC's review of climate impact studies suggests large differences in the vulnerability of water resource systems to climate variables. Isolated single-reservoir systems in arid and semiarid areas are extremely sensitive; they lack the flexibility to adapt to climate impacts that could vary from decreases in reservoir yields in excess of 50% at one extreme to increased seasonal flooding at the other. In contrast, highly integrated systems are inherently more robust. A set of studies undertaken largely in the United States and based on the most recent transient GCM simulations suggests that most of these integrated systems are sufficiently robust and resilient and possess adequate institutional capacity to

Table 1. Impacts of Hypothetical Climatic Changes on Mean Annual Runoff in Mountainous River Basins.

Precipitation change	River basin	Temperature change (%) +2 °C	+ 4 °C	Precipitation change	River basin	Temperature change (%) +2 °C	+ 4 °C
−25%	Carson	−25%	−25%		White River	−4%	−8%
	American	−51%	−54%		East River	−9%	−4 to −16%
−20%	Upper Colorado	—	−41%		Animas River	−2 to −7%	−14%
	Animas River	−26%	−32%	+10%	Great Basin Rivers	+20 to +35%	—
	White River	−23%	−26%		Sacramento River	+12%	+7%
	East River	−19%	−25 to −30%		Inflow to Lake Powell	+1%	−10%
	Sacramento	−31%	−34%		White River	+7%	+1%
−12.5%	Carson	−24%	−28%		East River	+1%	−3%
	American	−34%	−38%		Colorado River	−18%	—
−10%	Great Basin Rivers	−17 to −28%	—		Animas River	+3%	−5%
	Sacramento River	−18%	−21%	+12.5%	Carson	+13%	+7%
	Inflow to Lake Powell	−23%	−31%		American	+20%	+19%
	White River	−14%	−18%	+20%	Upper Colorado		+2%
	East River	−19%	−25%		Animas River	+14%	+5%
	Upper Colorado	−35%	—		East River	+12%	+7 to 23%
	Lower Colorado	−56%	—		White River	+19%	+12%
	Colorado River	−40%	—		Sacramento	+27%	+23%
	Animas River	−17%	−23%	+25%	Carson	+39%	+32%
0	Sacramento River	−3%	−7%		American	+67%	+67%
	Inflow to Lake Powell	−12%	−21%				

Note: Entries are drawn from a variety of studies. Ranges reported here in some cases reflect the results of more than one study of the same basin.

— indicates that no figure was reported in the study being cited.

Source: Table 2 in Frederick and Gleick (see Suggested Reading).

adapt to likely changes not only in the climate but in such factors as economic and population growth. However, much of the world's water is managed through single-source, single-purpose systems.

Uncertainties regarding how the climate and hydrology of a region will change in response to global greenhouse warming are enormous. However, one of the more likely impacts involves areas where precipitation currently comes largely in the form of winter snowfall and streamflow comes largely from spring and summer snowmelt. Warming would likely result in a distinct shift in the relative amounts of snow and rain and the timing of snowmelt and runoff. A shift from snow to rain could increase the likelihood of flooding early in the year and reduce the availability of water during periods of peak demand, espe-

cially for irrigation. Many of the basins in the western United States are vulnerable to such changes.

Sea Level Rise

Sea levels rising in response to the thermal expansion of the oceans and increased melting of glaciers and land ice also affect water availability. The global sea level increased about 18 cm (7 in.) during the past century. The 1996 IPCC results suggest average sea level might rise another 15–95 cm (6–37 in.) by 2100, with a best guess of about 50 cm (20 in.).

Higher sea levels and increased storm surges could adversely affect freshwater supplies in coastal areas. Saltwater in river mouths and deltas would advance inland and coastal aquifers would face an increased threat of saltwater intrusion, jeopardizing the quality of water for many domestic, industrial, and agricultural users. For example, sea level rise would aggravate water supply problems in several coastal areas in the United States, including Long Island, New York; Cape Cod, Massachusetts; New Jersey shore communities; and the Florida cities of Miami, Tampa, and Jacksonville.

Rising sea level also threatens critical freshwater supplies in California. The Sacramento–San Joaquin Delta, for example, which is already under stress, is a major source of water for the farms and cities of southern California and the San Joaquin Valley. It is also the habitat for scores of fish species, several of which have been so weakened by the diversion and degradation of delta water that they have either been granted protection or are being considered for listing under the federal Endangered Species Act. Saltwater intrusion from San Francisco Bay threatens the delta's ecology as well as its use as a freshwater source. Over the last century, sea level rise in conjunction with ground subsidence in the delta has exacerbated these water supply and environmental problems.

Limiting the intrusion of saltwater depends on sufficient freshwater flows from the delta to the bay and on maintaining the levees that protect the more than 500,000 acres of islands within the delta. These islands are now rich farmlands created out of the marshland that originally characterized much of the delta. Gradual compaction of the delta's peat soils has caused many of the islands to fall well below sea level. When a levee breaks, as happens on average about twice a year, freshwater (that would otherwise help prevent saltwater from entering into the delta) floods onto the land. Any wide-scale failure of these levees thus would increase salinity levels, threatening the ecosystem and water for the farms and cities to the south. As the level of the sea rises, additional scarce freshwater supplies are required to prevent saltwater intrusion into the delta; maintaining the more than 1,100 miles of levees becomes increasingly difficult and expensive as well.

CO$_2$ Effects

A growing body of research suggests that atmospheric CO$_2$ levels may affect water availability

through its influence on vegetation. Controlled experiments indicate that elevated CO$_2$ concentrations increase the resistance of plant "pores" (stomata) to water vapor transport. Experiments suggest that a doubling of CO$_2$ would increase stomatal resistance and reduce the rate of transpiration—the passage of water vapor from plants—by about 50% on average. The resulting decrease in transpiration would tend to increase runoff. On the other hand, CO$_2$ also has been demonstrated to increase plant growth, leading to a larger area of transpiring tissue and a corresponding increase in transpiration. Other factors that might offset increases in plant water-use efficiency associated with a CO$_2$-enriched atmosphere are a potential increase in leaf temperatures caused by reduced transpiration rates and species changes in vegetation communities. The net effect of opposing influences on water supplies would depend on the type of vegetation and other interacting factors, such as soil type and climate.

Climate Impacts on Water Demand

Precipitation, temperature, and CO$_2$ levels can affect the demand for water as well as the supply.

Irrigation

Irrigation, the most climate-sensitive use of water, accounts for 41% of all water withdrawn from ground and surface sources in the United States and 81% of consumptive use (that is, water withdrawn that is evaporated, transpired, incorporated into crops, or otherwise removed from the immediate water supply). In the 17 water-scarce western states, these percentages increase to 77% and 85%, respectively.

The yields and profitability of irrigated relative to dryland farming tend to increase as conditions become hotter and drier. Consequently, in areas with available and affordable water supplies, hotter and drier conditions would increase both the land under irrigation and the amount of water applied per irrigated acre. Increased water use efficiency attributable to higher atmospheric CO$_2$ levels would tend to counter the tendency to apply more water as temperatures increase.

McCabe and Wolock (see Suggested Reading) of the U.S. Geological Survey used an irrigation model to simulate the effects of hypothetical changes in temperature, precipitation, and stomatal resistance (to illustrate the effects of changes in atmospheric CO_2 concentrations) on irrigation demand. Their results, which are based on annual plant water use in a humid temperate climate, suggest that increases in mean annual irrigation demand are strongly associated with increases in temperature and less strongly associated with decreases in precipitation. When temperature and precipitation were the only changes, irrigation demand increased with a 2 °C (3.6 °F) warming, even with 20% more precipitation. Plant water use is even more sensitive to changes in stomatal resistance than to temperature. For instance, a 20% increase in stomatal resistance reduced irrigation demand with a 2 °C (3.6 °F) warming and no change in precipitation.

Domestic Use

Water for normal household purposes—drinking, preparing food, bathing, washing clothes and dishes, flushing toilets, and watering lawns and gardens—accounts for 8% of withdrawals and 6% of consumptive use in the United States. Water demands for gardening, lawn sprinkling, and showering are the most sensitive of these uses to climate changes.

Aggregate annual domestic water use is not very sensitive to changes in temperature and precipitation; estimates suggest that a 1% increase in temperature would increase use from 0.02% to 3.8%, and a 1% decrease in precipitation would increase residential water use from 0.02% to 0.31%. Nevertheless, because they are likely to be greatest during the seasons and years when supplies are under the most stress, climate-induced increases in domestic demand can aggravate the problems of balancing supplies with demands during drought.

A study of urban water use in four mountainous counties of Utah (see Hughes and others in Suggested Reading) illustrated how climate variables can increase domestic water demands when supplies are likely to be scarcest. It found that potential evapotranspiration and rainfall best explain changes in residential water use attributable to the climate. Higher

evapotranspiration attributable to a temperature increase of about 2.2 °C (4 °F) increased residential water demand by an estimated 2.8% during the summer season and by as much as 8% during the month of June, when supplies in the region are likely to be in short supply. A temperature increase of 4.4 °C (8 °F) increased demand by 5% in the summer and as much as 16% in June.

Industrial and Thermoelectric Power Uses

Industrial use of water—which includes purposes such as processing, washing, and cooling in manufacturing facilities—accounts for 7% of withdrawals and 4% of consumptive use in the United States. Thermoelectric power use—which includes water for cooling to condense the steam that drives the turbines in the generation of electric power with fossil fuel, nuclear, or geothermal energy—accounts for 39% of all withdrawals but only 4% of consumptive use in the United States.

An increase in water temperature would reduce the efficiency of cooling systems, contributing to an increased demand for cooling water. Because more than 95% of the freshwater withdrawn for industrial and thermoelectric power use is now returned to groundwater and surface water sources, this increased demand would not represent a major increase in consumptive use. However, if aquatic ecosystems were threatened by higher water temperatures resulting from either global warming or from warmed returnflows of water used for cooling, then these uses might be subjected to more stringent environmental regulations.

Stricter regulations on returnflows might prompt a switch from once-through cooling systems to cooling towers and ponds that return little or no water to the source. Such a shift would reduce withdrawals and returnflows but have little effect on consumptive water use. Evaporative losses occur on site with cooling towers and ponds. In a once-through system, more of the evaporation occurs off-site and is attributable to the increased temperature of the receiving water body.

Global warming would also have indirect effects on industrial and thermoelectric water use. For instance, summer energy use for air conditioning would increase, and winter demand for space heating would decline. Changes in the temporal and perhaps the spatial demand for energy would alter the demand for cooling water.

Instream Uses

Changes in the quantity, quality, and timing of runoff stemming from greenhouse warming would affect instream water uses such as hydroelectric power generation, navigation, recreation, and the maintenance of ecosystems. These changes might also affect instream water demands, directly or indirectly. For example, changes in streamflows would alter actual and potential hydroelectric power generation, which in turn would affect the demand for alternate sources of electricity. Because thermoelectric cooling is one of the largest withdrawal uses of water, shifts in hydroelectric power production could have a significant impact on the demand for water within a watershed.

Warming would increase the potential length of the navigation season on some northern lakes and rivers that typically freeze in winter. To the extent that lake depth and river flow are constraints on navigation, demand could increase for water to facilitate navigation during the extended ice-free period. Similarly, seasonal water demands associated with recreational uses such as swimming, boating, and fishing might increase.

During the last quarter century, public policy has shifted from one that encouraged withdrawal uses at the expense of aquatic ecosystems to one that has emphasized protecting instream flows and recovering some of the recreational and environmental benefits that had been sacrificed under the earlier policy. Aquatic ecosystems and the benefits they provide are vulnerable to hydrological shifts that could result from greenhouse warming, especially if the major burden of adaptation falls on streamflows. Maintaining minimum instream flows to protect an endangered species or recreation benefits when supplies become scarce requires major adjustments in the use of water. On the other hand, protecting offstream uses could threaten the sustainability of some aquatic ecosystems. Tradeoffs between instream and withdrawal uses would increase if water supplies

became scarcer or more variable as a result of climate change.

Socioeconomic Impacts and Policy Implications

Climate is only one of many factors that influence the future supply of and demand for water. Indeed, population, technology, economic conditions, social and political factors, and the values society places on alternate water uses probably will have more of an impact on the future availability and use of water than changes in the climate. Even in the absence of human-induced changes in the climate and hydrological cycle, there is cause for concern over the adequacy of water supplies. Demand is outpacing supply, water costs are rising sharply, and current uses are depleting or contaminating some valued resources. Climate change has the potential to either aggravate or alleviate an area's water situation. On balance, however, the impacts probably will be adverse because the existing water infrastructure and use are based on an area's past climate and hydrology.

During most of this century, dams, reservoirs, pumps, canals, and levees provided the primary means of adapting to climate and hydrological variability and meeting the growing demands for water. Although the focus was on supply-side solutions, institutions that establish opportunities as well as incentives to use, abuse, conserve, or protect water resources were slow to adapt to the challenges of growing scarcity, rising instream values, and the vulnerability and variability of supplies. However, in recent decades, the high financial and environmental costs of water projects, along with limited opportunities for building additional dams and reservoirs to develop new water supplies, have shifted the focus away from new construction toward the improved management of existing supplies and facilities as well as toward demand management.

New infrastructure may, in some instances, eventually prove to be an appropriate response to climate-induced shifts in hydrological regimes and water demands. But it is difficult to plan for and justify expensive new projects when the magnitude, timing, and even the direction of the changes at the basin and regional levels are unknown. Narrowing the range of uncertainty for improved water planning depends on a better understanding of

- processes that govern global and regional climates,
- links between climate and hydrology,
- climatic effects on unmanaged ecosystems,
- impacts of ecosystem changes on the quantity and quality of water, and
- effects of increased atmospheric CO_2 on vegetation and runoff.

Meanwhile, the possibility that warming could result in greater hydrological variability and storm extremes should be considered in evaluating margins of safety of long-lived structures such as dams and levees that are under consideration anyway. In particular, low-cost structural and managerial modifications that ensure against the possibility of a range of climate-induced impacts should be sought.

More importantly, the prospect that global warming will alter local and regional supplies and demands in unknown ways reinforces the need for institutions that facilitate adaptation to whatever the future brings and promote more efficient water management and use. Unlike the structural supply-side approach, demand management that introduces incentives to conserve and opportunities to reallocate supplies as conditions change does not require long lead times, large financial commitments, or accurate information about the future climate.

Although likely to increase in the future, the magnitude and nature of water costs will be determined by the policies adopted to deal with all of these challenges. Critical determinants of future water costs will be the efficiency with which supplies are managed, how supplies are allocated among competing uses, and the effectiveness and costs of efforts to protect aquatic environments and drinking water quality. Integrated management of existing supplies and infrastructure at the river basin and watershed levels offers a potentially cost-effective means of increasing reliable supplies and resolving water conflicts in many regions. Providing appropriate incentives to conserve and protect the resource as well as

opportunities to voluntarily reallocate supplies (subject to consideration of third-party impacts) in response to changing conditions would ensure supplies for those purposes we value most highly. Measures required to achieve this objective include eliminating subsidies for the use of federally supplied irrigation water; pricing water at its social cost; establishing well-defined, transferable property rights in water; and creating water banks and other institutions to facilitate voluntary water transfers.

Although the prospect of climate change adds another element of uncertainty to the challenge of matching future supplies with demands, it does not alter what needs to be done to ensure that water is managed and apportioned wisely.

Suggested Reading

Frederick, Kenneth D. 1991. Water Resources: Increasing Demand and Scarce Supplies. In *America's Renewable Resources: Historical Trends and Current Challenges,* edited by K. D. Frederick and R. A. Sedjo. Washington, DC: Resources for the Future.

Frederick, Kenneth D., and Peter H. Gleick. 1999. *Water and Global Climate Change: Potential Impacts on U.S. Water Resources.* Arlington, VA: Pew Center on Global Climate Change.

Frederick, Kenneth D., David C. Major, and Eugene Z. Stakhiv (eds.). 1997. Water Resources Planning Principles and Evaluation Criteria for Climate Change. *Climatic Change* 37(September): 291–313.

Hughes, T., K.M. Wang, and R. Hanson. 1994. Impacts of Projected Climate Change on Urban Water Use: An Application Using the Wastch Front Water Supply and Demand Model. Provo, UT: U.S. Bureau of Reclamation.

Intergovernmental Panel on Climate Change. 1996. *Climate Change 1995: The Science of Climate Change.* Contribution of Working Group I to the Second Assessment Report of the Intergovernmental Panel on Climate Change. New York: Cambridge University Press.

————. 1996. *Climate Change 1995: Impacts, Adaptations and Mitigation of Climate Change: Scientific-Technical Analyses.* Contribution of Working Group II to the Second Assessment Report of the Intergovernmental Panel on Climate Change. New York: Cambridge University Press.

McCabe, Gregory, and David Wolock. 1992. Sensitivity of Irrigation Demand in a Humid-Temperate Region to Hypothetical Climatic Change. *Water Resources Bulletin* 28(3): 535–43.

Waggoner, Paul E. (ed.). 1990. *Climate Change and U.S. Water Resources,* New York: John Wiley and Sons.

U.S. Congress, Office of Technology Assessment. 1993. *Preparing for an Uncertain Climate* (Volume 1). OTA-O-567. Washington, DC: U.S. Government Printing Office.

7 Forests and Climate Change

Roger A. Sedjo and Brent Sohngen

Experts agree generally (however, not universally) that increased concentrations of greenhouse gases (GHGs) in the atmosphere will result in changes in Earth's climate. There is much less agreement about how such climate change could affect its forests. The question is important because forests are an important global source of both valuable market goods (especially timber) and environmental benefits (such as species habitat, biodiversity, and soil and climate stabilization). Additionally, the Kyoto Protocol requires significant reductions in net GHG emissions over the next 10 years and provides carbon sink credits only for deliberate human actions in forests that sequester carbon.

In this chapter, we identify potential sources of forest damage from climate change and evaluate the socioeconomic consequences. We conclude that the effects of climate change on forests in general and timber harvests in particular probably will be positive, which means that previous research that warned of severe consequences has overstated the risk. We also conclude that effects of climate change on ecological values associated with forests are a source of concern in some places but need not be great overall, especially if climate change is relatively gradual and adaptation to climate change is enhanced. (The sink function of forests is addressed in Chapter 13, Carbon Sinks in the Post-Kyoto World.)

Forest–Climate Interactions

Forests thrive in many climatic conditions, from wet tropical forests to the forests of dry boreal (high latitude) regions. The vegetative transition from deserts to grasslands to forests is commonly determined by moisture conditions. Under extreme conditions, the vegetative transition reverses course, and forests are displaced by grassland. Forests generally flourish in warm, wet environments and do progressively less well as temperature and moisture decrease. Thus, forests would be expected to respond to changes in temperature and precipitation brought on by climate change. The kinds of trees growing at a given site might give way to those more suited to the new climatic conditions, or a forest might cease to exist altogether. However, the socioeconomic impact will depend on not only the disruptions that occur in the established relationships between forests and climates but also how humans adapt to the consequences.

It is now generally agreed that forests would grow somewhat more rapidly under most climate change scenarios and that the composition of forests is likely to change. Ecological predictions of the effect that climate change might have on forests are generally based on the outputs of general circulation models (GCMs). Most GCMs predict increased precipitation overall as a consequence of warming, but the pattern of regional increases and decreases in precipitation is more complex (for example, mid-continental areas could experience significantly less precipitation as a consequence of warming). Moreover, the amount of temperature change is likely to be lowest at the equator and progressively higher toward the poles. The most recent GCMs tend to point to an expansion of forest area as a result of global warming, in contrast to 1980s studies that suggested a contraction.

Nevertheless, the climate projections of the various GCMs vary greatly. For example, the *CCC model* (developed by the Canadian Centre for Climate Modeling) predicts generally higher temperatures and lower precipitation than many of the other models. By contrast, the *Hadley model* (of the Hadley Centre for Climate Prediction and Climate Research in Great Britain) tends to predict lower temperature increases and more precipitation. Both were combined with two vegetation models to simulate transition climate scenarios for the U.S. National Climate Assessment. Under the CCC scenario (the more severe of the two), the model predicts the conversion of southern forest to savanna, with no reversion to forest over time. Under the more benign Hadley scenario, the same vegetation model shows no conversion of southeastern forest to grassland (see MARA in Suggested Reading).

More generally, the highly respected Vegetative/ Ecosystem Modeling and Analysis Project (VEMAP) project has combined biogeochemical and biogeographic models to allow the integrated simulation of change in the vegetation structure, productivity, and carbon storage. This approach provided estimates that the U.S. forest area in the lower 48 states could vary from a slight decline (–0.7%) to substantial increases (+39.7%) (see Malcolm and Pitilka in Suggested Reading).

In light of the expected pattern of temperature changes, tropical forest areas would be less affected than boreal forests in higher latitudes. If higher moisture levels were to accompany the higher temperatures in general, as the GCMs now suggest, then the boreal forests would be expected to exhibit increased growth and productivity as well as expansion into the tundra of the northern latitudes. The forest type and mix of species that now exist in temperate regions would give way to types and species better adapted to the new climate.

Climate change could have significant negative effects on existing forests if a decrease in soil moisture is widespread. It could occur where climate change caused precipitation to decline (or where warming occurred in conjunction with constant or only slightly elevated precipitation). Where a drier climate ensued, existing forests would give way to ones more suited to the new conditions—or to other vegetation altogether, such as grasses or shrubs. However, increased carbon dioxide can improve the water use efficiency of individual trees, thereby offsetting the effects of the drier climate to some extent by making forests more tolerant to drought.

A modeling study of the prairie–forest interface in the U.S. Midwest illustrates the sensitivity of forests to climate (see Bowes and Sedjo in Suggested Reading). Results showed that the condition of natural forest was critically dependent on precipitation. Where warming and drying coincided, forest productivity declined, and some of it was overtaken by prairie. However, where warming coincided with increased precipitation and thus decreased stress on trees, forest productivity increased, and trees displaced some grass. In the short term (by forest standards—30 years), the species mix of the forest changed only minimally. In the long term, however, it underwent considerable change as trees suited to the new climatic conditions displaced those that could not adapt.

Other studies have shown that a higher level of atmospheric carbon dioxide generates a "fertilization effect" that causes an increase in biological growth rates for certain types of plants, including trees. However, the evidence on carbon fertilization is not complete. Moreover, carbon is often not the "limiting factor" in forest growth. If some other nutrient such as nitrogen were the limiting factor, then productivity gains would be minimal. Species competition also might constrain overall productivity, so total forest biomass at maturity might not increase.

Concerns about Forest Impacts

Some analysts are concerned that climate change would mean a global environment less hospitable to forests, one where conditions would be ripe for "dieback"—a high incidence of decline and individual tree death—because the change in climatic conditions would make forests vulnerable to disease and insect predation. Alternatively, weakened or non-adaptive species might simply be overwhelmed by competition from tree species or vegetation more suited to the site in the wake of climate change. A related concern is that any transition from one kind of forest to another—or from forest to another kind of vegetation—could be difficult and disruptive, perhaps leaving a site barren for years.

Even without climate change, a forest may not be able to migrate effectively over a landscape. Even though forests can migrate and have migrated (pollen evidence provides some information about past forest migrations), the rate at which they do so is not well understood. This capacity appears to vary with tree species. In addition, the heavy impact of human activities that involve the conversion and fragmentation of land, such as agriculture and urbanization, could hinder transition and replacement. If the climate change were modest and gradual, then species more suited to a site would probably infiltrate relatively easily and gradually displace forest types that could no longer thrive under the new conditions. However, should the change be rapid, greater transitional problems might be expected. Replacement species would have difficulty migrating fast enough to replenish a dying site. Dead and ailing forest areas might not be overtaken, and land with the look of a moonscape might appear.

However, the likelihood of this latter scenario is fairly small. Mature trees can survive in inhospitable habitats outside their normal range, although often they cannot procreate. Such stands of trees might well endure until more appropriate species replaced them at a more or less normal rate of mortality. Grasses and shrubs also might fill the space during the interim. Moreover, most of the updated GCMs suggest that temperature increases are likely to be

smaller and more gradual than anticipated by earlier GCMs. This possibility bodes well for a relatively orderly natural transition.

Some ecological models predict a difficult transition because of the unavailability of seeds from appropriate species. However, most forests consist of many species that overlap each other's natural range. Thus, whereas climate change may seriously affect *some* species, it is highly unlikely that *all* the forest's species will be negatively affected, and seed sources may not be a problem. Additionally, human intervention could make a big difference in ameliorating this problem when and if it did occur, either through low-cost aerial seeding or costlier on-site sowing and replanting. Our own recent research suggests that in many cases, economic incentives would be sufficient to justify various kinds of forest regeneration investments during climate change. Typically, these investments would occur after timber stands were harvested or logs salvaged where forests were experiencing dieback.

Among the many unanswered questions is whether forests would act more as a sink or a source of carbon as the global climate changed. Models demonstrate that forests have the potential to be either, depending on the underlying ecological scenario. Some early analyses suggest that forests would be sources, emitting large quantities of carbon in the transition period. However, more recent analyses suggest that forests would expand and thus absorb more carbon in the process, particularly when human activities in response to anticipated climate change are taken into account.

Socioeconomic Consequences: Timber Harvest and Nonharvest Values

Economic studies of U.S. and global timber markets indicate that global forests—both natural and cultivated—can be expected to expand with climate change on the basis of many alternative assumptions about climate and ecology. The transition to new kinds of forests could occur through interspecies competition or dieback. Foresters are presumed to invest in reforestation where it is economically justified. In either case, future timber supplies would be larger and timber prices lower than in the absence of climate change.

This result contrasts sharply with some earlier studies. One reason is that more recent ecological models indicate less stress from climate change, as discussed above. More recent analyses also take into account the possibility of long-term human intervention and adaptation in forest management.

Forests are valued not only for their timber production but also for their ecological services and recreational values. If forest dieback were to occur so rapidly that new forest had difficulty replacing existing ecosystem services (such as water protection and erosion control), then the damage could be substantial. However, such a result appears unlikely. When forests experience catastrophic damage from natural disasters (such as major wildfires, volcanic eruptions, or serious pest infestations), natural systems typically respond with resilience. Because tree and plant species have different climatic ranges, the absence of one or several species need not condemn a site to desolation. The horse chestnut tree, for example, was common in the eastern United States until the end of the nineteenth century, when disease eradicated it. Yet the forest continued to perform its ecosystem functions unhindered as other tree species (especially hickory) filled in for what was lost. Similarly, a forest's recreational services—providing places for hunting, fishing, hiking, bird watching, and skiing—presumably could continue as the forest cover gradually changed.

In the case of wildlife, perhaps the major challenge would not be climate change per se but the need to preserve a landscape that provided for adaptation. The migratory capacities of many species allow for relatively easy adaptation. The important caveat is that the necessary habitat areas and migratory pathways have to remain available.

Perhaps the most challenging problem associated with climate change and forests is the probable loss of biodiversity. Even without climate change, there is concern that substantial numbers of genes, species, and ecosystems are being lost in ongoing deforestation. Climate change probably would add to the problem by disrupting certain delicate relationships within forests.

Some experts argue that more species on a site is better, because the site is filled more completely, and that the redundancy provides for rapid adaptation to external disturbances and changes. Other experts disagree, countering that beyond a certain number, species add nothing to the productivity of a site. They point to the fact that nature limits to a very few the number of species on certain sites (for example, eastern salt marshes and boreal forests) and conclude that neither evolutionary theory nor empirical studies present convincing evidence that species diversity and ecosystem function are consistently and causally connected.

Regardless of whether less diversity is a problem, a forest that is adapting to climate change is likely to experience losses. Endemic species in particular area could disappear, because they are not widely distributed and may not migrate. However, tropical areas—which are the richest in biodiversity—are predicted to experience the least amount of climate change.

Conclusions

Although climate change is not necessarily a pleasant prospect to consider, its overall effect on global forests probably would be modest. However, its effect on individual forests could be substantial while vegetation adapted to its new climatic conditions. New forests might rise up in the tundra. Others might wane in places where moisture levels declined. In general, the effects probably would be greatest in the higher latitudes, where more warming is expected.

The effect of climate change on the industrial wood supply probably would be positive. The negative effects on most ecological, environmental, and recreational services would not be great—provided that climate change occurred gradually, as most of the recent GCMs now predict. The major negative effect probably would be on biodiversity, particularly on endemic species that would have difficulty migrating. Mitigating the negative effects of climate change on forests will depend on enhancing their capacity for adaptation.

Suggested Reading

General

Grime, J.P. 1997. Biodiversity and Ecosystem Function: the Debate Deepens. *Science* 277(29): 1260–1.

IPCC. 1996. *Climate Change 1995: Impacts, Adaptations, and Mitigation of Climate Change: Scientific-Technical Analysis.* Contribution of Working Group II to the Second Assessment Report of the Intergovernmental Panel on Climate Change. New York: Cambridge University Press.

MARA (Mid-Atlantic Regional Assessment Team). 2000. Forests. In *Preparing for a Changing Climate: The Potential Consequences of Climate Variability and Change—Mid-Atlantic Foundations*, Chapter 5. August. Public comment draft report prepared for the U.S. Global Change Research Program, Office of Research and Development, U.S. Environmental Protection Agency. http://www.essc.psu.edu/mara/results/index.html (accessed November 20, 2000).

Terborgh, John. 1992. *Diversity and the Tropical Rain Forest.* New York: Scientific American Library.

Technical

Bowes, M., and R.A. Sedjo. 1993. Impacts and Responses to Climate Change in Forests of the MINK Region. *Climatic Change* 24(June): 63–82.

Joyce, Linda A., John R. Mills, Linda S. Heath, A. David McGuire, Richard W. Haynes, and Richard A. Birdsey. 1995. Forest Sector Impacts from Changes in Forest Productivity Under Climate Change. *Journal of Biogeography* 22: 703–13.

King, G.A., and R.P. Neilson. 1992. The Transient Response of Vegetation to Climate Change: A Potential Source of CO_2 to the Atmosphere. *Water, Air, and Soil Pollution* 94(7): 365–83.

Malcolm, Jay R., and Louis F. Pitilka. In press. *Ecosystems and Global Climate Change: A Review of the Potential Impacts on U.S. Terrestrial Ecosystems and Biodiversity.* Arlington, VA: Pew Center on Global Climate Change.

Sedjo, R.A., and A. Solomon. 1989. Climate and Forests. In *Greenhouse Warming: Abatement and Adaptation*, edited by Norman J. Rosenberg, William E. Easterling III, Pierre R. Crosson, and Joel Darmstadter. Washington, DC: Resources for the Future, Chapter 8.

Shugart, H.H.; M.Y. Antonovsky, P.G. Jarvis, and A.P. Sandford. 1986. CO_2, Climatic Change, and Forest Ecosystems. In *The Greenhouse Effect, Climatic Change, and Ecosystems*, edited by Bert Bolin, Bo R. Doos, Jill Jager, and Richard A. Warrick. Chichester, U.K.: Wiley, 475–521.

Smith, T.M., and H.H. Shugart. 1993. The Transient Response of Terrestrial Carbon Storage to a Perturbed Climate. *Nature* 361: 523–26.

Sohngen, Brent, and Robert Mendelsohn. 1998. The U.S. Timber Market Impacts of Climate Change. In *The Economic Impacts of Climate Change on the U.S. Economy*, edited by R. Mendelsohn and J. E. Neumann. Cambridge, U.K.: Cambridge University Press, Chapter 5.

Sohngen, Brent, Robert Mendelsohn, and Ronald Neilson. 1998. Predicting CO_2 Emissions from Forests during Climate Change: A Comparison of Natural and Human Response Models. *Ambio* 27(7): 509–13.

Sohngen, Brent, Robert Mendelsohn, and Roger Sedjo. 2000. Measuring Climate Change Impacts with a Global Timber Model. Working Paper. Columbus, OH: The Ohio State University, Department of Agricultural, Environmental, and Development Economics.

Solomon, A.M. 1986. Transient Response of Forest to CO_2-Induced Climate Change: Simulation Modeling Experiments in Eastern North America. *Oecologia* 68: 567–79.

8

"Ancillary Benefits" of Greenhouse Gas Mitigation Policies

Dallas Burtraw and Michael A. Toman

To a large extent, policies for limiting the emission of greenhouse gases (GHGs) have been analyzed in terms of their costs and their potential for reducing the rate of increase in atmospheric concentrations of these gases. However, actions to slow atmospheric GHG accumulation could have several other impacts, such as a reduction in conventional environmental pollutants. The benefits (or costs) that result are often referred to as *ancillary* to the benefits and costs of GHG abatement (although controversy surrounds this terminology and the underlying concepts, as we discuss below).

A failure to adequately consider ancillary benefits and costs of GHG policy could lead to an inaccurate assessment of the overall impacts of mitigation policies. In particular, not accounting for ancillary benefits and costs would lead to an incorrect identification of a "no regrets" level of GHG mitigation. It also could lead to the choice of an unnecessarily expensive policy because of its failure to fully exploit potential ancillary benefits.

In this chapter, we first discuss in broad terms the concept of ancillary benefit. The concept turns out to be surprisingly difficult to define precisely. What is considered an ancillary benefit depends on the scope of policies being considered, the policy objectives being pursued, and the identity of the interests being served. That said, however, we describe what we believe is a serviceable definition of ancillary ben-

efits from the perspective of evaluating GHG mitigation policies within the Annex I countries who would have emission limitation obligations under the Kyoto Protocol. We focus on mitigation while acknowledging that adaptation policies also could have ancillary effects (for example, improved surveillance of tropical diseases could yield immediate health dividends; protection of coastal lands could harm wetland habitats in the more immediate term).

Having established a workable definition, we then turn to issues related to measuring ancillary benefits. To illustrate these issues, we consider how lower GHG levels resulting from less fossil fuel use could also reduce various conventional air pollutants (so-called criteria pollutants, as defined in the U.S. Clean Air Act). Reductions in premature mortality from reduced exposure to various forms of air pollutants (mainly particulates) typically account for about 75–85% of *all* estimated benefits (not only health benefits) in economic assessments of improved air quality in the United States and other developed countries (see Lee and others and the European Commission in Suggested Reading). Thus, focusing on this category of ancillary benefits probably will provide a fairly reliable picture of total ancillary benefits, even though controversy remains regarding the magnitude of nonhealth effects.

Nevertheless, estimates of ancillary health benefits are quite variable. Ancillary benefits could offset a significant fraction of the costs of reducing carbon emissions in some cases. Thus, they should figure prominently in estimating the overall costs and benefits of GHG policies. However, the considerable uncertainty about the size of ancillary benefits precludes identification of a single "best estimate" of their magnitude. And for various reasons we explain below, we have much more confidence in more conservative estimates of ancillary benefits compared to estimates which equal or exceed the costs of GHG control.

What Is an Ancillary Benefit or Cost?

An *ancillary benefit* of a GHG mitigation policy is understood by many analysts to refer to a benefit derived from GHG mitigation that is reaped in addition to the benefit targeted by the policy, which is a reduction in the adverse impacts of global climate change. An *ancillary cost* would be a negative impact experienced in addition to the targeted benefit. The key elements of these definitions—and the sources of much of the controversy surrounding the notion of "ancillary"—are "in addition to" and "targeted."

In the context we have used for defining ancillary benefits and costs, the principal policy goal is GHG mitigation to reduce the adverse impacts of climate change. Asserting that ancillary benefits are additional to the benefits of reducing climate change does not mean that these benefits are necessarily less important or that other policy goals are less important than reducing the negative effects of climate change. Benefits that are ancillary to those targeted by climate change policy could be bigger in magnitude and more salient for the affected citizens and their decisionmakers. Our definition simply puts ancillary benefits in a certain policy context.

This policy context can be and is debated. Developing countries have argued with justification that they have more immediate developmental and environmental needs than reducing their GHGs. This sentiment is reflected in the upcoming Third Assessment Report of the Intergovernmental Panel on Climate Change (IPCC), which emphasizes the notion of integrating climate change considerations into a broader context of sustainable development. In this broader policy context, what we call "ancillary bene-

fits" could be considered "co-benefits" of policies designed to promote various objectives. Whatever context is used, it needs to be clearly stated so that users of information about ancillary benefit and cost information can understand which effects of the policy in question are viewed to be "targeted" and which are "additional." Our view is that when discussing climate change policies, the benefits and costs targeted by the policies should be considered as those associated with GHG mitigation and reducing the risk of climate change; other benefits and costs should be treated as ancillary in the sense defined earlier, but not given short shrift.

Some more specific but related considerations that arise in defining ancillary benefits and costs involve the scope of what is included in the calculation and the perspective of the decisionmaker evaluating the ancillary benefits and costs. Many kinds of impacts can be considered. Much of the emphasis in these calculations has been on near-term health effects in relatively close proximity to the GHG mitigation (for example, reduced incidence of lung disease in the same area as a coal-fired power plant if that plant is used less as a consequence of GHG mitigation measures), but various other effects also could be important.

Ecological systems could be affected by reductions in the flow of conventional pollutants (for example, less fossil fuel use could mean less nitrogen oxide deposition into water bodies). Decreased emission of pollutants also could reduce some direct costs, such as the cost of infrastructure maintenance and pollution-related losses. Traffic accidents could be reduced by less driving or slower traffic speeds. Less traffic could lower road maintenance costs. Increased forest areas could increase recreational opportunities and reduce erosion. GHG policies also could stimulate technical innovation.

Ancillary costs can be incurred if energy substitution leads to other health and environmental risks (from nuclear power; uncontrolled particulate emissions from biomass combustion; or the use of diesel fuel in lieu of gasoline, because diesel fuel emits less carbon but more of other pollutants). Better building insulation can add to indoor air pollution, such as radon, and switching from coal to gas raises the specter of fugitive emissions of methane, a more potent GHG than carbon dioxide (CO_2). Policies that promote reforestation also could encourage the destruction of old-growth natural forests because younger forests can store more carbon. GHG mitigation policies could mainly redirect innovation efforts away from other productive activities, rather than increasing it. In addition, relatively expensive GHG mitigation policies could have some negative side effects on health by reducing the resources available to households for other health-improving investments.

An economic perspective on ancillary benefits sees them as part of a larger concern with economic efficiency, as typically expressed in measures of aggregate benefits and costs. From this perspective, it is important not to isolate ancillary benefit and cost information from other relevant benefit and cost information associated with GHG policy. Ancillary benefits of a policy could be substantial, but they are nonetheless a questionable achievement if the cost of garnering them is high. Often ancillary benefits are expressed in terms of a monetary measure per ton of carbon not emitted to the atmosphere as a consequence of the mitigation policy. Expressed this way, ancillary benefits (and costs) can be compared with the cost of mitigation. This comparison is usually meaningful and useful, because ancillary benefits often but not always occur on the same relatively short-term time scale as mitigation costs, whereas the benefits of reducing climate change will be realized in the more distant future.

Although the economic focus is largely on some aggregation of individual benefits and costs, it is important to recognize that as with any policy, some actors may benefit more than others. With ancillary costs, there can be losers as well as winners. These distributional effects are not in themselves ancillary benefits and costs in the way benefits and costs are typically used in economic assessments of policies, because we lack any agreed-upon monetary metric for evaluating distributional impacts. Nonetheless, these effects are an important component of assessing the ancillary impacts of GHG policies and should receive careful consideration.

A final related point is that the scope and magnitude of ancillary benefits and costs depends on the perspective of the decisionmaker as to what constitute policy-relevant impacts. From the perspective of a hypothetical global decisionmaker concerned with global social well-being, ancillary benefits and costs are important wherever they are incurred. From this perspective, it thus is important to consider how a redistribution in the location of GHG mitigation could affect ancillary benefits and costs.

In particular, policy mechanisms such as international emissions trading or the clean development mechanism (see Chapter 20, Policy Design for International Greenhouse Gas Control) will redistribute ancillary impacts toward those countries undertaking more GHG mitigation. Efforts by Annex I countries alone to mitigate GHGs could have collateral effects in developing countries not bound by quantitative emissions limits, in that lower energy prices in international markets will stimulate some additional energy use and associated local environmental effects. On the other hand, for an Annex I decisionmaker evaluating the benefits and costs of GHG mitigation policies in his or her own jurisdiction, the relevant ancillary benefits and costs are likely to consist primarily of those that affect individuals in that political jurisdiction. Cross-boundary spillovers are relevant for the Annex I decisionmaker only to the extent that a sense of ethical responsibility or altruism motivates a broader concern for the spillovers.

Still another perspective would be adopted by developing country decisionmaker contemplating involvement in the Clean Development Mechanism. In this case, the primary benefits in terms of importance for the developing country considering hosting a GHG-reducing investment are likely to be the benefits that are ancillary to the GHG control, according to our definition of the term.

Empirical Challenges in Assessing Ancillary Benefits and Costs

Having discussed some of the key conceptual issues surrounding ancillary benefits and costs, we turn next to some of the key problems in measuring these values. Figure 1 is a simple but useful illustration of how an integrated assessment framework can be

Figure 1. An Integrated Assessment Representation of GHG and Conventional Pollutant Interrelationships.

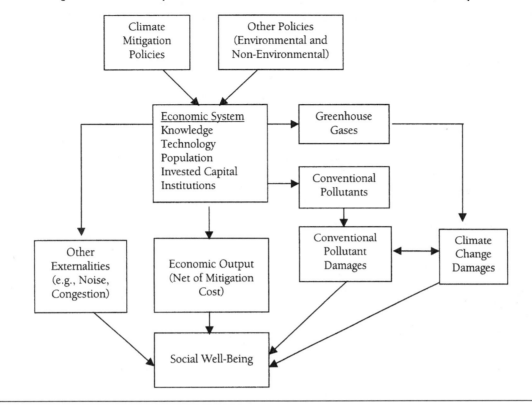

used to describe links among climate and other policies, the economic system, health and environmental impacts, and social well-being (a simpler example of such an approach in presented in Chapter 4, How Much Climate Change Is Too Much?). The diagram shows an economic system whose key elements are the population, endowments of other production inputs (capital), knowledge embodied in technology, and institutions. The overall output of this economic system is affected by the application of GHG mitigation policies and other policies (environmental and nonenvironmental). Specifically, these policies reduce standard economic output as reflected in mitigation costs; however, they also reduce GHG emissions, other pollutants, and nonenvironmental impacts such as traffic congestion.

We can use this diagram to highlight some of the key challenges that arise in operationalizing this framework to develop empirical measures of ancillary benefits and costs. To calculate ancillary benefits and costs over time, two hypothetical situations must be compared. The first is a baseline scenario without any modification of GHG mitigation policy, sometimes referred to as "business as usual"; however, this term is somewhat misleading because over time, the status quo can change even without modification of GHG policies. The baseline is compared with an even more hypothetical scenario that involves changing the current and future "state of the world" by modifying GHG mitigation. To carry out this exercise in practice means addressing several challenges.

How the baseline is defined crucially affects the magnitude of ancillary benefits and costs generated by a change in GHG mitigation policy. One important influence on the baseline is the status of nonclimate policies. It can be vividly illustrated with two environmental examples. Suppose that even in the absence of climate policy, conventional air pollutants are expected to drop sharply because of trends in policies for the regulation of conventional pollutants. (Such a trend requires not only tougher standards over time but also a maintained or increased degree of compliance with those standards.) In this case, we would expect the incremental benefits from a reduction of conventional air pollutants in the wake of tougher GHG controls to be smaller than if the increased GHG controls were being applied to a dirtier baseline environment.

The second example involves the establishment of total emission caps for conventional pollutants, such as the cap on sulfur dioxide (SO_2) from power plants in the United States. If such a cap is imposed, then a stronger GHG mitigation policy will not have an effect on the total emissions of conventional pollutants unless a much tougher GHG policy is imposed—so tough that it leads to polluters reducing conventional emissions below the legal cap. In less stringent cases, the location of the conventional emissions would be affected, and the total cost of meeting the conventional emissions cap will be lower because GHG policy will pay for achieving some of the conventional pollutant goal. However, this example illustrates the need for careful cost and benefit accounting when calculating ancillary benefits and costs.

Aside from the interaction of GHG policies and conventional pollutant policies over time, several other elements are important in specifying the baseline. All the factors driving the evolution of the economic system are included in the list. The state of technology will affect the energy and emissions intensity of economic activity. The size and location of the population as well as the volume and location of total economic output will affect both the scale of physical impacts on the environment and the risks posed to the population. Finally, the status of natural systems is also part of the baseline; it indicates the sensitivity of humans and ecological resources to changes in conventional pollutants.

Another important set of influences on estimates of ancillary benefits and costs includes the scale of analysis, the level of aggregation, and the stringency of the GHG policy being considered. As discussed later, we find that estimates of ancillary health benefits from reduced conventional air pollutants (expressed as dollars per ton of carbon release avoided) tend to get smaller when the analysis shifts from an aggregate perspective to one that considers more carefully the effects of GHG policies on specific sectors at specific locations. These latter analyses appear to be better able to model the distribution of gains and losses as well as the behavioral responses to GHG policies. As for the stringency of GHG policy, we would expect that a stronger GHG program will generate successively smaller increments in ancillary benefits and more ancillary costs as other risks decline relative to baseline levels.

A final point involves the assessment of the ancillary impacts themselves. In the area of conventional air pollutants and human health, which has received more research support than others, considerable uncertainty about how a change in ambient environmental conditions will affect health end points (for example, how many fewer cases of disease will result from somewhat cleaner air) and how much society values these changes nonetheless continues. The effects of these uncertainties are illustrated later. The uncertainties are especially acute and troubling when one tries to use studies of impacts and valuations from developed countries to assess ancillary benefits in developing countries with lower incomes, different health status and infrastructure, and different cultural norms. Other health and nonhealth ancillary environmental benefits and costs are even less well researched or understood.

Illustration: Adverse Human Health Effects of Conventional Air Pollutants

An extensive scientific literature exists on the adverse human health effects of exposure to criteria air pollutants. These effects are widely seen as significant; however, their magnitude depends on the amount and duration of exposure to specific pollutants, the nature of the exposed population, and other factors. Effects include the initiation or aggravation of various pulmonary disorders and cardiovascular problems that result in illness and sometimes premature mortality.

The pollutants described as particulates seem to have the greatest impact on public health, as already noted. They take several forms and arise from many sources. Particulates include soot emitted directly from the combustion process, soil dust (often mobilized in the air by human activities), and secondary pollutants such as sulfate and nitrate aerosols that form chemically in the atmosphere from SO_2 and nitrogen oxides (NO_x). The proportion of particulate substances varies from area to area, and particulates have different degrees of impact on human health. Of the various kinds of particulates, sulfate and nitrate aerosols and their potency raise particular concern.

Another secondary pollutant that impairs human respiration is ground-level ozone (O_3), which is formed from the mixing of NO_x and volatile organic compounds (VOCs) such as hydrocarbons in sunlight. Although some short-term adverse health effects can result from increased ozone concentrations (the magnitude of which continues to be debated), little evidence indicates that ozone is associated with long-term illness or premature mortality for most of the population. Consequently, ozone receives much less weight than particulates in economic analysis. Carbon monoxide (CO), which is fatal at high concentrations, has much more limited health effects (primarily related to the cardiovascular system) at ambient exposure levels normally encountered. Moreover, CO emissions are decreasing over time as new vehicles with low emissions replace older vehicles.

Health effects predominate in assessments of environmental benefits from reduced emissions, but there are other impacts, too. At high enough concentrations, criteria air pollutants can damage ecosystems. NO_x and SO_2 are precursors to acidic deposition (commonly referred to as acid rain), which has adverse effects on some forest species and aquatic wildlife. Atmospheric nitrogen deposition is a potentially significant contributor to an increase in damaging algal blooms in certain estuaries (for example, the Chesapeake Bay in the United States). Both SO_2 and ground-level ozone can damage the foliage of several crops and trees; ozone is responsi-

ble for agricultural yield losses in the United States valued at several billion dollars each year, and the damages to forests and other ecosystems are still being assessed.

Criteria air pollutants also impair visibility and damage materials, affecting both aesthetic and property values. Airborne sulfur and nitrogen compounds, particulates, and ground-level ozone, for example, tend to impair visibility. Particulates soil buildings, statues, and monuments; acid rain accelerates their decay.

The vast majority of the U.S. economy's GHG emissions arise from fossil fuel combustion (see Chapter 3, The Energy–CO_2 Connection). Natural gas (methane) is the least carbon-intensive fossil fuel per unit of energy content. Methane also is a relatively clean fuel with respect to conventional pollutants. The main pollutants resulting from its combustion are relatively small amounts of NO_x. Petroleum products have intermediate carbon intensity and can generate significant emissions of particulates, SO_2, NO_x, VOCs, and (in older cars) CO in the absence of effective emissions controls. Coal is the most carbon-intensive fossil fuel, and its combustion also generates relatively high emissions of criteria pollutants (especially SO_2, NO_x, and particulates) in the absence of effective emissions controls. However, current regulatory standards require very stringent controls on gases, dust, and soot from stationary sources such as power plants, and these controls reduce emissions substantially.

The most likely sources of particulate reductions to be had from GHG policies that affect energy use are controls on emissions of SO_2 and NO_x (which form sulfate and nitrate aerosols) and on emissions from diesel engines (which produce soot). Previous studies indicate that significant reductions in NO_x and CO are possible as a result of policies aimed primarily at reducing carbon emissions. Reductions in VOCs or direct particulate emissions from carbon policies may be much smaller than the NO_x and CO reductions, depending on the policies pursued. The actual extent of emissions reduction depends critically on both the kind of fuel used and the technologies used for combusting fossil fuels and trapping pollutants in the waste gas stream.

Many factors that will influence the size of ancillary human health benefits from GHG policies also reduce conventional air pollutants. One important influence is the prospect of future tightening of U.S. pollution standards, for example, as reflected in proposed new air quality standards for ozone and particulates (which, as of this writing, are the subject of continued legal wrangling). Future air quality improvements will reduce the ancillary benefits actually achieved by climate policies compared with projections that fail to take into account future abatement measures for conventional pollutants. However, tighter standards with respect to conventional pollutants also are likely to raise the relative cost of using more carbon-intensive fuels. This means that tougher conventional standards will lower the opportunity costs of GHG emissions reduction (for example, through fuel switching) in the future as well.

Another important interaction between GHG mitigation and conventional pollutant policies arises in considering the effects of GHG policies on U.S. SO_2 emissions. With the cap on SO_2 emissions from electric utilities in the United States, aggregate SO_2 emissions from electric utilities (the major source category in the country) are essentially independent of the amount of GHG emissions reduction, up to the point where SO_2 emissions become so small that the cap is not binding. This means that ancillary health benefits from SO_2 reductions as a consequence of small-to-moderate GHG initiatives will arise only from a spatial redistribution of SO_2 emissions, and these effects in turn are likely to be very modest. However, GHG policies could lower the cost of complying with the SO_2 cap by reducing the use of coal and thus the demand for SO_2 emission allowances.

More generally, the ancillary economic benefits of GHG emissions reduction depend critically on geographic location. Differences in air quality imply different benefits from pollution mitigation. Population density also affects total benefits. For example, far more people are affected by emissions from a power plant located in New York than one in New Mexico. Failure to account for growth in population or migration that increases the number of exposed individuals leads to understatement of the ancillary benefits of GHG mitigation through the reduction of conventional air pollutants.

Several other factors discussed earlier in connection with baselines influence the scale of ancillary benefits. For example, continued technical innovation that improves energy efficiency and encourages the use of cleaner fuels will reduce baseline emissions and thus reduce ancillary benefits. Finally, the scale of ancillary benefits will depend on the scale of GHG mitigation—larger GHG mitigation should generate more ancillary benefits, though we would expect the incremental benefits to decrease. (The possibility of increased health benefits as GHG controls become so strict as to drive SO_2 emissions below the current cap is a counterexample to this point.)

Table 1 summarizes various ancillary benefit estimates, expressed in the common metric of dollars per ton reduction of carbon emissions. (*Note:* References for all estimates are given in Source Information at the end of this chapter.) In every case, the original studies that had produced the data identified a wide range of possible estimates around the midpoint estimate that we report. Lower and upper bounds for each estimate vary from the midpoints by factors of 2 to 10 or more.

Table 1 indicates a large variation even among the midpoint estimates in previous studies. Several differences in the analyses help to explain the different results, which include the modeling of criteria pollutant emissions reductions from GHG abatement, the estimation of health impacts from the criteria pollutant changes, and the evaluation of these impacts. One reason for the variation among studies is differences in the coverage of sectors, pollutants, and impacts. For example, one study considers a small voluntary program, whereas others consider the entire electricity sector or the economy as a whole. Some studies include a few of the larger health effects, and others attempt a more comprehensive accounting of ancillary benefits.

Another important element here is the treatment of locational differences (or lack thereof). More aggregated analyses calculate total emissions changes and apply a single unit value to value the avoided

Table 1. Comparisons of Estimates of Ancillary Benefits.

Model[a]	Targeted sectors, pollutants, and policy	Average ancillary benefit per ton of carbon reduction ($)[b]
HAIKU/ TAF	• Nationwide carbon tax of $25/ton of carbon in electricity sector, analyzed at state level • Only health effects from NO_x changes valued, including secondary particulates, excluding ozone effects • Range of estimates with and without NO_x SIP call[c] reductions included in baseline	2–5
ICF/ PREMIERE	• Nationwide Motor Challenge—voluntary program (industry), analyzed at regional level • Only health effects from NO_x changes valued, including secondary particulates, excluding ozone effects	3
Dowlatabadi and others/ PREMIERE	• Nationwide seasonal gas burn in place of coal, analyzed at regional level • Health effects from NO_x changes valued using PREMIERE, including secondary nitrates, excluding ozone effects	3
EXMOD	• Reduced use of existing coal steam plant at a suburban New York location • Only PM, NO_x, and SO_2 (under emission cap) changes valued (based on 1992 average emissions), including secondary particulates and ozone effects • All quantifiable health, visibility, and environmental effects included	26
Coal/ PREMIERE	• Equal percentage reduction in use of all existing (1994) U.S. coal plants analyzed at state level • Only health effects from NO_x changes valued, including secondary particulates and excluding ozone	8
Coal/ PREMIERE/RIA	• Equal percentage reduction in use of all existing (1994) U.S. coal plants analyzed at state level • Only NO_x-related mortality changes valued, including secondary particulates and excluding ozone, using new EPA RIA estimates of impacts and valuations	26
Abt and Pechan	• Carbon taxes of $30 and $67/ton of carbon • Modeled changes in conventional emissions and concentrations of particulates (no ozone) and changes in health status, visibility, and materials damages • Estimates of avoided abatement costs for NO_x and SO_2 • Cost savings in attainment areas, air quality improvements in nonattainment areas • NO_x SIP call reductions in baseline for all scenarios • Estimates of outcomes with and without reductions in SO_2 below 1990 Clean Air Act, based on size of carbon tax (high tax leads to net SO_2 reductions)	8 and 68
Goulder/ Scheraga	• Economy-wide carbon tax of $144/ton of carbon with stabilization at 1990 levels in 2000 • Human health effects calculated from reduced total emissions of all criteria pollutants, no secondary particulates or ozone	32
Boyd and others	• Economy-wide carbon tax of $9/ton of carbon • Human health and visibility effects calculated from reduced total emissions of all criteria pollutants	39
Viscusi and others	• Equal percentage reduction in use of existing (1980 average) U.S. coal steam plants nationwide • Human health and visibility effects calculated from reduced total emissions of all criteria pollutants	86

Notes: NO_x, nitrogen oxides; PM, particulate matter; SO_2, sulfur dioxide; EPA, U.S. Environmental Protection Agency; RIA, regulatory impact analysis.

[a] See Source Information for an explanation of the abbreviations used.

[b] Prices are expressed in 1996 dollars, rounded to the nearest dollar.

[c] Revision of State Implementation Plans (SIP) to accommodate EPA's proposed reduction in NO_x emissions, to take effect in 2004.

health impacts. In contrast, more disaggregated models can more precisely model the location of emissions, their transport through the atmosphere, and the exposure of affected populations. These analyses show that benefits do not have a simple proportional relationship to reduced emissions. Sensitivity analyses show that the above-mentioned aspects are important influences on ancillary benefits, so the greater precision with which they are calculated in disaggregated models gives us greater confidence in these results.

Moreover, long-term changes in pollution standards are not accounted for in any of the studies for assessing GHG policies that we discuss below (including our own). As a practical matter, our estimates of ancillary benefits should be considered more reliable for near-term GHG policies than for policies that are actually implemented during the 2008–12 commitment period identified in the Kyoto Protocol. Other things equal (which in practice is not the case), we would expect progress toward improved air quality in the United States to reduce ancillary benefits below the amounts shown in Table 1.

Treatment of the aggregate cap on SO_2 emissions created under Title IV of the 1990 Clean Air Act Amendments presents another important distinction among the studies. The avoided SO_2 abatement costs when emissions are lowered as a consequence of GHG policy likely are considerably smaller than the additional health benefit that would accrue if total SO_2 emissions were reduced below the cap. This aspect of ancillary benefits estimation is addressed in only a few of the studies in Table 1. Similar issues would emerge if EPA increased its use of economic incentive approaches, such as cap-and-trade regulation of NO_x to cut other pollutants.

Yet another factor is the uncertainty surrounding the economic valuation of avoided adverse impacts. For instance, one recent analysis suggests that the value of reducing premature mortality, when considered in the context of reduction in conventional air pollutants, is significantly lower than the usual estimates applied in all of the studies reported here (see Krupnick and others in Suggested Reading). On the other hand, some evidence points to a stronger link

between ozone concentrations and premature mortality than is represented in the existing studies considered here.

Firm conclusions are all but impossible to draw from the welter of estimates in Table 1, given the current state of knowledge. There is no best estimate of benefits per ton of carbon reduced for any particular GHG limitation, let alone for all possible GHG limitations.

We believe that modest but important ancillary benefits per ton of carbon reduction would result from a modest level of GHG control and that the benefits might be more substantial in certain locations (for example, those with denser populations and greater exposures to damaging criteria pollutants). The benefits per ton of carbon reduction would be larger with a greater degree of GHG control; however, it is difficult to gauge by how much. Moreover, the literature provides little in the way of estimates for ancillary benefits other than those associated with the electricity sector. A more reliable and comprehensive set of estimates must await analysis of how GHG abatement policies would affect other emissions sources, among other advances in knowledge.

Having said this, if the goal is to identify the ancillary benefit per ton of carbon reductions for a modest carbon abatement program, then we have greater confidence in the first five estimates in Table 1, all of which reflect the impact of GHG reductions in the electricity sector. These estimates reflect the most detailed methodologies, including locational differences in emissions and exposures, and they take into account the role of the SO_2 cap in limiting ancillary benefits. Note that these estimates suggest modest (less than \$10/ton) benefits on average for the United States as a whole, although benefits could be significantly higher in certain areas. The higher sixth estimate in the table reflects alternative assumptions about the scale of health impacts, the role of nitrates, and the economic valuation of impacts. The difference illustrates that ancillary benefits are sensitive to such assumptions, but given the controversy surrounding these specific assumptions, we put less stock in it.

However, the applicability of all these results is necessarily limited. Specific utility-sector policies for

CO_2 reduction may have different effects in different geographic areas than assumed in these estimates and may include changes not anticipated in the use of other technologies besides coal-fired plants. For example, an energy efficiency policy could reduce the use of low-emitting gas as well. Moreover, health is not the only environmental benefit. GHG policies affecting other sectors—notably, transportation—could also generate ancillary environmental benefits not captured in the utility-sector analyses. Finally, benefits would be larger than estimated if nonmarginal GHG mitigation policies were imposed, especially those that drive SO_2 emissions below the regulatory cap.

It may be tempting to embrace the last three studies in Table 1 that attempt to describe the effects of nonmarginal, economy-wide GHG reductions and include various pollutants and effects. However, the methodologies in these studies simply compute a total economic benefit from a national reduction in criteria pollutant emissions. They lack attention to locational differences in emissions and exposures, and they inherently overestimate the total ancillary benefits from SO_2 reduction by failing to take into account the effect of the SO_2 cap. Moreover, the assessments and valuations of health impacts in these studies are based on literature from the 1980s, but the field has developed rapidly in recent years. Finally, the ancillary benefits from a comprehensive carbon tax may not reflect the benefits generated by other, less comprehensive and cost-effective policies.

Our focus in the discussion thus far has been on estimating ancillary benefits and determining health-based benefits in the United States. We conclude this section with some brief comments on studies for other countries. Some efforts at ancillary benefits estimation also have been undertaken for Europe, particularly in the United Kingdom (see the papers by Ekins and by Barker and Rosendahl in Suggested Reading). They tend to be much larger than even the larger U.S. figures. Several reasons seem to explain the difference. Population concentrations are higher in Europe than in the United States, and wind patterns tend to direct more European emissions over populated areas, whereas more U.S. emissions are blown out over the Atlantic Ocean. In addition, the studies include a range of ecological as well as health effects, and the unit values used in these calculations tend to be substantial. The high valuation of ecological impacts (for example, acid precipitation damage to forests) could be attributed to a high European willingness to pay for ecological protection. On the other hand, there is reason to think the studies overestimate health benefits compared with the most recent literature on the subject. Moreover, many of the estimates accord substantial ancillary benefit to SO_2 and NO_x reduction that will already have taken place under new European regulations such as the Second Sulphur Protocol. In other words, the studies reflect a wrongly specified regulatory baseline.

Ancillary benefit studies also are being undertaken for developing countries. Whereas that literature is beyond the scope of this chapter, it will be reviewed in the forthcoming Third Assessment Report of the IPCC. In general, studies of ancillary benefits in developing countries are few and generate highly variable conclusions. The estimates are fraught with uncertainty for several reasons. For example, detailed modeling of how emissions disperse in the atmosphere is rarely available, and detailed emission inventories are rare, so studies often have simply applied "unit values" that express a change in health status resulting from a change in emissions without modeling emissions diffusion, population exposure, and health responses. Even when these intermediate steps are modeled, studies have used relationships from the United States and elsewhere that may not be applicable because of other important influences on health status, including differences in expected lifetimes and other risk factors. Without a doubt, much potential exists for health improvements in developing countries. Just how much GHG policy could contribute to this is uncertain, and the concern is that GHG policy is not only an indirect way to achieve such goals but potentially a more expensive way than direct interventions in local environmental problems.

Lessons for Policy

One important application of information about ancillary benefits and costs is in gauging the overall costs and benefits of GHG controls given the presumed baseline conditions and, in particular, to evaluate what degree of GHG mitigation might be "no regrets." The uncertainties that plague current estimates of ancillary benefits make it impossible to confidently answer this question. However, our analysis (the first row in Table 1; see Burtraw and others in Source Information) leads us to conclude that at least for relatively modest GHG control levels, ancillary benefits may be a significant fraction of costs. The marginal costs of small initial reductions are likely to be fairly low; indeed, there is reason to think they would be close to zero (some would even argue less than zero, but we remain skeptical). Compared with such a low cost, ancillary environmental benefits of even $3/ton of carbon reduced, let alone $7–10/ton, could have a significant effect on the volume of no-regrets emissions reduction.

However, we emphasize that large uncertainties in these estimates are related to the measurement of health benefits, the valuation of those benefits, the magnitude of nonhealth benefits, and policies that will change baseline air quality and conventional pollutant control costs in the future. Our analysis and review of the literature does not support the conclusions of some other ancillary benefits studies—that these benefits can offset much, all, or more than the cost of GHG abatement, especially for nonmarginal GHG policies that involve significant GHG controls.

Some insights can be derived from our analysis and applied to the design of policy, but they must be interpreted with care. Ancillary benefits may be larger for GHG policies that heavily target coal use, but the reason has at least as much to do with the continued use of old, relatively polluting boilers as with the coal itself. And GHG abatement policies that have relatively greater effects and impose greater costs on new plants will have the perverse effect of creating a new bias against construction of new facilities, resulting in continued use of old facilities and low ancillary benefits.

Compared with other options, GHG mitigation that occurs in areas especially conducive to the formation of secondary pollutants (ozone and secondary particulates) will confer larger ancillary benefits. Similarly, GHG mitigation that occurs at sources whose emissions affect large populations and ones that are subject to high levels of pollution also will tend to confer larger benefits than where populations are small and pollution is lower.

The possible trend in ancillary benefits over time also is of interest. It is often argued that abatement costs associated with a goal such as GHG emissions stabilization will rise over time because of growing energy demand, even though technical progress and ultimately a transition to noncarbon backstop energy resources should ease the trend. The ancillary benefits of GHG control also will rise over time, as a result of growth in population density and congestion and growth in income that can be expected to increase public willingness to pay for environmental protection. However, ancillary benefits will trend downward to the extent that more ambitious national goals for conventional pollutant control are set and achieved.

Last but not least, it is important to be cautious about the implications of ancillary benefits with regard to the desired level of GHG control. Put simply, when ancillary benefits are taken into account, the GHG policies that have the lowest net cost to society do not necessarily target the least expensive sources for reducing carbon emissions. Nor do they necessarily maximize the ancillary benefits of GHG control. Ancillary benefits are important enough that they should be considered jointly with the costs of carbon reduction to identify the preferred policies for society. At the same time, the choice of policies can have important distributional effects, in both economic costs and ancillary benefits that must be considered. These distributional issues are an important topic for additional research.

Suggested Reading

Barker, Terry, and K.E. Rosendahl. 2000. Ancillary Benefits of GHG Mitigation in Europe: SO_2, NO_x, and PM_{10} Reductions from Policies to Meet the Kyoto Targets Using the E3ME Model and EXTERNE Valuations. Paper presented at the IPCC Expert Workshop on Assessing the Ancillary Benefits and Costs of Greenhouse Gas Mitigation Policies, March 27–29, 2000, Washington DC.

Burtraw, Dallas, Alan Krupnick, Erin Mansur, David Austin, and Deirdre Farrell. 1997. The Costs and Benefits of Reducing Air Pollutants Related to Acid Rain. *Contemporary Economic Policy* 16: 379–400.

Davis, Devra Lee. 1997. Short-Term Improvements in Public Health from Global-Climate Policies on Fossil-Fuel Combustion: An Interim Report. *Lancet* 350: 1341–49.

Ekins, Paul. 1996. How Large a Carbon Tax Is Justified by the Secondary Benefits of CO$_2$ Abatement? *Resource and Energy Economics* 18: 161–87.

European Commission. 1995. ExternE: *Externalities of Energy*. For the Directorate General XII. Prepared by Metroeconomica, CEPN, IER, Eyre Energy-Environment, ETSU, Ecole des Mines. Luxembourg: Office of Official Publications of the European Communities.

Krupnick, Alan J., and Dallas Burtraw. 1997. The Social Costs of Electricity: Do the Numbers Add Up? *Resources and Energy* 18(4): 467–90.

Krupnick, Alan, Dallas Burtraw, and Anil Markandya. 2000. The Ancillary Benefits and Costs of Climate Change Mitigation: A Conceptual Framework. Paper presented at the IPCC Expert Workshop on Assessing the Ancillary Benefits and Costs of Greenhouse Gas Mitigation Policies, March 27–29, 2000, Washington DC.

Krupnick, Alan, Anna Alberini, Maureen Cropper, Natalie Simon, Bernice O'Brien, Ron Goeree, and Martin Heintzelman. 2000. Age, Health, and the Willingness to Pay for Mortality Risk Reductions: A Contingent Valuation Survey of Ontario Residents. RFF Discussion Paper 00-37. Washington, DC: Resources for the Future.

Lee, R., A.J. Krupnick, D. Burtraw, and others. 1995. *Estimating Externalities of Electric Fuel Cycles: Analytical Methods and Issues, and Estimating Externalities of Coal Fuel Cycles*. Washington, DC: McGraw-Hill/Utility Data Institute.

Morgenstern, Richard D. 2000. Baseline Issues in the Estimation of the Ancillary Benefits of Greenhouse Gas Mitigation Policies. Paper presented at the IPCC Expert Workshop on Assessing the Ancillary Benefits and Costs of Greenhouse Gas Mitigation Policies, March 27–29, 2000, Washington DC. (Several important influences on the baseline are identified in this paper.)

Munasinghe, Mohan, and Rob Swart (eds.). 2000. *Climate Change and its Linkages with Development, Equity, and Sustainability*. Geneva, Switzerland: Intergovernmental Panel on Climate Change.

U.S. DOE/EIA (Department of Energy, Energy Information Administration). 1997. *Annual Energy Outlook 1997*. Washington, DC: U.S. EIA.

U.S. EPA (Environmental Protection Agency). 1996. *The Benefits and Costs of the Clean Air Act, 1970–1990*. Washington, DC: U.S. EPA.

Source Information

HAIKU/TAF

Burtraw, Dallas, Alan Krupnick, Karen Palmer, Anthony Paul, Mike Toman, and Cary Bloyd. 1999. Ancillary Benefits of Reduced Air Pollution in the U.S. from Moderate Greenhouse Gas Mitigation Policies in the Electricity Sector. RFF Discussion Paper 99-51. September. Washington, DC: Resources for the Future.

ICF/PREMIERE

Holmes, Rebecca, Doug Keinath, and Fran Sussman. 1995. *Ancillary Benefits of Mitigating Climate Change: Selected Actions from the Climate Change Action Plan*. Final report prepared for Adaptation Branch, Climate Change Division, Office of Policy, Planning and Evaluation, U.S. Environmental Protection Agency. Contract No. 68-W2-0018. March 31. Washington, DC: ICF Incorporated. (To generate the entries for Table 1, results from this study were combined with analysis using the PREMIERE model [see later citation under Coal/PREMIERE].)

Dowlatabadi and others/PREMIERE

Dowlatabadi, Hadi, F. Ted Tschang, and Stuart Siegel. 1993. Estimating the Ancillary Benefits of Selected Carbon Dioxide Mitigation Strategies: Electricity Sector. August 5. Unpublished report prepared for the Climate Change Division. Washington, DC: U.S. Environmental Protection Agency. (To generate the entries for Table 1, results from this study were combined with analysis using the PREMIERE model [see later citation under Coal/PREMIERE].)

EXMOD

Hagler Bailly Consulting, Inc. 1995. *Human Health Benefits Assessment of the Acid Rain Provisions of the 1990 Clean Air Act Amendments*. Unpublished Final Report prepared by Hagler Bailly Consulting, Inc., Boulder, CO, under subcontract to ICF Incorporated, Fairfax, VA, for the U.S. Environmental Protection Agency, Acid Rain Division.

Coal/PREMIERE

Bloyd, Cary, and others. 1996. *Tracking and Analysis Framework (TAF) Model Documentation and User's Guide*. ANL/DIS/TM-36. December. Argonne, IL: Argonne National Laboratory.

Palmer, Karen L., and Dallas Burtraw. 1997. Electricity Restructuring and Regional Air Pollution. *Resources and Energy* 19(1–2): 139–74.

Coal/PREMIERE/RIA: same as above, plus

U.S. EPA (Environmental Protection Agency). 1996. *Regulatory Impact Analysis for Proposed Particulate Matter National Ambient Air Quality Standard* (Draft). December. Unpublished report prepared by Innovative Strategies and Economics Group, Office of Air Quality Planning and Standards. Research Triangle Park, NC: U.S. EPA.

Abt and Pechan

Abt Associates and Pechan-Avanti Group. 1999. Co-Control Benefits of Greenhouse Gas Control Policies. February. Unpublished report prepared for Office of Policy. Washington, DC: U.S. Environmental Protection Agency.

Goulder/Scheraga

Goulder, Lawrence H. 1993. Economy-Wide Emissions Impacts of Alternative Energy Tax Proposals. July. Unpublished report submitted to Adaptation Branch, Climate Change Division. Washington, DC: U.S. Environmental Protection Agency.

Scheraga, Joel D., and Susan S. Herrod. 1993. Assessment of the Reductions in Criteria Air Pollutant Emissions Associated with Potential CO_2 Mitigation Strategies. August. Unpublished report prepared in the Office of Policy, Planning and Evaluation, Climate Change Division. Washington, DC: U.S. Environmental Protection Agency.

Scheraga, Joel D., and Neil A. Leary. 1993. Costs and Side Benefits of Using Energy Taxes to Mitigate Global Climate Change. *Proceedings 1993 National Tax Journal,* 133–38.

Boyd and others

Boyd, Roy, Kerry Krutilla, and W. Kip Viscusi. 1995. Energy Taxation as a Policy Instrument to Reduce CO_2 Emissions: A Net Benefit Analysis. *Journal of Environmental Economics and Management* 29(1): 1–24.

Viscusi and others

Viscusi, W. Kip, Wesley A. Magat, Alan Carlin, and Mark K. Dreyfus. 1994. Environmentally Responsible Energy Pricing. *The Energy Journal* 15(2): 23–42.

B

Appendix
Climate Change, Health Risks, and Economics

Alan J. Krupnick

Changes in greenhouse gas (GHG) concentrations in the atmosphere may result in warmer air and ocean temperatures. In addition, there is speculation that such changes could lead to increased frequency and severity of weather disturbances and of oscillations, such as the El Niño Southern Oscillation.

If these changes occur, they could significantly affect death rates by increasing heat-related deaths and the incidence of infectious disease. For instance, one researcher predicts that heat-related deaths could double by 2020 and increase even more by 2050 on the basis of about a doubling of very hot summer days predicted by one of the climate change models. For instance, the Atlanta, Georgia, population of 3 million experiences 78 deaths related to excess heat each year. This total could rise to 191 in 2020 and 293 in 2050 if the modeled temperature increases take place (see the 1996 IPCC report in Suggested Readings) and behavior and technology remain the same.

Not surprisingly, the size of such effects and even their direction remain highly uncertain. In addition, the cost of policies or programs to change behavior to reduce such effects may not be high. Economic

This essay is based in part on material presented at Emerging Public Health Threats and the Role of Climate Change, a conference sponsored by the U.S. Environmental Protection Agency held in March 1998 in Atlanta, Georgia.

analysis provides a means for both estimating such costs and developing policies to reduce them.

Nature of the Threat

Previous heat waves have been associated with significantly increased death rates in areas where the population is not acclimatized to such temperature variability and lives in housing that is not designed to protect its occupants from temperature spikes (for example, windows may not open, or rooms may not be ventilated). In July 1995, 500 deaths in excess of normal mortality rates were recorded in Chicago during a three- to five-day hot spell. Recent research identifies certain synoptic air masses (such as hot, moist, tropical air) as being particularly highly correlated with increased summer death rates. Hot nighttime temperatures have been identified as a particular concern.

One might think that warmer temperatures would result in fewer deaths overall simply because more deaths occur in the winter than in the summer months (10–25% more in temperate regions). If the effect of GHGs on temperature and the effect of temperature on health were proportional and operated symmetrically with respect to high and low temperatures, then the net effect of warming would be to reduce deaths. However, because respiratory infections account for a large fraction of winter deaths and no correlation has been observed between small changes in winter temperatures and outbreaks of these infections, the effects are likely to be asymmetric, and the net effect will be a further increase in deaths beyond the status quo.

Warmer temperatures are also likely to affect the concentration of air pollutants. Exposure to air pollutants can have serious effects on human health. No conclusive results show the effects of climate change on air pollutants. As ground level ozone increases with higher temperatures, it would be reasonable to expect to observe higher concentrations of this pollutant and its associated health damages. However, these damages by and large are limited. Particulate concentrations—particularly those that form through chemical processes in the atmosphere—have been more strongly linked to effects on morbidity and increases in premature mortality. However, these chemical reactions are not particularly sensitive to temperature (but relative humidity may play a significant role).

Warm weather combined with increased precipitation also can be a potent engine for spreading various vectors (such as mosquitoes) that in turn spread infectious diseases. The World Health Organization (WHO) has identified several diseases that are "likely" or "highly likely" to spread with warmer average air temperatures: malaria, lymphatic filiariases, onchocerciasis, schistosomiasis, African typanosomiasis, Dengue fever, yellow fever, and Japanese encephalitis. Waterborne infections (such as cholera) and effects associated with warmer water temperatures (such as red tides) also can be expected to worsen.

Malaria is probably the most serious concern, because it is rated by the WHO as highly likely to spread with warm temperatures and already affects 270 million people worldwide. In addition, vectors (mosquitoes) and disease carriers (blood-borne parasites) are becoming increasingly resistant to insecticides and antimalarial drugs, respectively, used in Africa and South East Asia.

Temperature change could affect malaria incidence through several channels. First, the parasite carrying the disease cannot reproduce below 16 °C, so the minimum daily temperature in a location is the critical variable. Second, at temperatures higher than 20 °C, the time between the mosquitoes ingesting infected blood and being able to transmit the parasite shortens, which means that the disease can spread more rapidly. Third, higher temperatures increase the breeding and feeding rates of mosquitoes.

Large cities such as Nairobi and Harare in developing countries are currently protected from major malaria incidence by their low minimum daily temperatures. However, these cities would be potentially vulnerable to malaria outbreaks after even a relatively small amount of warming.

Warmer ocean temperatures are of concern because they may lead to ecosystem changes that affect health. (In the following discussion, I do not address the effects of increased frequency or severity of natural disasters on health.) Some of the most serious issues appear to be those associated with El Niño, the periodic change in ocean temperature gradients that warms parts of the Pacific Ocean and thereby affects global weather patterns. Recent evidence suggests this effect could become more frequent and more severe by climate change. For example, high rainfall during a recent El Niño period in the southwestern United States resulted in an unusually large increase in rodent populations, which carry the hantavirus. Increased rodent populations have been linked to an unusually large number of hantavirus infections in the Southwest. Changes in temperature patterns also can affect disease rates. Some studies have shown an association between malaria rates in various countries and temperature swings associated with El Niño episodes, but the evidence is far from conclusive.

Recent research links cholera outbreaks to El Niño and other events that produce warmer sea surface temperatures. Warmer temperatures increase algae growth, which in turn leads to growth in the zooplankton that feed on the algae. Some species of zooplankton were found to harbor bacteria that cause cholera. The six-week lag between algal blooms and zooplankton growth and the appearance of cholera outbreaks where the blooms were occurring suggests a link between sea surface temperatures and cholera.

Uncertainties and Adaptation Potential

The key word in many of these dire predictions is "may." Uncertainties exist at every step in the chain from GHG emissions to outbreak of disease. For example, the coincident introduction of cholera bacteria from ship bilge and through other means unaffected by El Niño can't be ruled out as a cause of recent outbreaks in Peru. More research also is needed to address the links between temperature spikes and increased mortality, such as links between the use of air conditioning and increased pollution, and the interaction of weather characteristics and pollution. It is not even clear that malaria rates will increase. Some of the many mosquito species that carry malaria may not adapt well to changes in temperature, humidity, or precipitation. Furthermore, whereas pools of standing water contribute to mosquito breeding, heavy rainfall just washes away the eggs. So, increasing precipitation could decrease malaria rates in certain regions.

Moreover, the conclusion that there is a link between climate change and increased incidence of disease does not automatically imply the need to devote significant resources to reducing GHGs. Changing lifestyles and behaviors—*adaptation*—is a key part of risk management. To reduce health risks, it may be more effective to invest in technologies and measures that improve the protection of public health through traditional means, such as improved sanitation, greater dissemination of bed nets to protect against mosquitoes, more vaccination programs, and the like. Greater attention to protecting people from heat may be far more effective than reducing the chance of temperature spikes by a small amount.

The Role of Economic Analysis

Economics can play a role first in helping people to understand the benefits and costs of reducing the incidence of deaths and diseases by assessing their trade-offs between reduced threats and other benefits. In this way, the value of reduced health threats can be estimated and compared with the costs of various risk-reducing actions by using a common metric. Such a comparison is important because trade-offs need to be made in the allocation of resources to meet climate change concerns versus meeting other needs. Economic analysis offers tools and a framework to investigate these trade-offs.

In a similar vein, economic analysis can inform a traditional debate in the public health area. This debate concerns the share of resources that should be directed toward improving public health through treatment (either before, as vaccinations, or after, with therapy); avoidance of exposure (by using mosquito nets and air conditioning, or by making other behavioral and technological changes); and mitigation of the conditions that raise exposure risks (such as reducing GHGs).

A more targeted illustration of the role for economic analysis is in developing better estimates of the costs of a disease outbreak that may be related to climate change. Although the direct medical and productivity costs of an outbreak are sometimes measured, indirect costs—such as those associated with the avoidance of particular areas or particular foods—are rarely considered quantitatively. A few cases of food poisoning discovered at a theme park can result in huge costs, for instance, if the park is shut down.

Another promising area for economic analysis is in contributing to the design and effectiveness of health and environmental surveillance and monitoring systems. As early as 1983, the Peruvian government started predicting rainfall patterns related to El Niño disturbances and passing on this information to its department of agriculture, which subsequently alerted farmers. The farmers were then assisted in deciding whether to plant rice, which thrives in wet weather, or cotton, which tolerates drier weather. Other countries have since used El Niño predictions to help make agricultural decisions. In addition, major efforts have been made to bolster ocean surveillance for El Niño events and to plan the development of integrated health and environmental monitoring systems to help understand and predict the linkages among climate change, weather, ecology, and health.

A "value of information" analysis can be used to help identify the potential gains from such information systems and the associated systems for human responses to better information. This approach examines the costs and benefits of better (more frequent, higher quality, more targeted) information. In the case of El Niño, it means better predictions about the amount of rainfall over particular areas in the coming year. The benefits of better information depend on the kind of information provided, to whom it is provided, how it is disseminated, the available actions to take (and their costs) on receiving the information, the risks and consequences of the predictions being wrong, and an estimate of the baseline (that is, what choices would have been made in the absence of the better information). These benefits can be compared with the costs to help determine which investments in early warning and response systems are most justified.

Such analyses seem particularly timely now, when much of the debate over climate change risks and response costs remains so inadequately informed. Analysts and policymakers need to take a broad, interdisciplinary approach to determine both the magnitude of the risks posed and the appropriate preventative and adaptive response measures to be taken.

Suggested Reading

IPCC (Intergovernmental Panel on Climate Change). 1996. *Climate Change 1995: Impacts, Adaptations, and Mitigation of Climate Change: Scientific-Technical Analysis.* Contribution of Working Group II to the Second Assessment Report of the Intergovernmental Panel on Climate Change. New York: Cambridge University Press. (Chapter 18 is a review of the health implications of climate change and provides numerous literature citations.)

————. 1998. *The Regional Impacts of Climate Change: An Assessment of Vulnerability.* New York: Cambridge University Press. (This report is a detailed discussion of regional impacts, including those on human health.)

The report by Working Group II in the Third IPCC Assessment, due to be published in 2001, will update and extend the Second Assessment Report.

Part 3

Policy Design and Implementation Issues

9

Choosing Price or Quantity Controls for Greenhouse Gases

William A. Pizer

Much of the debate surrounding climate change has centered on verifying the threat of climate change and deciding the magnitude of an appropriate response. After years of negotiation, this effort led to the 1997 signing of the Kyoto Protocol, a binding commitment by industrialized countries to reduce their emissions of carbon dioxide (CO_2) to slightly below 1990 recorded levels. Without approving or disapproving of the response effort embodied in the Kyoto Protocol, I believe that an important element has been ignored. Namely, should we specify our response to climate change in terms of a quantitative target?

The appeal of a quantitative target is obvious. A commitment to a particular emissions level provides a straightforward measure of environmental progress as well as compliance. Commitment to an emissions tax, for example, offers neither a guarantee that emissions will be limited to a certain level nor an obvious way to measure a country's compliance (when other taxes and subsidies already exist). Yet, this concern points to an important observation.

Quantity targets guarantee a fixed level of emissions. Emission taxes guarantee a fixed financial incentive to reduce emissions. Both can be set at aggressive or modest levels. Aside from the appeal of the known and verifiable emissions levels that quantity targets can ensure, might there be other important differences between price and quantity controls? Economists would say "Yes." With uncertain outcomes and policies that are fixed for many years, it is important to carefully consider both the costs and benefits of alternate price and quantity controls to judge which is best. My own analysis of the two approaches indicates that price-based greenhouse gas (GHG) controls are much more desirable than quantity targets, taking into account both the potential long-term damages of climate change and the costs of GHG control. This can be argued on the basis of both theory and numerical simulations. On the basis of the latter, I find that price mechanisms produce expected net gains five times higher than even the most favorably designed quantity target.

To explain this conclusion, I first characterize the differences between price and quantity controls for GHGs. I then present both theoretical and empirical evidence that price-based controls are preferable to quantity targets on the basis of these differences. Finally, I discuss how price controls can be implemented without a general carbon tax. This point is particularly salient for the United States, where taxes are generally unpopular. The "safety valve," as it is often called, involves a cap-and-trade GHG system accompanied by a specified fee or penalty for emissions beyond the initial cap.

How Do Quantity- and Price-Based Mechanisms Work?

A quantity mechanism—usually referred to as a permit or cap-and-trade system—works by first requiring individuals to obtain a permit for each ton of CO_2 they emit, and then limiting the number of permits to a fixed level. (CO_2 emissions from fossil fuel sources constitute the bulk of GHG emissions and are the general focus of most policy discussions. However, the arguments made in this context apply equally well to the regulation of GHG emissions more broadly defined.) This kind of system has been used with considerable success in the United States to regulate sulfur dioxide and lead. The permit requirement could be imposed on the individuals who release CO_2 into the atmosphere by burning coal, petroleum products, or natural gas. However, unlike the emissions of conventional pollutants, which depend on various other factors, CO_2 emissions can be determined very accurately by the volume of fuel being used. Rather than requiring *users* of fossil fuels to obtain permits, we could therefore require *producers* to obtain the same permits. This method has the advantage of involving far fewer individuals in the regulatory process, thereby reducing both monitoring and enforcement costs (see Chapter 10, Using Emissions Trading to Regulate National Greenhouse Gas Emissions).

One key element in a permit system is that individuals are free to buy and sell existing permits in an effort to obtain the lowest cost of compliance for themselves, which in turn leads to the lowest cost of compliance for society. In particular, when individuals observe a market price for permits, those who can reduce emissions more cheaply will do so to sell excess permits or to avoid having to buy additional ones. Similarly, those who face higher reduction costs will avoid reductions by buying permits or by keeping those they already possess. In this way, total emissions will exactly equal the number of permits, and only the cheapest reductions are undertaken.

A price mechanism—usually referred to as a carbon tax or emissions fee—requires the payment of a fixed fee for every ton of CO_2 emitted. Like the permit system, this fee could be levied upstream on fossil fuel producers or downstream on fossil fuel consumers. Either way, we associate a positive cost with CO_2 emissions and create a fixed monetary incentive to reduce emissions. Such price-based systems have been used in Europe to regulate a wide range of pollutants (although the focus is usually revenue generation rather than substantial emissions reductions).

Like a tradable permit system, price mechanisms are cost-effective. Only those emitters who can reduce emissions at a cost below the fixed fee or tax will choose to do so. Because only the cheapest reductions are undertaken, we are guaranteed that the resulting emission level is obtained at the lowest possible cost.

The important distinction between these two systems is how they adjust when costs change unexpectedly. A quantity or permit system adjusts by allowing the permit price to rise or fall while holding the emissions level constant. A price or tax system adjusts by allowing the level of total emissions to rise or fall while holding the price associated with emissions constant. Ignoring uncertainty and assuming that we know the costs of controlling CO_2, both policies can be used with the same results. Consider the following example:

Suppose we know that with a comprehensive domestic CO_2 trading system in place in the United States by 2010, a permit volume of 1.2 gigatons (billion tons) of carbon equivalent emissions (GtC) will lead to a $100 permit price per ton of carbon. (U.S. emissions of carbon from fossil fuels were estimated at 1.5 GtC for 1998.) In other words, faced with a price incentive of $100 per ton to reduce emissions, regulated firms in the United States will find ways to reduce emissions to 1.2 GtC. Then, the same outcome can be obtained by imposing a $100 per ton carbon tax.

Uncertainty about Costs

In reality, we have only a vague idea about the permit price that would occur with emissions of 1.2 GtC or any other emission target. These costs are hard to pin down for three reasons. First, little evidence exists concerning reduction costs. There are no recent examples of carbon reductions on a substantial scale from which to base estimates. In the 1970s, energy prices doubled and encouraged increased energy efficiency, but these events occurred in a context of considerable uncertainty about the future and alongside many other confounding factors (such as increased environmental regulation). Alternatively, engineering studies provide a bottom-up approach to estimating costs. However, comparisons of past engineering forecasts with actual implementation costs suggest that forecasts are inaccurate at best.

A second source of uncertainty arises because we need to forecast compliance costs in the future. This task involves difficult predictions about the evolution of new technologies. Proponents of aggressive policy argue that reductions will be cheap as new low-carbon or carbon-free energy technologies become available. Proponents of more modest policies argue that these are unproven, pie-in-the-sky technologies that may never be practical.

Finally, it is impossible to know how uncontrolled emission levels will change in the future. That is, to achieve 1990 emission levels in 2010, it is unclear whether reductions of 5%, 25%, or even 50% will be necessary. The Intergovernmental Panel on Climate Change (IPCC), the international agency charged with studying climate change, gives a range of six possible global emission scenarios in 2010 that include a low of 9 GtC and a high of 13 GtC. My own simulations suggest a broader possible range, 7–18 GtC.

The low end of both ranges reflects the possibility that population and economic growth may slow in the future and the energy intensity of production may fall. The high end reflects the opposite possibility, that growth remains high and energy intensity rises. Figure 1 shows the distribution of uncontrolled emissions arising from my simulations of 1,000 possible outcomes in 2010 alongside the six IPCC scenarios. (For details about the model, see Pizer in Suggested Reading.)

In summary, we have only vague ideas about the cost of alternative emission targets for two important reasons. First, there is little historic evidence about costs. Second, as we examine policies 10 or more years in the future, it is unclear how baseline emissions and available technologies will change between now and then. Figure 1 indicates that global emissions could be anywhere from 7 to 18 GtC in 2010. The cost associated with a target of 8.5 GtC (1990 level) will be uncertain because the necessary reduction is uncertain—somewhere between 0 and 10 GtC—and because costs are difficult to estimate, even knowing the reduction level.

Effects of Price and Quantity Controls with Cost Uncertainty

When the cost of a particular emission target is uncertain, price and quantity controls will have distinctly different consequences for the actual level of emissions as well as the overall cost of a climate policy. Even if both policies are designed to deliver the same results under a best-guess scenario, they will necessarily behave differently when control costs deviate from this best guess. These differences arise because a price policy provides a fixed incentive (dollars per ton of CO_2 emissions), regardless of the emission level, and a quantity policy generates whatever incentive is necessary to strictly limit emissions to a specified level.

Figure 2 illustrates these differences by showing the emission consequences in 2010 associated with two policies that are roughly equivalent under a

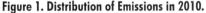

Figure 1. Distribution of Emissions in 2010.

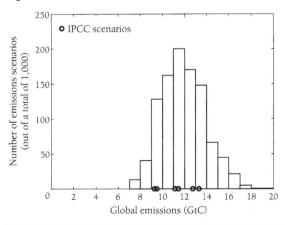

Figure 2. Effect of Price and Quantity Controls on Emissions in 2010.

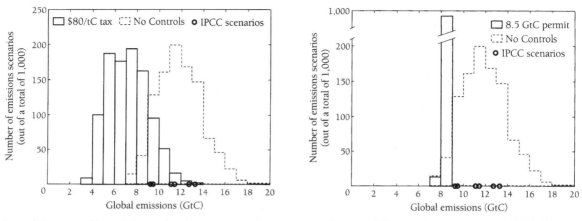

Two policies are roughly equivalent under a best-guess scenario: a carbon tax of $80/ton (left) and a quantity target of 8.5 GtC (right).

best-guess scenario: a quantity target of 8.5 GtC and a carbon tax of $80/ton. Using the same 1,000 emission scenarios shown in Figure 1, simulations are used to calculate the effect of these two policies for each outcome. With a carbon tax, emissions are below 8.5 GtC in more than 75% of the outcomes. In other words, on average the carbon tax achieves more reductions than a quantity target of 8.5 GtC. Sometimes, the reductions are much more; emissions may be as low as 3 GtC. Yet, the carbon tax fails to guarantee that emissions will always be below any particular threshold.

The quantity target, in contrast, never results in emission levels above 8.5 GtC. Because some emission outcomes in the absence of controls were rather high, on the order of 18 GtC, we would expect that the cost of this policy could be quite high. At the other extreme, the quantity policy would be costless if uncontrolled emissions were unexpectedly low.

These data suggest that the cost associated with quantity controls will be high or low depending on future reduction costs as well as the future level of uncontrolled emissions. In contrast, price controls create a fixed incentive to reduce each ton of CO_2 regardless of the uncontrolled emission level. Therefore, costs under a carbon tax should fluctuate much less than costs under a quantity control.

Figure 3 shows the estimated cost consequences of both policies. The range of costs associated with the quantity target is quite wide, as we suspected. The estimates extend from 0 to 2.2% of global gross domestic product (GDP), almost four times higher than the highest cost outcome under the carbon tax. In fact, the cost associated with emission reductions under a carbon tax are concentrated entirely in the range 0.2–0.6% of GDP. Because the carbon tax always applies the same per ton incentive to reduce emissions, the cost outcomes are more narrowly distributed than those occurring under a quantity target.

Choosing between Price and Quantity Controls

So far, the discussion has been limited to the different emission and cost consequences of alternative price and quantity controls. Choosing between them, as well as choosing the appropriate stringency of either policy, requires making judgments about climate change consequences as well as control costs. To understand when one policy instrument probably will be preferred to the other, it is useful to consider two extreme cases.

First, imagine that there is a known climate change threshold. When CO_2 emissions are below this threshold, the consequences are negligible.

Figure 3. Distribution of 2010 Costs Associated with Price and Quantity Controls.

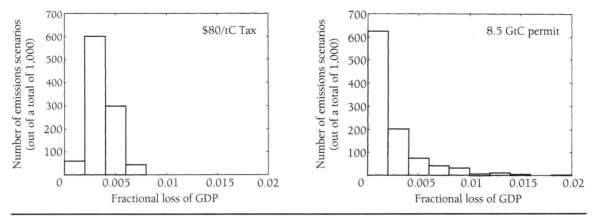

Above this threshold, however, damages are potentially catastrophic. For example, research suggests that the process by which CO_2 is absorbed at the surface of the oceans and circulated downward could change dramatically under certain circumstances. If we also believe that these changes will have severe consequences and that we can identify a safe emission threshold for avoiding them, then quantity controls seem preferable. Quantity controls can be used to avoid crossing the threshold, and in this case, large expenditures to meet the target are justified by the dire consequences of missing it.

Now, imagine instead that every ton of CO_2 emitted causes the same incremental amount of damage. These damages might be very high or low, but the key is that each ton of emissions is just as bad as the next. Such a scenario is also plausible, as indicated by a survey of experts including both natural and social scientists who do research on global warming. Their beliefs suggest that the damage caused by each ton of emitted CO_2 may be quite high but that there is no threshold: Damages are essentially proportional to emissions. Each additional ton is equally damaging, whether it is the first ton emitted or the last.

In this case, it makes sense to use a price instrument. Specifically, a carbon tax equal to the damage per ton of CO_2 will lead to exactly the right balance between the cost of reducing emissions and the resulting benefits of less global warming. Every time a

firm decides to emit CO_2, it will be confronted with an added financial burden equal to the resulting damage. It will lead to reduction efforts as well as investments in new technology that are commensurate with the alternative of climate change damage. In this scenario, little emphasis is placed on reaching a particular emission target because there is no obvious quantity target to choose. This argument applies even if we are uncertain about the magnitude of climate damage per unit of CO_2.

Arguments for Price Policies

Given this characterization of circumstances under which alternative price and quantity mechanisms are preferred, we can make the argument for price controls. This argument hinges on two basic points. The first point is that climate change consequences generally depend on the stock of GHGs in the atmosphere, rather than annual emissions. GHGs emitted today may remain in the atmosphere for hundreds of years. It is not the level of annual emissions that matters for climate change but the total amount of CO_2 and other GHGs that have accumulated in the atmosphere. The second point is that although scientists continue to argue over a wide range of climate change consequences, few advocate an immediate halt to emissions. For example, the most aggressive stabilization target discussed by the IPCC is a 450-ppm concentration in the atmosphere (roughly 1,035 GtC), a level that we will not reach before 2030, even in the absence of emission controls (see the Technical Summary of the IPCC report in Suggested Reading).

If only the stock of atmospheric GHGs matters for climate change, and if experts agree that the stock will grow at least in the immediate future, then there is almost no rationale for quantity controls. The fact that only the stock matters should first draw our attention away from short-term quantity controls for emissions and toward long-term quantity controls for the stock. It cannot matter whether a ton of CO_2 is emitted this year, next year, or in 10 years if all we care about is the total amount in the atmosphere. Taking the next step and presuming that the stock will grow over the next few decades, this approach suggests that there is some room to rearrange emissions over time and that a short-term quantity control on emissions is unnecessary.

Quantity controls derive their desirability from situations where strict limits are important, when dire consequences occur beyond a certain threshold. Such policies trade off low expected costs in favor of strict control of emissions in all possible outcomes. However, under the assumption that it is acceptable to allow the stock of GHGs to grow in the interim, there is no advantage to such strict control. We give up the flexible response of price controls without the benefit of an avoided catastrophe.

Even for those who believe the consequences of global warming will be dire and that current emission targets are not aggressive enough, price policies are still better. An aggressive policy designed to stabilize the stock eventually does not demand a strict limit on emissions before stabilization becomes necessary. Additional emissions this year are no worse than emissions next year. Why not abate more when costs are low, less when costs are high—exactly the outcome under a price mechanism? When we eventually move closer to a point where the stock must be stabilized, a switch to quantity controls will be appropriate.

In addition to these theoretical arguments, integrated assessment models can provide support. To this end, I have constructed an integrated model of

the world economy and climate based on the dynamic integrated climate-economy (DICE) model developed by William Nordhaus (see Suggested Reading). In contrast to the DICE model, I simultaneously incorporate uncertainty about everything from growth in population and energy efficiency to the cost of emission reductions, the sensitivity of the environment to atmospheric CO_2, and the damages arising from global warming.

The results of these simulations indicate the price-based mechanisms can generate overall economic gains (expected benefits minus expected costs) that are *five times higher* than even the most prudent quantity-based mechanism. These results are robust. Even allowing for catastrophic damages beyond 3 °C of warming, price mechanisms continue to perform better. This robustness can be explained in two ways. First, the catastrophe—if it exists—is in the future. Before we reach that point, it is desirable to have some flexibility in emission reductions. Specifically, we will want to delay those reductions if the costs are unexpectedly high in the short run, provided those reductions can be obtained more cheaply in the future but before the catastrophe.

Second, unlike the stylized description in which climate consequences depended directly on CO_2 concentrations presented earlier, in this model, damages depend on temperature change. In reality, damages probably depend on an even more complex climatic response. Either way, the links between CO_2 emissions, concentrations, temperature change, and other climatic effects are not precisely known. Therefore, a quantity control on *emissions* is not equivalent to a quantity control on *climate change*. Both price and quantity controls will lead to uncertain climate consequences. Therefore, the advantage of the quantity control—namely, its ability to avoid with certainty the threat of climate catastrophe—is substantially weakened.

Combined Price and Quantity Mechanisms

Even if a carbon tax is preferable to a cap-and-trade approach in terms of social costs and benefits, this policy obviously faces steep political opposition in the United States. Businesses oppose carbon taxes because of the transfer of revenue to the government. Under a permit system, there is a hope that some, if not all, permits would be given away for free. Environmental groups oppose carbon taxes for an entirely different reason: They are unsatisfied with the prospect that a carbon tax, unlike a permit system, fails to guarantee a particular level of emissions. Such antagonism from both sides of the debate makes it unlikely that a carbon tax will become part of the U.S. response to the Kyoto Protocol.

However, the advantages of a carbon tax can be achieved without the baggage accompanying an actual tax. In particular, a combined mechanism (often referred to as a hybrid, or a safety valve) can obtain the economic advantages of a tax while preserving at least some of the political advantages of a permit system.

In such a scheme, the government first distributes a fixed number of tradable permits—freely, by auction, or both. The government then provides additional permits to anyone willing to pay a fixed ceiling or "trigger" price. The initial distribution of permits allows the government the flexibility to give away a portion of the right to emit CO_2, thereby satisfying concerns of businesses about government revenue increases. The sale of additional permits at a fixed price then gives the permit system the same compliance flexibility associated with a carbon tax.

With a combined price/quantity mechanism, it will be necessary to consider how both the trigger price and the quantity target should evolve over time. One possibility is to raise the trigger price over time to guarantee that the quantity target is eventually reached. A second possibility is to carefully choose future trigger prices as a measure of how much we are willing to pay to limit climate change. As we learn more about the costs of future emission reductions, however, this distinction between price and quantity controls will diminish. That is, after uncertainty about future compliance costs is reduced through experience, then price and quantity controls can be used to obtain similar cost and emission outcomes.

Operationally, when this safety valve is used in conjunction with international emissions trading, as the Kyoto Protocol allows, problems potentially arise. In general, there would be a need for either harmonization of the trigger price across countries or restrictions on the sale of permits from those countries with low trigger prices. Otherwise, there would be an incentive for countries with a low trigger price to simply print and export permits to countries with higher permit prices. This action would not only effectively create low trigger prices everywhere; it also would create large international capital flows to the governments of countries with the low trigger prices.

Instead of harmonizing trigger prices, the trigger price could be set low enough to avoid the need for international GHG trades. This may be a desirable end in light of concerns about the indirect economic consequences of large volumes of international GHG trade flows.

Finally, if we find it desirable to raise the trigger price rapidly, it will be necessary to limit the possibility that permits can be purchased now and held for long periods of time. Otherwise, there will be a strong incentive to buy large volumes of cheap permits now to sell them at high prices in the future. This problem is easily addressed by assigning an expiration date for permits as they are issued, for perhaps one or two years in the future.

Building Domestic and International Support for a Price-Based Approach

Although the safety valve approach is potentially appealing to businesses concerned about the uncertainty surrounding future permit prices, environmental groups will be wary of giving up the commitment to a fixed emission target. Such a commitment is already an integral part of the Kyoto Protocol. However, a strict target policy ultimately may lack political credibility and viability. Although a low trigger price would clearly rankle environmentalists as an undesirable loosening of the commitment to reduce emissions, a higher trigger price could allay those fears while still providing insurance against high costs.

Perhaps more controversial than the concept of a safety valve is the fact that a hybrid policy requires setting a trigger price. It extends the debate over targets and timetables to include perceived benefits on the basis of the trigger price. Business interests undoubtedly will seek a low trigger price and environmental groups a high trigger price. I believe this conflict is desirable. The debate will focus on the source of disagreement between different groups—namely, the value placed on reduced emissions. Rather than leaning on rhetoric that casts reduction commitments as either the source of the next global recession (according to businesses) or the costless ushering in of a new age of cheaper and more energy-efficient living (according to environmentalists), it will be necessary to decide how much we are realistically willing to spend to deal with the problem.

Although seemingly provocative in its challenge of the core concept of targets and timetables embedded in the Kyoto Protocol, some concept of the safety valve is already part of many countries' notion of their commitments to the protocol. European countries that are likely to implement carbon taxes must have some idea how they will handle target violations if their tax proposals fail to sufficiently reduce emissions before the end of the first commitment period. Likewise, other countries that are considering either a quantity or command-and-control approach must envision a way out if their actual costs begin to surpass their political will to reduce emissions.

Among the many implicit safety valve possibilities, one could imagine a more flexible interpretation of existing provisions, such as the clean development mechanism or the use of carbon sinks. Alternatively, Article 27 specifies that parties can withdraw from the protocol by giving notice one year in advance. A country that foresaw difficulty in meeting its target in the first commitment period could serve notice that it wished to withdraw before the commitment period ended.

Therefore, flexibility in meeting current commitments already exists implicitly. Countries can choose to massage their commitments using existing provisions, violate their targets and risk penalties (which have yet to be defined), or simply withdraw. In these cases, however, the outcome and consequence are unclear. The advantage of a price mechanism is that it makes the safety valve concept explicit and transparent. Establishing a price trigger for additional emissions allows countries, and private economic decisionmakers in turn, to approach their reduction commitments with greater certainty about the future. This method not only improves the credibility of the protocol but also its prospects for future success in reducing GHG emissions.

Conclusions

The considerable uncertainty surrounding the cost of international GHG emission targets means that price- and quantity-based policy instruments cannot be viewed as alternative mechanisms for obtaining the same outcome. Price mechanisms will lead to uncertain emission consequences, and quantity mechanisms will lead to uncertain cost consequences. Economic theory as well as numerical simulations indicate that the price approach is preferable for GHG control, generating five times the net expected benefit associated with even the most prudent quantity control. The essence of this result is that a rigid quantity target over the next decade is indefensible at high costs when the stock of GHGs is allowed to increase over the same horizon.

Importantly, a price mechanism need not take the form of carbon tax. The key feature of the price policy is its ability to relax the stringency of the target if control costs turn out to be higher than expected. Such a feature can be implemented in conjunction with a quantity-based mechanism as a safety valve. A quantity target is still set, but with the understanding that additional emissions (beyond the target) will be permitted only if the regulated entities are willing to pay an agreed-upon trigger price.

This approach can improve the credibility of the protocol and its prospects for successful GHG emission reductions. The last point is particularly relevant for ongoing climate negotiations. Should the emission incentives and consequences remain ambiguous and uncertain, or should they be made explicit and transparent? Specifying a price at which additional, above-target emissions rights can be purchased provides a transparent incentive; the current approach does not. Although ambiguity may prove to be the easier negotiating route, it also may be a disincentive for true action.

Suggested Reading

General

Fischer, Carolyn, Suzi Kerr, and Michael Toman. 1998. Using Emissions Trading to Regulate U.S. Greenhouse Gas Emissions. RFF Climate Issues Brief 10-11. Washington, DC: Resources for the Future.

Harrington, Winston, and Richard Morgenstern. 1998. On the Accuracy of Regulatory Cost Estimates. RFF Discussion Paper 99-18. Washington, DC: Resources for the Future.

McKibbin, W., and P. Wilcoxen. 1997. A Better Way to Slow Global Climate Change. Brookings Policy Brief 17. Washington, DC: Brookings Institution. http://www.brookings.edu.

Newell, R.N., and W.A. Pizer. 1998. Stock Externality Regulation under Uncertainty. RFF Discussion Paper 99-10. Washington, DC: Resources for the Future.

Nordhaus, W.D. 1994. Expert Opinion on Climate Change. *American Scientist* 82: 45–51.

Nordhaus, W.D. 1994. *Managing the Global Common.* Cambridge, MA: MIT Press.

Parry, Ian. 1997. Revenue Recycling and the Costs of Reducing Carbon Emissions. RFF Climate Issues Brief 2. June. Washington, DC: Resources for the Future.

Pizer, W.A. 1998. Prices versus Quantity Revisited: The Case of Climate Change. RFF Discussion Paper 98-02. Washington, DC: Resources for the Future.

Technical

Broecker, Wallace S. 1997. Thermohaline Circulation, the Achilles Heel of Our Climate System: Will Man-Made CO_2 Upset the Current Balance? *Science* 278 (November 28): 1582–88.

Enting, I.G., T.M. Wigley, and M. Heimann. 1994. Future Emissions and Concentrations of Carbon Dioxide: Key Ocean/Atmosphere/Land Analyses. CSIRO Division of Atmospheric Research Technical Paper 31. Aspendale, Australia: Commonwealth Scientific and Industrial Research Organisation. http://cdiac.esd.ornl.gov.

Fischer, Carolyn, Suzi Kerr, and Michael Toman. 1998. Using Emissions Trading to Regulate U.S. Greenhouse Gas Emissions: An Overview of Policy Design and Implementation Issues. *National Tax Journal* 51(3): 453–64.

IPCC (Intergovernmental Panel on Climate Change). 1996. *Climate Change 1995: The Science of Climate Change.* Contribution of Working Group I to the Second Assessment Report of the Intergovernmental Panel on Climate Change. New York: Cambridge University Press.

Nordhaus, William D. 1992. The "DICE" Model: Background and Structure of a Dynamic Integrated Climate-Economy Model of the Economics of Global Warming. Discussion Paper 1009. New Haven, CT: Yale University, Cowles Foundation for Research in Economics.

Parry, Ian, Roberton Williams III, and Lawrence Goulder. 1999. When Can Carbon Abatement Policies Increase Welfare? The Fundamental Role of Distorted Factor Markets. *Journal of Environmental Economics and Management* 37(1): 52–84.

Roughgarden, Tim, and Stephen Schneider. 1999. Climate Change Policy: Quantifying Uncertainties for Damages and Optimal Carbon Taxes. *Energy Policy* 37: 415–29.

10 Using Emissions Trading To Regulate National Greenhouse Gas Emissions

Carolyn Fischer, Suzi Kerr, and Michael A. Toman

The 1997 Kyoto Protocol would limit average annual emissions of greenhouse gases (GHGs) of Annex I industrialized countries (as identified in the 1992 U.N. Framework Convention on Climate Change) to levels that average about 5% below 1990 emissions during 2008–12. Participants in the climate policy debate are expressing considerable and growing interest in the use of an emissions trading policy to achieve these limits. Through the trading of emissions permits, society as a whole benefits from having environmental goals achieved at lower cost than with a command-and-control system. Over time, GHG controls implemented by permit trading will induce investments in the development and diffusion of new technologies (such as renewable energy sources and enhanced energy efficiency) for reducing GHG emissions. Several recent experiments with emissions trading to control other pollutants—notably, the phaseout of lead in gasoline and the ongoing reduction of sulfur dioxide (SO_2) emissions from electric utilities under Title IV of the U.S. Clean Air Act—suggest that an emissions trading policy is very promising. However, considerable debate and some confusion remain about how a GHG emissions trading program would be organized and operated in practice.

In this chapter, we first address several basic questions that must be considered in designing and implementing a domestic GHG trading system: who and what is covered by regulation (which sources and gases), the commodity to be traded and the nature of the trading system, how rights to GHG permits are defined, and how to deal with different kinds of uncertainties. We then explore additional questions related to the initiation of the system, intertemporal flexibility, and interactions of permit trading with the tax system. Throughout, we consider the overall cost-effectiveness of alternative approaches (including administrative costs), their distributional implications, and political economy considerations. Our focus is exclusively on domestic policy; we do not consider how emissions trading or other GHG control policies should function in an international context.

Our fundamental conclusions are the following:

- The best overall economic performance will be achieved with an upstream program that caps total CO_2 emissions through tradable permits for fossil fuel inputs to the economy (that is, a cap-and-trade system), with other GHG types and sources added to the extent possible consistent with a cost-effective and administratively efficient system.
- Distributing bankable carbon permits through a periodic government auction has attractive economic and equity features, provided the govern-

ment uses the resulting revenues to reduce other distorting taxes in the economy and to provide carefully targeted assistance to those most adversely affected by GHG restrictions.

- The rules for altering permit allocations should create the proper incentives for the private sector to respond to risk while limiting opportunistic government behavior.
- Allowing permits to be borrowed from future commitment periods has the potential to substantially reduce the cost of compliance. However, to implement such an approach, it is necessary to determine how tighter future constraints would be made credible for current decisionmakers, a challenging problem that has not yet been solved.
- The efficiency of permit markets will be reduced somewhat by broader distortions in the economy related to the tax system's treatment of capital gains. These distortions will be smaller if permits are auctioned than if they are distributed free of charge. If permits are issued free of charge, then they should be taxed on their market value.

Basic Economics of GHG Trading

The fundamental idea behind emissions trading is that any particular environmental target can be achieved at lower total cost to society if those responsible for achieving the environmental target can exploit gains from trade. In this context, "gains from trade" refers to the possibility of emitters with higher costs of abatement paying those with lower costs of abatement to undertake more emission control. Those with higher costs benefit by reducing their net cost of compliance; they pay others to undertake emission control but also avoid the higher costs of controlling their own emissions. Emitters with lower costs benefit because they voluntarily enter into transactions that yield revenues at least as large as the extra control costs they assume. And society benefits by having a more cost-effective control program—fewer real resources devoted to the achievement of the environmental goal.

These concepts underlie both the SO_2 control program in the United States and previous programs that involved more informal trading of emission reduction credits. Under the SO_2 program, the government established emission limits for each covered source but made the allowed emissions tradable through the issuance of homogeneous (not source-specific) emission allowances that can be freely bought and sold. Those with higher abatement costs buy allowances from those with lower costs, who must then undertake additional emissions control to remain in compliance. Emissions reduction credits, in contrast, arise from the action of an individual emitter in a specific location who reduces emissions below the required level and thereby creates a credit that can be sold to other emitters (often restricted to nearby locations) who seek to cover excess emissions relative to their standards. Credits thus are not the homogeneous "emissions scrip" characteristic of a cap-and-trade program. Markets for credits often tend to be "thinner" and less well developed, with more transaction costs.

These examples illustrate the application of trading systems to the direct control of unwanted emissions. Generally, to be most efficient, environmental policy needs to be focused directly on the unwanted "bad," the emissions. Policies that seek instead to indirectly regulate emissions through the control of inputs to production or consumption—for example, by taxing coal based on its sulfur content—are less cost-effective because they single out some paths for control at the expense of others, whereas all options for control need to be on a level playing field to achieve the lowest costs. In SO_2 control, a tax on the sulfur content of coal would encourage the substitution of lower-sulfur fuels, but it would not make the best use of end-of-pipe controls (scrubbers).

However, this argument does not apply to CO_2, at least not until such time as technology for large-scale scrubbing and storage of waste CO_2 becomes economical. In the absence of such technology, there is a one-to-one correspondence between the carbon content of fossil fuels going into combustion and the CO_2 emitted. It means that it is possible to regulate CO_2 emissions by limiting fossil fuel use. This theory is the basis for a tax based on the carbon content of fossil fuels to restrict CO_2 emissions. As described later, the same idea underlies an upstream approach to GHG trading that places quantitative restrictions on fossil fuel flows through the economy.

Although there are various ways that GHG trading can be implemented, it is crucial to note that *any* efficient approaches (taxes or trading) that comparably restrict fossil fuel use or GHG reductions have essentially the same effects on energy costs, given markets that function fairly well (with reasonable levels of competition and limited regulatory distortions). Whether fossil fuel supplies are restricted or demand is comparably restricted through limits on users' emissions, whether permits are distributed free of charge or sold by the government, the effects on energy prices are the same. In particular, even if carbon allowances are given away, firms will value them at their market price, and the prices of energy will adjust accordingly to reflect the increased scarcity of primary fuel inputs. In the past, regulation of the electric utility industry resulted in differences in the price consequences of different systems, because regulators determined electricity prices based on formulas that reflected actual expenditures. But with progressive deregulation of the utility sector and greater reliance on market pricing, the market price of carbon permits—rather than the accounting value—will determine the price of electricity.

The fact that different methods of restricting GHG emissions will have comparable effects on energy prices and other prices throughout the economy means that the first-order effects of GHG limits on downstream energy purchasers (including households) and the effects on employment and capital investment are the same across different trading approaches. The principal differences among the trading approaches relate to administrative efficiency (which includes the monitoring of diverse energy suppliers and users) and to who obtains the "scarcity rent"—the revenues generated when final energy prices rise as a consequence of regulatory-induced scarcity. The latter issue raises questions related to both fairness and politics.

Design of a GHG Trading System

Direct regulation of all or even most GHG emissions is close to impossible in practice. Fossil fuel emissions of CO_2—the most important GHG—stem from scores of millions of sources. In addition, highly precise methods for directly measuring CO_2 emissions are economically practical only for large boilers that can be equipped with continuous emission monitors.

One alternative is to control fossil fuel inputs to the economy. Because the CO_2 emitted from fossil fuel is perfectly correlated with the carbon content of the fuel, an accurate measure of emissions can be obtained by keeping track of quantities of fossil fuel used and their correlating carbon content. For example, it would be possible to control natural gas entering interstate pipelines, crude oil (domestic and imported) entering refineries, and coal sales from mines or processing plants. Such a system would require the oversight of fewer than 2,000 actors.

One also might consider various hybrids that combine the regulation of downstream CO_2 emissions from large boilers (as in the U.S. SO_2 program) with upstream regulation of other fossil fuel flows. This approach requires more complex record keeping and enforcement and entails greater risk of sources "leaking" from the system. For example, it would be necessary to draw a distinction between natural gas flows to utilities (the downstream component) and gas flowing to household and commercial users for space heating (the upstream component). As noted above, this approach does not lead to a fundamentally different outcome in the economy that might justify the extra administrative burden and potential loss of efficiency. The main distinguishing feature of the hybrid approach seems to be that it could be used to allocate free emission permits to large sources of emissions, as was done in the SO_2 program. It raises some challenging questions related to equity (discussed in the next section).

Several adjustments and additions would be needed to increase the coverage and efficiency of a CO_2 program. The basic upstream system would not capture the relatively small amount of energy used in producing and initially transporting the fossil fuels to the regulatory control points. A rough adjustment might be made for that energy by requiring permits in excess of the actual carbon content of the fuel. It also would be necessary to adjust for noncombustion uses of fossil fuel inputs (such as chemical feedstocks; some of the carbon in these materials eventually may escape into the environment as wastes are burned). Biomass energy supplies could be included by requiring biomass supplies to be permitted for their carbon content but then providing credits for the carbon sequestered in the growing of the feedstock. Similar crediting provisions would be needed for other deliberate carbon sequestration activities—notably, reforestation.

Some arguments have been made for including producers of energy-using capital equipment in a trading system (for example, by requiring vehicle manufacturers to hold permits equal to the expected lifetime emissions of their vehicles). However, this approach controls only the performance of new equipment, has no effect on the utilization of equipment, and creates a bias that encourages the uneconomic life extension of old, inefficient equipment. It also can lead to double regulation of emissions from energy-using capital equipment. Over the long term, price signals provided by a fuel-based program will effectively guide equipment purchase and use decisions. It is certainly possible that market failures associated with energy-using equipment (such as information unavailability, regulatory rigidities in the utility sector that limit the transmission of effective price signals) need to be addressed to improve the efficiency of capital purchase decisions. But these problems are best attacked directly where they occur through targeted reforms in regulation, the provision of better information to consumers, and so forth.

A cap-and-trade program for CO_2 would involve issuing a quantity of homogeneous (not source- or sector-specific) permits equal to the total target level of emissions. Individuals would be free to buy and sell permits subject to whatever bookkeeping requirements are needed for ensuring compliance. The commodity traded is the opportunity to emit a unit of carbon (measured in terms of carbon content released) once, rather than defining the commodity as a stream of emissions over time. This approach to defining the commodity, which is used already to regulate U.S. utilities' SO_2 emissions, provides maximum flexibility and liquidity in the permit market.

Transaction costs (that is, the costs of identifying trading partners and effectuating trades) can be very low, and derivative transactions such as forward, futures, and options contracts can develop.

Other human-induced GHG emissions (such as methane from coal mines and landfills, or hydrofluorocarbons [HFCs] now used in lieu of chlorofluorocarbons [CFCs] in air conditioners) could be added to the cap-and-trade system (on the basis of their calculated relative contributions to global warming) depending on the capacity to measure or infer emission balances. Sources, sinks, and gases not included directly (because of difficulties in tracking aggregate balances) could be incorporated in ancillary project-specific efforts to create credits that could be "imported" into the core program.

Each permit can be used after a given date. For example, permits for the first budget period would be dated 2008, and those for the second period, 2013. Permits could be used or banked for future use. Banking is a key feature of an efficient trading system, because it allows for efficient arbitrage of marginal costs that rise over time as emission targets tighten or energy demand grows. Regulators also could issue or sell permits ahead of their "use after" dates, and emitters and fuel suppliers could engage in forward contracting to assemble portfolios of permits. In particular, it may be useful to initiate provision or sale of permits for the initial commitment period (2008–12) some time beforehand, to allow price information to develop and to build confidence in the trading and monitoring institutions. Experience with the U.S. SO_2 program suggests that confidence is enhanced by establishing in advance how trading will work and how results will be monitored. One concern with banking is that if a large stock of permits were banked and then used quickly, climate change damages might be exacerbated. However, the scale of this risk is uncertain. To limit such a risk, policy targets would have to be negotiated so as to not accelerate too rapidly over time, because rapid increases in target stringency are what give rise to incentives for large-scale permit hoarding.

After a capacity for measuring or reliably estimating individual actors' CO_2 emissions or fossil fuel sales is established (a feasible but not trivial task), it is necessary only to keep track of covered sources' permits relative to fuel flows or emissions. There would be no need to keep track of specific transactions in the trading registry, other than the creation of project-specific non-CO_2 emission credits imported into the core cap-and-trade program. Any shortfall between required and actual permits would be deterred by stiff financial penalties and a requirement to cover the shortfall. Permits would be fully fungible within the five-year commitment period, but each year firms would be required to show they hold permits equal to or greater than their emissions. This annual reporting would avoid the risk of large carbon debts when firms go bankrupt and could provide useful information to the permit market.

How GHG Permits Are Allocated

Before any trading program could commence, the permits would have to be allocated. The government could simply require prospective permit holders (and anyone else who so desired) to bid for permits in an auction. Unless foolishly prohibited by government, an active secondary market could and would evolve that would allow sources to adjust their permit holdings to changes in circumstances between auctions. Auctions could be held periodically—for example, quarterly. The government could auction more permits than are expected to be used within the immediate period to provide some degree of intertemporal flexibility within the overall commitment period. The government also could auction in advance permits for use later in the commitment period (for example, in subsequent quarters) to help permit markets to develop.

A great deal of research and a reasonable amount of practical experience is available for designing an auction. One option with strong potential for efficiency is a sealed-bid auction with a uniform price rule. Buyers would submit bids, the auctioneer would find the price that clears the market, and those who bid at least as much as the clearing price would receive permits at that price. Uniform pricing encourages participation by small bidders, because it is strategically simple. The usual concern with mar-

ket power in uniform price auctions does not arise in this competitive market. Ascending auctions are more complex approaches that can yield some efficiency improvements over sealed bids, an advantage that must be balanced against their complexity. The repeated bidding process in an ascending auction reveals information about participants' valuations of permits, which improves the bidders' own valuation estimates and hence the efficiency of the final permit allocation.

An auction would raise considerable revenue for the government. To illustrate, U.S. emissions of carbon from fossil fuel burning in 1990 were calculated to be 1,374 million metric tons. Reducing total emissions to 7% below the 1990 level by 2008 would imply emissions of about 1,280 metric tons. (The United States is obliged to reduce all GHGs to a level that averages 7% below 1990 levels by 2008–12.) If the price of a GHG permit were $50/metric ton of carbon emission, then 1,280 million metric tons of permits in the United States would be worth $64 billion/year.

This revenue could be used to mitigate the adverse effects of GHG control for certain groups of business owners or workers, to help finance innovation in technologies for GHG reduction (such as new energy sources or carbon storage technologies) and for adaptation to climate change, or to benefit taxpayers as a whole through reductions in other taxes. This last use of revenues has the potential to substantially reduce the net cost of GHG control, assuming the tax cuts are broad-based and increase efficiency. The reason is that any tax on consumption goods or income lowers the effective real wage and reduces the incentive for workers to increase the labor effort they provide. Any environmental policy that raises the cost of carbon-intensive goods exacerbates the preexisting distortions in the labor market that result from current income taxes.

The alternative to an auction is some form of no-charge allocation. Such an allocation could occur in many ways. One often-discussed strategy is a form of grandfathering, whereby regulated emission sources or fuel suppliers are given permits in proportions equivalent to their historical emissions or fossil fuel sales. Another approach is to divide up permits in proportion to market output shares, which could be updated over time (for example, in a rolling average); however, this approach would act as an efficiency-reducing subsidy on the outputs of the permit recipients (because expanded output is rewarded with increased availability of free permits). Allocations can involve some or all of the total permit supply.

No-charge allocation provides a tool for distributing the scarcity rents created by limits on fossil fuel use or GHG emissions. It reduces the resistance of those who would have to pay for all their carbon or emission permits in an auction scheme. However, the beneficiaries from a political decision to allocate permits to certain parties may not correspond to those who ultimately face the most adverse economic effects from imposition of the program. As already noted, the ultimate price and allocative effects (aside from interactions with the tax system) will be the same with any permit allocation or with a permit auction. It means that, for example, employees who are unable to easily change jobs in response to lower labor demand in specific sectors (such as coal) and households that cannot easily reduce energy demand in response to higher energy prices will suffer under any system. On the other hand, with free allocation of all permits, the rents will flow to the shareholders of the favored firms, and there will be no capacity to reduce the direct costs of GHG control with reductions in other taxes. Business owners suffer one-off costs of CO_2 control buffered by receipt of the rents, and households suffer potentially substantial ongoing costs. It is possible that these adverse distributional impacts can be softened considerably (and at low cost in terms if economic efficiency) if only a relatively small part of the total permit supply is allocated (see Chapter 12, Confronting the Adverse Industry Impacts of CO_2 Abatement Policies).

Finally, no-charge allocation also may reduce somewhat the incentive for technical innovation. If an innovator is motivated by anticipated cost savings due to permit price reductions, these savings will be smaller with a no-charge allocation.

Grandfathering is sometimes criticized as being biased against new sources, which must pay for their emissions permits while existing sources ob-tain theirs for free. However, unless there are deeper competition problems, existing firms will have neither the motivation nor the capacity to discriminate against new entrants. For their part, entrants can and should come into an industry if and only if they are more efficient than the incumbents, taking into account the opportunity cost of GHG permits. It is possible that small new sources could be disadvantaged because of imperfect capital markets that limit their access to finance. But this problem is best rectified by addressing the sources of any capital market distortions. No-charge allocations other than grandfathering also create inefficiencies; for example, allocating permits based on changes in market share acts as a de facto subsidy to the output of the favored firms and distorts product purchasing decisions.

We believe auctioning is the best approach. Because it may be problematic politically to start with an auction for all GHG permits, it would be possible to start with a mixture of allocations and auctions and gradually phase out the allocation. In light of the argument in the previous section, the allocation should be of upstream fossil fuel carbon permits, *not* emission permits. No-charge allocation of some permits to downstream actors could be undertaken; the permits would be bought back by upstream actors that directly face regulation (for example, refineries), making the flows of rents fairly transparent. Rent transfers also could be achieved in more direct and more transparent ways than through the allocation of free permits—in particular, through transfers of revenues from a permit auction to whichever groups of shareholders or consumers are deemed worthy of receiving them (however, again, care is needed to avoid inventing new forms of output subsidies).

Dealing with Uncertainties

A GHG trading program also must deal with various uncertainties over time. External factors reflect changes in aggregate national GHG emissions ceilings due to such international influences as changes in the perceptions of the risks of climate change and the ongoing process of negotiation of global emissions targets and national obligations under the

U.N. Framework Convention on Climate Change. For example, a discovery that climate change is more threatening than it seems today could cause national GHG budgets to be reduced to phase down emissions more quickly. Internal uncertainties reflect the government's own political or revenue-based motivations for changing the allocation of permits. The possibility that a future government may decide not to honor its Kyoto Protocol commitments is one major source of uncertainty. Other sources of uncertainty arise even if the Kyoto Protocol targets are met. One example is the possible reallocation of free permits among downstream sectors (that is, reducing permits given to sectors that have shown a particular capacity to reduce costs and increasing allocation to those that do not). Another example is unannounced movements from no-charge allocations to auctioned allocations to increase government revenue.

These uncertainties call for responses that are, to some extent, at cross purposes. Putting the external risk on the private sector (by designating allowances as a share of total national targets as set in advance in international agreements rather than as a fixed quantity of carbon) creates uncertainties in their investment planning. However, these risks are unavoidable, and emitters—not the government—are in the best position to bear them through a portfolio of investments in existing and new technologies that provide flexibility if quotas fall (and options to emit more if they should happen to rise). On the other hand, internal risks unnecessarily increase investment uncertainty and can undercut the incentives for permit market participants to undertake desirable investments in lowering the cost of emissions control. By their nature, hybrid upstream/downstream programs with no-charge allocation seem to be more prone to internal risks than an upstream auctioned approach, because the government has more control over the initial allocations of permits. Policy needs to spell out carefully the circumstances under which emissions allocations can be altered, to limit opportunistic regulatory behavior while recognizing the sovereignty of the government over GHG emissions and its need to respond to ongoing international discussions on GHG reductions.

Flexibility in the Timing of Emissions Reductions

Banking unused permits can improve cost-effectiveness, especially when regulatory targets are such that marginal control cost is rising faster than the relevant rate of interest over time (so that the present worth of future permits is higher than their current value). Under such circumstances, permits are more valuable when used to offset future abatement costs than if they are used today. This kind of arbitrage improves market efficiency in the same way that intertemporal contracting and inventories improve the efficiency of conventional commodity markets.

By the same token, long-term borrowing against future allocations to meet near-term targets will be attractive to those facing GHG constraints if the price of permits otherwise is expected to rise more slowly than the rate of return on other assets, indicating a disproportionate short-term burden of controlling GHG emissions relative to the long term. This scenario could take place because the composition of energy-using capital is more flexible in the long term than in the short term and because tougher short-term requirements provide relatively less opportunity to embed technical improvements over time. In this situation, all regulated entities—not only those in one country—would have an economic incentive to borrow if they could until the shift in permits from future to present eliminates the arbitrage opportunity by depressing current permit prices and raising future permit prices. One alternative to such aggregate borrowing would be an international regime in which stocks of permits covering longer periods of time were provided, and permit holders were allowed to freely allocate them over time as well as trade contemporaneously with each other. Neither approach is consistent with the Kyoto Protocol, which establishes a single set of quantitative emissions targets for a single period (which means that there is no future commitment against which to borrow) and treats the targets as binding.

Modeling studies suggest that, in principle, long-term atmospheric concentrations of GHGs could be stabilized at far lower costs (savings on the order of

50%) if emitters could maintain significantly higher emissions than are allowed under the Kyoto Protocol during the next couple of decades, then reduce emissions to a greater extent later. In such a situation, short-term costs from rapid capital obsolescence could be avoided and there would be more opportunity for incorporating new technical innovations into investment decisions over time. In particular, during the initial couple of decades, emitters would begin to install high-efficiency capital that will allow much lower emissions subsequently. One key element of these models is that they assume emitters or fossil fuel sellers view future constraints on GHGs as credible and that, in anticipation of these future constraints, emitters rationally begin to adjust their capital stocks in advance.

Critics of these analyses have raised several objections. One is that the postponement of short-term emissions reductions, by lowering energy prices below what would have occurred with tighter standards, also will retard some induced innovation and "learning by doing" that proponents assume to be available in the future. Innovation will not be neglected entirely if participants in the GHG market do anticipate tighter future controls. However, to the extent that innovation is slower than is assumed in the above-mentioned modeling studies, reaping the cost savings pointed to by advocates of borrowing might require increased government commitments to research and development (R&D). (A more complex question, which we do not address here, is what level of public and private R&D actually is desirable, and the extent to which a desirable level will be undertaken when a policy such as emissions trading is imposed.)

Critics of borrowing or long-term allocation approaches also have raised more fundamental questions related to policy consistency and credibility over time. What is to stop emitters from rapidly depleting their endowments of permits in anticipation that today's regulators will not be willing to impose sanctions in a few years? Moreover, if future regulatory actions to enforce repayment of borrowed permits are not seen as credible, then some needed investments in long-lived energy-efficient capital assets and some R&D will not be undertaken. In addition,

what would prevent future regulators, who themselves will have incentives to defer control costs, from allowing higher emissions than necessary to repay the debt in response to the higher future control cost burdens that a current decision to defer abatement would place on them? If emitters today anticipate this possibility, then the long-term credibility of the program is further undermined.

In lieu of borrowing, governments could attempt to negotiate a more intertemporally cost-effective time path of goals with looser short-term targets and tighter emission limits over time. However, this approach, too, has a credibility problem related to whether governments will follow through on a commitment to tighten emissions limits. However, at the same time, the higher cost of strict near-term emission controls also creates a credibility problem in that it deters compliance from the outset. Unless society has an underlying long-term commitment to reducing GHG emissions and eventually stabilizing their atmospheric concentrations, *no* time path of emissions controls will be credible, and paths with higher costs because of excessively stringent early targets arguably will be even less credible.

These problems might be reduced if new ways to limit GHG emissions with significantly lower costs are discovered and diffused, perhaps with government support, so that postponement is less attractive. Another possibility is that if permits are sold well in advance (but cannot be used in advance), then private holders will have an incentive to resist subsequent efforts by the government to add permits to the system and create capital losses on banked holdings.

Permit Trading and the Tax System

Any national GHG permit system has to operate within the context of the country's existing tax system. Each system will have an effect on the efficiency of the other, raising important design issues for GHG trading. Of particular concern is the tax treatment of banked permits over time, as general inefficiencies in the taxation of capital income raise some specific issues for intertemporal emissions abatement choices.

Permits are potential assets as well as inputs to production. This dual nature poses a challenge in choosing proper tax treatments and achieving efficiency in the permit market. When market conditions favor banking of permits (that is, when abatement costs are expected to rise sufficiently in the future to warrant holding permits for later use), the taxation of capital gains can introduce distortions in permit markets. One distortion is a lock-in effect that stimulates the excessive banking of permits. Given the choice of (a) selling a permit and paying tax on capital gains accrued to date and then investing the sale proceeds or (b) holding the permit in expectation of further capital gains, which will be taxed only when the permit finally is sold, existing permit holders tend to defer permit sales that might otherwise be economically efficient.

The tax system also creates a kind of reverse lock-in: Allowance purchase costs cannot be deducted as inputs until use, which discourages the purchase of allowances in advance. However, because of the first lock-in effect, the deferral of deductibility is crucial to deter inefficient tax avoidance. Without this rule, taxes could be saved by purchasing and immediately deducting allowance costs as well as deducting interest in funds borrowed to finance the purchase; tax would be paid on the accrued return only when the asset is sold. Still, the same kind of tax arbitrage can occur when permit buyers purchase and fully deduct high-cost permits while holding permits with a lower cost basis for accrual of gains. This particular scenario can be avoided by a type of first-in, first-out rule.

Inefficiencies arise when permit owners face different effective tax rates. For example, an owner of a permit would sell rather than hold the asset if there were a lower rate of return to holding permits. But with different cost bases (the acquisition cost or original market value against which profits are reckoned for tax purposes) as a consequence of changes in market conditions, different owners will require different returns to current sales to overcome the particular lock-in they face. On the other hand, a buyer willing to purchase and hold a permit must obtain a higher rate of return, because the cost cannot be deducted until later. No permit price path can simultaneously make all these actors indifferent to buying, selling, and holding.

No-charge allocation creates more of a bias toward excessive permit banking, the first lock-in effect. It will tend to inefficiently tilt the permit price path such that permits today will be more expensive than without lock-in, and the volume of banked permits will lower future price expectations. The tilting of the price path in turn will shift abatement activities inefficiently toward the present. On the other hand, with auctioned permits the bias is against holding permits, because the deduction must be deferred. An efficient amount of permit banking will occur only if the price path is inefficiently tilted up; in practice, too few permits will be banked, and some abatement that should be done immediately will be inefficiently deferred.

Ultimately, reducing these distortions would require a broader reform of capital gains taxation. Simply eliminating the taxation of capital gains on permits would not eliminate the problems; as long as permits are deductible, gains must be taxed. Otherwise, two permit users could swap permits, each could deduct the full value, and neither would be taxed on the proceeds. Exempting permits from taxation completely would treat firms symmetrically and get rid of this incentive to "churn." However, differences in income tax brackets among individuals and corporations mean that different investors still will face different trade-offs between permits and other assets, and distortions will remain. If permits were tax-exempt, then people with high marginal tax rates would invest in them, driving down the rate of return to the low after-tax rate of return these individuals would reap on other assets; consequently, the lower rate of return would cause excessive amounts of early abatement effort. Furthermore, exempting permits from taxation would tend to narrow the tax base, leaving other forms of income-generating activities to shoulder more of the distortionary tax burden.

How substantial are distortions to abatement decisions and the potential loss of tax revenues through arbitrage activity from these features of the tax system? Recent evidence in the literature on capital gains taxation suggests that when people choose to realize capital gains is not very sensitive to tax rates over the long term, which implies a relatively small lock-in effect. However, free permits pose a larger potential lock-in effect than most assets: Because the cost basis for free permits is zero, in effect, they instantaneously accrue the market value. The incentive to defer gains by holding onto permits is not only the postponement of taxes on the price increases but also the postponement of taxes on the initial permit price itself. Short of auctioning permits, this problem could be mitigated by taxing the recipient on the market value of the allocation, thereby allowing the initial basis to be the market rate and not zero. Firms then would deduct that market value for the permits they use, so at tax time they would effectively only be liable for tax on the no-charge permits they did not use. (In contrast, in the U.S. SO_2 program, allocated permits are given a zero-cost basis; tax is paid on the entire sale price when a permit is sold, but only when the sale takes place, and the cost of purchased permits is deducted when such permits are used.)

As noted previously, a permit system also can interact with the tax system through the generation of auction revenue that can be used to lower existing distortionary income taxes (see Chapter 11, Revenue Recycling and the Costs of Reducing Carbon Emissions). Thus, coupling an auctioned permit system with broad-based income tax reform would not only minimize additional labor market inefficiencies but also lessen tax distortions to the permit system itself.

Concluding Remarks

Our conclusions that the permit system should be upstream and auctioned and that auction revenue should be recycled may be controversial. However, if a long-term national commitment to the potentially costly control of GHG emissions is assumed, then regulatory institutions have a fair amount of time to evolve and gain public acceptance—and we must choose the best institutions possible. Without such a commitment, which is currently lacking in the political arena, any discussion of regulatory options is somewhat beside the point. Only time will tell whether our proposal gains currency in the years

leading up to the start of binding commitments under the Kyoto Protocol, if indeed the protocol is ultimately implemented. Many important practical questions need to be addressed before a national-scale GHG trading system can be constructed and implemented in the United States and in other industrialized countries. However, the deployment of a relatively efficient and transparent system seems within our grasp, especially after key questions related to who benefits from the revenues created by such a system are answered. In turn, a well-developed comprehensive domestic trading system would provide a good foundation for taking advantage of the potential opportunities afforded by GHG trading on an international scale.

Suggested Reading

General

Burtraw, D. 1996. The SO_2 Emissions Trading System: Cost Savings without Allowance Trades. *Contemporary Economic Policy* 14(1): 79–94.

Festa, D. 1998. *U.S. Carbon Emissions Trading System: Some Options that Include Downstream Sources.* Airlie Paper. Washington, DC: Center for Clean Air Policy.

Fischer, C. Forthcoming. Rebating Environmental Policy Revenues: Output-Based Allocations and Tradable Performance Standards. Discussion Paper. Washington, DC: Resources for the Future.

Fischer, C., S. Kerr, and M. Toman. 1998. Using Emissions Trading to Regulate U.S. Greenhouse Gas Emissions: An Overview of Policy Design and Implementation Issues. *National Tax Journal* 51(3): 453–64.

Fisher, B.S., and others. 1996. An Economic Assessment of Policy Instruments for Combatting Climate Change. In *Climate Change 1995: Economic and Social Dimensions of Climate Change*, edited by J. Bruce and others. Cambridge, U.K.: Cambridge University Press.

Ha-Duong, M., M.J. Grubb, and J.-C. Hourcade. 1997. Influence of Socioeconomic Inertia and Uncertainty on Optimal CO_2 Emission Abatement. *Nature* 390: 270–73.

Hahn, R., and G. Hester. 1989. Marketable Permits: Lessons for Theory and Practice. *Ecology Law Quarterly* 16: 361–406.

Hahn, R., and R. Stavins. 1995. Trading in Greenhouse Permits: A Critical Examination of Design and Implementation Issues. In *Shaping National Responses to Cli-*

mate Change, edited by H. Lee. Washington, DC: Island Press.

Hargrave, T. 1998. U.S. Carbon Emissions Trading: Description of an Upstream System. Airlie Paper. Washington DC: Center for Clean Air Policy.

Kerr, S. (ed.) 2000. Domestic Greenhouse Regulation and International Emissions Trading. In Global Emissions Trading: Key Issues for Industrialized Countries. Cheltenham, U.K.: Edward Elgar.

Smith, A.E., A.R. Gjerde, L.I. DeLain, and R.R. Zhang. 1992. CO₂ Trading Issues (Volume 2). Choosing the Market Level for Trading. Report to the U.S. Environmental Protection Agency. Washington, DC: Decision Focus, Inc.

Stavins, R. 1998. What Can We Learn from the Grand Policy Experiment? Positive and Normative Lessons from SO₂ Allowance Trading. Journal of Economic Perspectives 12(3): 69–88.

Stewart, R., J. Wiener, and P. Sands. 1996. Legal Issues Presented by a Pilot International Greenhouse Gas Trading System. Geneva, Switzerland: United Nations Conference on Trade and Development.

Swift, B., and others 1997. Implementing an Emissions Cap and Allowance Trading System for Greenhouse Gases: Lessons from the Acid Rain Program. September. Washington, DC: Environmental Law Institute.

Swisher, J., and others. 1997. Analysis of the Potential for a Greenhouse Gas Trading System for North America. Unpublished report. Boulder, CO: Econergy International Corporation.

Tietenberg, T., and D. Victor. 1994. Possible Administrative Structures and Procedures. In Combatting Global Warming. Geneva, Switzerland: United Nations Conference on Trade and Development.

Technical

Auerbach, A.J. 1992. On the Design and Reform of Capital Gains Taxation. American Economic Review 82(2): 263–67.

Burman, L.E., and W.C. Randolph. 1994. Measuring Permanent Responses to Capital Gains Tax Changes in Panel Data. American Economic Review 84: 794–809.

Cramton, P., and S. Kerr. 1998. Tradable Carbon Permit Auctions: How and Why to Auction Not Grandfather. Discussion Paper 98-34. Washington, DC: Resources for the Future.

Fischer, C. 1999. Rebating Environmental Policy Revenues: Output-Based Allocations and Tradable Performance Standards. FEEM, IDEI, and INRA Conference on Environmental, Energy Uses and Climate Change, Toulouse, France, June 1999.

Goulder, L.H. 1995. Effects of Carbon Taxes in an Economy with Prior Tax Distortions: An Intertemporal General Equilibrium Analysis. Journal of Environmental Economics and Management 29: 271–97.

Gravelle, J. 1991. Limits to Capital Gains Feedback Effects. Tax Notes (April 22).

Grubb, M. 1997. Technologies, Energy Systems and the Timing of CO₂ Emissions Abatement: An Overview of Economic Issues. Energy Policy 25(2): 159–72.

Jaffe, A.B., and R.N. Stavins. 1994. The Energy Paradox and the Diffusion of Conservation Technology. Resource and Energy Economics 16(2): 91–122.

Kerr, S., and D. Maré. 1998. Transaction Costs and Tradable Permit Markets: The United States Lead Phasedown. Unpublished report. New Zealand: Motu Economic Research.

Manne, A.S., and R. Richels. 1996. The Berlin Mandate: The Cost of Meeting Post-2000 Targets and Timetables. Energy Policy 24(3): 205–10.

Manne, A.S., and R. Richels. 1997. On Stabilizing CO₂ Concentrations—Cost-Effective Emission Reduction Strategies. Environmental Modeling & Assessment 2(4): 251–65.

Parry, I.W.H. 1995. Pollution Taxes and Revenue Recycling. Journal of Environmental Economics and Management 29: S64–77.

Parry, I.W.H., R.C. Williams III, and L.H. Goulder 1997. When Can Carbon Abatement Policies Increase Welfare? The Fundamental Role of Distorted Factor Markets. Journal of Environmental Economics and Management 37: 52–84.

Richels, R., and J. Edmonds. 1995. The Economics of Stabilizing Atmospheric CO₂ Concentrations. Energy Policy 23(4/5): 373–78.

Toman, M.A., R.D. Morgenstern, and J. Anderson. 1999. The Economics of "When" Flexibility in the Design of Greenhouse Gas Abatement Policies. Annual Review of Energy and the Environment 24: 431–60.

Zodrow, G.R. 1995. Economic Issues in the Taxation of Capital Gains. Canadian Public Policy 21(Supplement): 27–57.

11

Revenue Recycling and the Costs of Reducing Carbon Emissions

Ian W.H. Parry

The continued accumulation of heat-trapping gases in the atmosphere raises the prospect of future global warming and other associated changes in climate. Carbon dioxide (CO_2), a by-product emission of the burning of fossil fuels, is the most important example of these gases. At a 1997 conference in Kyoto, Japan, developed countries agreed on a timetable to begin reducing the emissions of CO_2 and other gases.

The benefits from reducing CO_2 emissions are difficult to assess. Enormous uncertainties surround the extent, global distribution, and economic impacts of possible changes in climate. Global warming may turn out to be a very serious problem, or it may not. Despite these uncertainties, economists such as William Nordhaus of Yale University have cautioned against drastic measures to control CO_2 emissions at this time. Others have expressed greater concern.

It is crucial to understand the economic costs that might be incurred by achieving the emissions targets agreed upon at Kyoto and additional emissions controls that might be imposed in the future. In particular, it is important to use policy instruments that minimize the economic costs associated with a given amount of emissions reduction. Economic research suggests that much will be at stake in this respect; even the costs of modest reductions in CO_2 emissions may differ substantially under different kinds of regulatory policies. Traditionally, economists have argued that the costs of reducing emissions are significantly lower under incentive-based policy instruments (such as emissions taxes and tradable emissions permits) than under more direct regulations (such as those that force all firms to adopt the same pollution abatement technologies).

Recent research has emphasized the difference between policy instruments for controlling CO_2 emissions that raise revenues for the government and those that do not. Revenue-raising instruments include taxes on the carbon content of fossil fuels and tradable CO_2 permits that are sold or auctioned by a regulatory agency. Non-revenue-raising instruments include tradable carbon permits that are distributed free of charge to firms.

This research has focused on the interactions between proposed carbon abatement policies and the tax system. Taxes distort economic behavior. Taxes on labor income reduce the overall level of employment in the economy below levels that would maximize economic efficiency. To the extent that employers must pay more for labor—because they have to pay Social Security taxes or because employees demand higher wages to compensate for taxes—the demand for labor will be lower. Also, to the extent that taxes reduce the effective take-home pay of employees, they reduce the supply of labor (for example, the partner of a working spouse may be discouraged from joining the labor force, an older worker may retire early, or a worker with one job may be discouraged from working additional hours at a second job). Taxes on the income from capital lead to "too much" current consumption and "too little" investment and savings. They include income and capital gains taxes that individuals pay on income from savings as well as taxes that corporations must pay on the income earned from expanding the size of their operations by investment.

Similarly, regulations that increase the costs to firms of producing output also tend to reduce the level of employment and investment in the economy and thereby add to the distortions created by the tax system. The costs of these spillover effects in labor and capital markets, caused by regulatory policies, are called the *tax interaction effect*. For example, a tax on the carbon content of fossil fuels drives up the cost to firms of producing electricity and gasoline, which tends to slightly reduce the overall level of economic activity and employment. In addition, this tax on fossil fuels reduces the incentives for capacity-enhancing investments in these industries. Alternative policies to reduce carbon emissions, such as tradable CO_2 permits, have the same impacts on fossil fuel industries and similar impacts on the overall level of employment and investment in the economy. Indeed, these kinds of regulations can be thought of as implicit taxes, because they raise the costs to firms of producing output in the same way that an explicit tax on the firm's activities would.

Taking into account the tax interaction effect raises the overall costs of regulations, possibly by a substantial amount. This effect was demonstrated in a recent collaborative work in which we estimated that the economic costs to the United States of using freely issued tradable permits to reduce CO_2 emissions by 10% below current levels is five times higher when interactions with the tax system are taken into account (see Parry and others in Suggested Reading).

On the other hand, some kinds of regulatory policies can raise revenues for the government. These policies include taxes on the carbon content of fossil fuels and tradable CO_2 permits that are sold by the government rather than distributed to firms free of charge. Significant economic gains are to be had from using these revenues to reduce other taxes that distort the level of employment and investment. Indeed, the benefits from this *revenue-recycling effect* can offset—perhaps more than offset—most of the costs of the tax interaction effect. Thus, our research suggests that the overall costs of reducing CO_2 emissions would be much lower under a policy that raised revenues (and used those revenues to cut other taxes) than under a permit policy with permits allocated free of charge.

Recently, considerable confusion has arisen about the implications of tax distortions in the economy regarding the costs of carbon abatement policies. In particular, several analysts have argued that there would be a double dividend from carbon taxes. They have—correctly—pointed to the potential benefits from recycling carbon tax revenues in other tax reductions. However, they have failed to recognize the adverse impact on employment and investment when the tax is initially imposed (before revenue recycling). In the rest of this chapter, I elaborate on some misperceptions behind the double dividend hypothesis and on other issues related to the interaction between CO_2 regulation and the tax system.

The Double Dividend Hypothesis

Goulder has identified several different notions of the double dividend hypothesis (see Suggested Reading). My preferred statement of the hypothesis is that environmental taxes can reduce pollution and reduce the overall economic costs associated with the tax system at the same time. At first glance, this hypothesis seems to be self-evident, if the revenues raised are used to reduce other taxes that discourage work effort and investment. In some European countries, where high taxes (among other factors) have contributed to double-digit unemployment rates, the double dividend hypothesis has been particularly appealing. It was thought that environmental taxes could be used to reduce unemployment and pollution simultaneously. More generally, it was argued that it is better to finance government spending by taxing economic "bads," such as pollution, rather than economic "goods," such as employment and investment.

Indeed, a stronger form of the double dividend hypothesis asserts that a carbon tax can produce net economic gains for society in addition to the benefits from reduced pollution emissions. It has been argued that this would occur if the benefit from the revenue-recycling effect outweighed the costs imposed on industries affected by an environmental tax. This notion is very appealing because the benefits from emissions reduction—particularly in the case of CO_2 emissions—often are very difficult to quantify. If the strong double dividend were possible, then introducing a carbon tax could produce a net gain for society, even if the benefits from reducing CO_2 emissions turned out to be very modest.

General agreement exists among economists that revenue recycling per se reduces the net cost of a carbon tax, and therefore a weak notion of the double dividend hypothesis is valid. However, recent analysis suggests that the overall impact of a carbon tax on the costs of the tax system is much more subtle than initially recognized.

The crucial flaw in the original double dividend argument was that it ignored the tax interaction effect. Environmental taxes tend to slightly reduce employment and investment (prior to revenue recycling), adding to the distortions created by the tax system, because they raise the costs to firms of producing output. For example, a tax levied on fossil fuels would increase the costs of purchasing energy. Bovenberg, Goulder, and others have demonstrated that these adverse employment and investment effects are not fully offset by using the environmental tax revenues to reduce other taxes, such as taxes on personal and corporate income. Therefore, in these more recent analyses, the tax interaction effect dominates the revenue-recycling effect, and the net impact of environmental taxes typically is to reduce the level of employment and investment in the economy.

Because environmental taxes have a relatively narrow base, we would expect them to be more distortionary than a broad-based tax such as the personal income tax. The base of a tax refers to the range of productive activities that the tax is levied on. The narrower the base, the greater the scope for firms

and consumers to substitute away from the tax (by reducing employment, increasing spending on other goods, and so on) and hence the greater the distortion created by the tax.

A tax on the carbon content of fossil fuels penalizes the industries that use fossil fuels intensively, particularly the electricity and transportation industries. It leads to two kinds of distortion (ignoring investment impacts): First, the overall level of employment tends to fall as these industries contract; second, the tax causes a shift in consumer spending away from goods produced by these industries—the prices of which are driven up—toward goods such as food, clothes, and entertainment, whose prices are not affected as much. In contrast, broad-based taxes, such as individual taxes on take-home pay, create only one kind of distortion. These taxes discourage employment, but they do not affect relative consumer prices and hence the pattern of consumer spending across different goods. The environmental tax policy discussed earlier effectively replaces revenues raised from broad-based taxes by revenues from a narrow-based tax. Tax economists have long recognized that this kind of policy change should increase the overall costs of the tax system, because it introduces a distortion in consumer choices among different goods.

It should be emphasized that the above discussion is concerned with only the *economic costs* of a carbon tax. One consequence of a carbon tax would be to shift spending away from activities that produce carbon emissions and into other activities. This result is exactly what is desired from an environmental perspective. If the environmental benefits from reducing carbon emissions exceeded the overall economic costs of the carbon tax, then society still would be better off with the tax.

Can There Ever Be a Double Dividend?

The answer to this question would be no, if the tax system were fully efficient—that is, if the composition of the tax system could not be changed so as to reduce the economic costs of the tax system for a given amount of revenue raised. In practice, tax systems are full of inefficiencies, which increase the possibility of a double dividend from environmental taxes under certain circumstances.

A key source of inefficiency in the case of the U.S. tax system is that certain kinds of spending—notably, mortgage interest on owner-occupied housing and employer-provided medical insurance—receive favorable tax treatment relative to other types of spending. Mortgage interest can be deducted from personal income tax liabilities, and medical insurance—unlike ordinary wage compensation—is not subject to personal income or Social Security taxes. Consequently, the tax system can induce too much spending on housing and health insurance relative to other spending (on food, clothes, and so on). These kinds of tax deductions raise the overall economic costs of the tax system, but they also increase the economic gains to be had from recycling carbon tax revenues in income tax cuts. In fact, under these circumstances, the revenue-recycling effect can easily outweigh the tax interaction effect, and the overall impact of a carbon tax can be to reduce the costs of the tax system, thereby producing a significant double dividend (see Parry and Bento in Suggested Reading).

Another source of inefficiency arises from capital being overtaxed relative to labor. Therefore, if the net effect of an environmental tax is to shift the tax burden away from capital and onto labor, the costs of the tax system will be further reduced. This prospect is more likely if the environmental tax revenues are used to cut taxes on capital (such as the corporate income tax) and if the industries affected by the environmental tax are relatively labor-intensive.

Presumably, shifting the burden of taxation away from capital is something the government could do directly. Similarly, the government could cut back on deductions for mortgage interest and so on. Why disguise such tax shifts by using an environmental tax? One possible reason is that direct tax reform can be politically difficult. For example, a cut in the rate of corporate income tax is perceived as benefiting higher income groups, and homeowners would oppose any phasing out of the mortgage subsidy. Bringing about some of these effects through environmental tax shifts is more opaque and possibly a little easier to obtain.

Alternative Means of Recycling Carbon Tax Revenues

Are there other ways that carbon tax revenues might be recycled to reap economic benefits besides cutting taxes? Yes, if the revenues were used to reduce the federal budget deficit. It would mean that in the future, fewer tax revenues would be required for interest payments and repayment of principal on the national debt. As a result, the distortion in the level of employment, investment, and consumer spending among different goods caused by future taxes would be lower. Thus, using environmental tax revenues for deficit reduction can also produce a revenue-recycling benefit for society, although this economic gain occurs in the future rather than the present.

If the revenues were used to finance additional public spending, then the answer to the same question would be, "It depends." The bulk of government expenditure in the United States consists of transfer payments, such as pensions, or expenditures that are close substitutes for private-sector spending, such as medical care and education. Loosely speaking (and ignoring distributional impacts), the benefit to people from a billion dollars of this kind of spending is a billion dollars. Suppose instead that the billion dollars were returned to the private sector by reducing taxes—say, the personal income tax. In this case, the benefits exceed a billion dollars. Not only do people get a billion dollars, but the lower tax rates will favorably alter relative prices in the economy. The rewards for working as opposed to not working and for saving as opposed to consuming are increased, thereby encouraging more employment and investment. In contrast, the increased public spending does not alter relative prices; the benefits from a billion dollars in additional public spending are generally less than those from a billion dollars of tax cuts.

Another component of government expenditure is on what economists call public goods. For various reasons, these goods may not be provided by the private sector: defense, crime prevention, and aid to needy families. People may (or may not) value a billion dollars of additional spending on these kinds of goods at more than a billion dollars. If so, then the economic benefits of a billion dollars from this kind of revenue recycling may be as large as (or even larger than) the benefits of a billion dollars in tax reductions.

Carbon Taxes versus CO_2 Permits

Another way to reduce CO_2 emissions is to require that firms have a permit for each unit of emissions that they produce. By controlling the total quantity of permits available to firms, a regulatory agency can limit total CO_2 emissions to a given target level. Allowing firms to trade these permits among themselves affords a lot of flexibility in achieving emissions reductions. For example, firms for whom it would be very costly to reduce emissions can purchase permits from firms that can reduce emissions at relatively low cost. The same flexibility is achieved under an emissions tax, because firms can choose to pay more taxes rather than reduce emissions. For this reason, economists have traditionally favored tradable emissions permits and emissions taxes over a more direct command-and-control type of regulation that might require all firms to reduce emissions by the same amount.

A regulatory agency could either auction off CO_2 permits or distribute them free of charge to existing firms. In the former case, the policy is essentially equivalent to a carbon tax (for a given amount of emissions reduction) because the policy raises revenues for the government that can be used to reduce other taxes in the economy. In the latter case, no revenue-recycling effect occurs, because no revenues are raised for the government. However, the policy produces the same tax interaction effect as would a carbon tax that produced the same reduction in emissions because it causes the same contraction in the industries that use fossil fuels.

Because a free CO_2 permits policy would not produce the revenue-recycling effect, it would be more costly than a carbon tax or a policy under which the permits were auctioned (but it still may be more cost-effective than command-and-control policies). Our estimates suggest that the difference can be striking. For example, we estimate that the overall economic cost to the United States of a 10% reduction in CO_2 emissions below current levels would be 300% greater under free CO_2 permits than if a carbon tax were imposed. The reason for this result is that the tax interaction and revenue-recycling effects are large compared with the economic costs of the regulation in the affected industries.

We estimate that the economic costs of a nonauctioned CO_2 permit program for the United States will exceed the environmental benefits—unless the benefits from reducing current carbon emissions are more than \$18/ton. Estimates of these benefits tend to be less than \$18/ton, except under more extreme scenarios for climate change (for examples of benefit estimates, see Parry and Toman, Nordhaus, and Roughgarden and Schneider in Suggested Reading). In contrast, Parry and others (Suggested Reading) estimate that a policy to reduce U.S. CO_2 emissions that produces the revenue-recycling effect can easily induce a net gain. It is important to remember that benefit estimates are preliminary and highly speculative at this stage. For example, they do not take into account the possibility of drastic changes in climate, should warming disturb some unstable mechanism within the climate system, nor do they take into account possibly adverse impacts on the distribution of world income arising from the greater vulnerability of poorer countries to climate change.

Moreover, other factors should be considered in choosing among policy instruments besides whether they raise revenue. For example, the affected industries might be less opposed to nonauctioned emissions permits than to an emissions tax because under the permit scheme, they would retain the rents from the policy. Other considerations might include the potential impact of a given instrument on private incentives to develop energy-saving technologies over the long run.

Conclusion

Carbon abatement polices are likely to add to the distortions in the economy created by the tax system. The added distortion raises the overall economic cost of reducing emissions by a potentially substantial amount. However, much of this added cost can be offset—perhaps more than offset—if the policy raises revenue for the government that are used to cut other taxes. Thus, a potentially important distinction exists between revenue-raising policies (such as carbon taxes) and non-revenue-raising policies (such as nonauctioned tradable CO_2 permits).

Suggested Reading

Bovenberg, Lans, and Lawrence Goulder. 1996. Optimal Environmental Taxation in the Presence of other Taxes: An Applied General Equilibrium Analysis. *American Economic Review* 86(4): 985–1000.

Bovenberg, Lans, and Ruud de Mooij. 1994. Environmental Levies and Distortionary Taxation. *American Economic Review* 84(4): 1085–89.

Goulder, Lawrence. 1995. Environmental Taxation and the Double Dividend: A Reader's Guide. *International Tax and Public Finance* 2(2): 157–83.

Nordhaus, William D. 1994. *Managing the Global Commons: The Economics of Climate Change.* Cambridge, MA: MIT Press.

Oates, Wallace. 1995. Green Taxes: Can We Protect the Environment and Improve the Tax System at the Same Time? *Southern Economic Journal* 61(4): 915–22.

Parry, Ian. 1995. Pollution Taxes and Revenue Recycling. *Journal of Environmental Economics and Management* 29(3, Part 2): 564–77.

Parry, Ian, and Antonio Bento. 2000. Tax Deductions, Environmental Policy, and the "Double Dividend" Hypothesis. *Journal of Environmental Economics and Management* 39: 67–96.

Parry, Ian, and Wallace Oates. 2000. Policy Analysis in the Presence of Distorting Taxes. *Journal of Policy Analysis and Management* 19: 603–14.

Parry, Ian, and Michael Toman. 2000. Early Emission Reductions Programs: An Application of CO_2 Policy. RFF Discussion Paper 00-26. Washington, DC: Resources for the Future.

Parry, Ian, Roberton Williams III, and Lawrence Goulder. 1999. When Can Carbon Abatement Policies Increase Welfare? The Fundamental Role of Distorted Factor Markets. *Journal of Environmental Economics and Management* 37(1): 52–84.

Roughgarden, Tim, and Stephen Schneider. 1999. Climate Change Policy: Quantifying Uncertainties for Damages and Optimal Carbon Taxes. *Energy Policy* 37: 415–29.

12 Confronting the Adverse Industry Impacts of CO₂ Abatement Policies
What Does it Cost?

Lawrence H. Goulder

Over the past decade, policy analysts have investigated several domestic policies to reduce carbon dioxide (CO_2) emissions and thereby address the risk of global climate change. One approach is a carbon tax. Another is a system of carbon permits or "caps," whereby supplies of carbon fuels or emissions of CO_2 are restricted to the amounts implied by the number of permits allocated to firms by the government. A third approach is the application of energy efficiency standards to industrial equipment and household appliances.

Economists tend to embrace the carbon tax, largely because it is seen as the most cost-effective approach to reaching the desired reductions in CO_2 emissions. Economists also find certain tradable carbon permits systems attractive in terms of cost-effectiveness—particularly systems under which permits are auctioned to firms rather than distributed free of charge. Yet these relatively cost-effective policies gain little political support. Indeed, carbon taxes and auctioned tradable carbon permits have been described as "political nonstarters."

The political resistance to these policies has many sources, but one key factor seems to be that their cost impacts are not spread evenly across the economy. These policies would place a large share of the economy-wide burden of regulation on the workers, managers, and stockholders of a few key energy industries. According to most studies, a carbon tax or system of auctioned carbon permits would significantly reduce profit and employment for producers of fossil fuels as well as industries that rely heavily on these fuels as inputs in the production process, such as electric utilities and petroleum refining. The uneven distribution of costs inhibits political feasibility, because the affected industries are well organized and can exert considerable political opposition.

Enhancing political feasibility might seem to require policies that avoid placing exceptionally large cost burdens on key energy industries. Yet, as discussed below, alternative policies that spread the burden more evenly can involve higher costs to the economy as a whole. Can CO_2 policies be designed that avoid placing especially large costs on key energy industries, without significant loss of cost-effectiveness, that is, without significant increases in overall costs to the economy? In this chapter, I examine this challenge, drawing on recent work that I have undertaken in collaboration with Lans Bovenberg of Tilburg University (the Netherlands). Our initial findings suggest that the challenge can be met.

Winners, Losers, and Political Feasibility

Policies to reduce CO_2 emissions produce "winners" as well as "losers." Under a carbon tax, for example, future generations would benefit to the extent that the tax helps reduce or prevent harmful changes to climate patterns around the globe. In addition, many producers, workers, and households in the current generation would benefit as well. For example, such a tax could boost employment and profits in industries that produce alternatives to conventional fossil fuels. Much of the rationale for a carbon tax derives from the idea that these benefits are likely to be greater in the aggregate than the costs that they might impose on others.

But the prospect of aggregate net benefits—what economists call an "efficiency improvement"—does not guarantee political feasibility. Even if benefits to winners exceed the costs to losers, the losers may be more mobilized politically and thus can dominate the outcome, thereby blocking political enactment. In particular, if the potential costs are concentrated (so that costs to each member of the losing group are substantial) while the potential benefits are diluted (so that gains to each member of the winning group are small), then the potential losers may be more likely to mobilize politically and thus may have a more powerful voice in the political process. In such circumstances, efficiency-improving policies will fail to be enacted.

This type of situation seems to apply to the carbon tax and to tradable carbon permits systems whereby the permits are initially auctioned rather than distributed free of charge. It may partly explain the significant resistance to these policies by the most politically mobilized stakeholders. How can alternative policies be designed to avoid this problem?

At first blush, the solution to this problem may seem simple: Simply offer financial compensation to address potential losses of profit or employment in the industries that might suffer serious hardship. Compensation could come, for example, in the form of tax cuts and transitional unemployment assistance. But compensation schemes can be costly. To the extent that the government uses up some of its revenues to pay compensation to certain industries, it will have to raise more revenue from ordinary taxes (such as income and sales taxes) to finance its other goods and services. These taxes create their own inefficiencies that reduce overall economic performance. Thus, offering compensation does not simply transfer wealth or resources from one group to an-

other; it often reduces the overall level of output or income in the economy. It thus becomes important to consider not only how policies can be designed to avoid serious adverse impacts on key industries but also whether avoiding these impacts comes at a very high price to the economy as a whole.

Our initial research on these questions has generated potentially encouraging results. We find that it is possible to design CO_2 abatement policies that avoid significant financial costs to key energy industries without adding significantly to economy-wide economic costs. Through partial exemptions to the carbon tax, a very modest amount of "grandfathering" (or free allocation) of CO_2 emissions permits, or relatively modest tax breaks to the most vulnerable industries, the government can promote desired reductions in CO_2 emissions without harming profits of the most vulnerable energy industries and without significantly raising the cost to the economy as a whole.

Ways to Avoid Adverse Impacts on the Most Vulnerable Industries

Most discussions of a carbon tax center on an "upstream" tax, that is, a tax imposed at the point where carbon first enters the economy. This type of carbon tax is levied on fossil fuel suppliers and importers of fossil fuels, and the tax rate is proportional to the carbon content of fossil fuels. Similarly, under an upstream carbon permit policy, permits are given out to domestic fossil fuel suppliers and importers of fossil fuels. Each permit confers the right to bring a certain amount of carbon into the economy. Because CO_2 emissions are proportional to the carbon content of fossil fuels, limiting the supply of carbon entering the economy ultimately limits the amount of CO_2 that results from the combustion of fossil fuels or fossil fuel products.

An upstream carbon tax raises costs to fossil fuel producers and, to the extent that these costs are shifted forward, leads to higher prices for these fuels. An upstream permits system also leads to higher fuel prices because it restricts the supply of these fuels. Higher fuel prices, in turn, encourage consumers and industrial users to shift their demands toward goods and services that rely less on fossil fuels in their production. This system promotes reduced combustion of fossil fuels or refined products from these fuels, which implies reduced emissions of CO_2 relative to what otherwise would occur.

Each of these policies would cause fossil fuel producers to lose profits; these losses can be substantial. However, alternative policies can ease or entirely eliminate these losses.

A Modified Carbon Tax

Under the standard carbon tax, fossil fuel suppliers pay a tax on every unit of fossil fuel supplied. If each producer were granted an exemption from the carbon tax for some fixed amount of supply of fuel, then the burden of the carbon tax on firms would be lower. (The size of the exemption could differ by firm or fuel, depending on historical supplies and other considerations.) As long as fossil fuel producers still face the tax "at the margin" (that is, for the last units supplied), the modified carbon tax will have the same effect on fossil fuel prices, fuel supplies, and CO_2 emissions as the standard carbon tax while reducing the burden on the fossil fuel suppliers. However, while this modified tax helps fossil fuel suppliers, it offers no relief to "downstream" users of carbon, who do not pay the carbon tax in the first place. Additional measures (such as targeted tax breaks) would need to be invoked to help vulnerable downstream producers, such as electric utilities and petroleum refiners.

Free Allocation of CO_2 Permits

An upstream CO_2 permit policy can be designed to avoid significant reductions in profit. Profit losses can be trimmed or avoided through the free allocation (as opposed to auctioning) of some or all of the permits. As with the standard carbon tax, under a permit policy in which all permits are auctioned, firms must pay for every unit of output supplied; they must purchase a permit that entitles them to each unit of output. In contrast, under a system of freely allocated permits, firms do not pay for each unit of output; instead, they pay (by purchasing additional permits) only for units of output beyond the level to which they were entitled in their initial permit allocation. Firms pay for output at the margin, but some "inframarginal" output or pollution can be supplied without charge. Hence, this policy imposes smaller costs on the regulated industries. Indeed, as discussed later, in some cases this policy could *benefit* the regulated industries by raising their profits substantially. Like the modified carbon tax, a carbon permit policy with free allocation to fossil fuel suppliers does not help downstream users of carbon. To help the most vulnerable downstream users, the government could offer these permits to such users (who would then sell the permits to upstream suppliers) or offer alternative forms of compensation, such as tax relief.

Corporate Tax Cuts

A third way to prevent concentrated costs on energy industries is to offer financial compensation through tax relief, such as corporate tax cuts. Such tax cuts could be directed to fossil fuel suppliers and to the most vulnerable downstream industries.

The Trade-off between Overall Economic Efficiency and a More Even Distribution of Costs

There are several ways to lessen or eliminate the adverse impacts on profits in key energy industries. But studies have suggested that avoiding these adverse effects on profits could raise the costs of CO_2 abatement considerably relative to the costs under the most cost-effective policies (the standard carbon tax or a system of auctioned carbon permits). The cost-effective policies bring in substantial revenue. In contrast, the modified carbon tax forgoes considerable revenue, and the system of freely allocated carbon permits brings in no revenue. Bringing in revenue enables the carbon tax to exploit a cost-reducing "revenue-recycling effect." This beneficial efficiency impact arises when revenues are used to finance cuts in rates of preexisting distortionary taxes such as income and sales taxes. The modified carbon tax exploits this effect less, and the policy of freely allocated carbon permits cannot exploit this effect at all. Thus, the government will be unable to exploit the revenue-recycling effect to the degree that it could under the standard policies, and the overall economic costs are higher.

The Good News

My recent work with Bovenberg does not contradict the idea of a trade-off: Our results support the idea that avoiding profit losses raises overall economic costs. However, we find that the cost increase is relatively small.

The basis for this finding is the observation that the potential revenues from CO_2 abatement policies are very large relative to the losses of profit that would occur in energy industries under the "standard" policies. Thus, only a small fraction of the potential revenues need to be sacrificed to fully compensate the most vulnerable firms. Under a modified carbon tax, the sacrifice is directly proportional to the size of the tax exemption for a portion of the fossil fuels supplied. Under a carbon permits scheme, the sacrifice occurs to the extent that some permits are given out free (or grandfathered) rather than auctioned. Under policies involving compensation through reductions in corporate tax rates, the sacrifice increases with the size of the tax cut. In all cases, the sacrifice is small, because the exemption, grandfathering, or corporate tax cut forgoes a small fraction (less than 15%) of the potential carbon revenues.

Table 1 offers a glimpse of how large the potential revenues from a carbon tax (or set of auctioned permits) might be compared with the potential losses of profit. (The values are intended to give a rough idea of the magnitudes involved; a more complex set of calculations is necessary to address the issues seriously. The numerical findings in the next section derive from a more comprehensive examination of the issues.) Gross output from the coal industry is about $29 billion. A carbon tax of $25/ton, if introduced in 2000, would bring in about $11 billion in revenues in that year. This estimate is based on the idea that the tax would raise prices by about 57% and that coal output would fall by about one-third. By comparison, if the tax reduced after-tax profits in the industry by 30%, then it would cost the industry about $90 million. This cost to the industry, while

Table 1. Potential National Carbon Revenues Compared with Potential Profit Losses in the Coal Industry: Illustrative Calculations for the United States in 2000.

Potential national carbon revenues (in billions of 1997 dollars)	
Gross output of coal industry under status quo[a]	29.28
Estimated gross output after $25/ton carbon tax (assumes 33% reduction)[b]	19.62
Estimated potential carbon tax revenues ($25 per ton carbon × 0.0228 tons of carbon per $1of fuel × $19.62 billion)	11.18
After-tax profits (in billions of 1997 dollars)	
After-tax dividends plus retained earnings under status quo[c]	0.33
Loss in earnings from $25/ton carbon tax (assumes 30% reduction)[d]	0.10
Loss in earnings as a percentage of potential carbon revenues	0.9

[a] Based on gross output in 1997, as published in Table 15 of U.S. Department of Commerce (1998). Gross output was projected to 2000 assuming a 2% real growth rate.

[b] Underlying assumptions: 55% price increase and a price elasticity of demand of 0.6.

[c] Based on average before-tax profits for 1990–97, as reported by U.S. Department of Commerce (1998). These figures were converted to after-tax values assuming an overall effective corporate tax rate of 35%.

[d] The detailed numerical model described in the text projects a loss of earnings of about 28%.

large in absolute terms, is less than 1% of the value of the potential carbon revenues. Thus, only a small fraction of the revenues that potentially could be collected from the coal industry would have to be forgone to preserve the industry's profits.

Why is the potential profit loss (or required compensation) small in relation to potential carbon revenues? Much of the cost of standard abatement policies—a carbon tax or auctioned carbon permits—is not borne by fossil fuel suppliers but instead is shifted to downstream firms (industrial users of fossil fuels) and to households that consume fossil-based consumer goods. In addition, the most vulnerable downstream firms, electric utilities and petroleum refiners, also shift a large share of their added costs to consumers. The cost borne by fossil fuel suppliers, electric utilities, and petroleum refiners is significant in absolute terms, but it appears to be small relative to the total amount of potential carbon revenue. Hence, the relative increase in cost to the economy is fairly small.

Numerical Findings

Bovenberg and I have been examining these issues using a general equilibrium numerical economic model of the U.S. economy with international trade. The model considers the interactions among the household, production, and government sectors as well as between various industries. The model generates paths of equilibrium prices, outputs, and incomes for the U.S. economy and the rest of the world under specified policy scenarios.

The model combines a fairly realistic treatment of the U.S. tax system and a detailed representation of energy production and demand. It enables us to examine interactions between energy-oriented environmental policies and the tax system. A distinctive feature of the model is its attention to the dynamics and rigidities associated with the installation and removal of industrial plants and equipment. This feature is crucial for understanding how new policies affect profits in various industries. (A detailed description of the model can be obtained from the author upon request.)

Table 2 displays results for policies that involve a $25/ton carbon tax or carbon permits that require

Table 2. Industry Impacts of CO₂ Abatement Policies, 2002 and 2025.

	Percentage changes from reference case			
	Standard carbon tax	*Carbon tax with industry-specific corporate tax cuts*	*Carbon permits, partial grandfathering*	*Carbon permits, 100% grandfathering*
Gross of tax output price				
Coal mining	54.5, 57.0	54.3, 55.9	54.5, 57.0	54.5, 57.0
Oil and gas	13.2, 8.3	13.2, 8.3	13.2, 8.3	13.2, 8.3
Petroleum refining	6.4, 5.1	6.3, 4.7	6.4, 5.1	6.4, 5.1
Electric utilities	2.5, 5.5	2.5, 5.1	2.5, 5.5	2.4, 5.5
Average for other industries	−0.6, −0.7	−0.6, −0.6	−0.6, −0.7	−0.6, −0.6
Output				
Coal mining	−19.1, −23.3	−18.9, −21.9	−19.1, −23.3	−19.0, −23.3
Oil and gas	−2.1, −4.4	1.5, −0.4	−2.1, −4.3	−2.0, −4.2
Petroleum refining	−7.8, −5.3	−7.8, −5.0	−7.8, −5.3	−7.8, −5.4
Electric utilities	−3.0, −5.4	−2.9, −5.0	−3.0, −5.4	−3.0, −5.5
Average for other industries	−0.1, 0.1	−0.1, 0.1	−0.1, 0.1	−0.1, −0.1
After-tax profits				
Coal mining	−32.3, −25.5	−19.9, −12.0	−16.6, −10.4	542.7, 526.9
Oil and gas	−2.3, −3.9	−6.6, −9.1	1.3, −1.8	21.4, 9.4
Petroleum refining	−9.1, −3.6	−5.5, −0.9	−9.1, −3.6	−9.1, −3.8
Electric utilities	−7.4, −4.8	−5.2, −2.7	−7.4, −4.8	−7.5, −5.0
Average for other industries	−0.7, −0.7	−0.7, −0.7	−0.7, −0.8	−0.7, −0.9

Note: The carbon tax rate applied under the first two policies is $25/ton. Under the permits policies (the remaining two columns), the level of abatement is such as to yield a permit price of $25/ton. All policies are revenue-neutral. Policy-generated revenues (after compensation via corporate tax cuts, if applicable) are recycled through reductions in marginal rates of the personal income tax.

Source: Author's model (see text).

approximately the same reductions in CO_2 emissions. It shows the effects on prices, output, and after-tax profits for 2002 (two years after implementation) and 2025.

Standard Carbon Tax

The impacts of a standard carbon tax (that is, a $25/ton carbon tax with recycling of the revenues through cuts in the marginal rates of personal income taxes) are shown in Table 2. This policy is functionally the same as a policy involving tradable CO_2 permits, whereby the permits are competitively auctioned and the number of permits issued is such as to yield a market price of $25/ton. This carbon tax (or auctioned permit) policy does not include any provisions to soften the impacts on energy industries. Under this policy, the coal industry experiences the largest percent changes in prices and out-

put. Coal prices rise by about 54% by the time the policy is fully implemented (2002), and the price increase is sustained at slightly above that level. The price increase implies an output reduction of about 23% in the long run. The other major impacts on prices and output are in the oil and gas, petroleum refining, and electric utilities industries. Although the carbon tax is imposed on the oil and gas industry, the price increase is considerably smaller than in the coal industry, reflecting the lower carbon content (per dollar of fuel) of oil and gas compared with coal. The petroleum refining and electric utilities industries also experience significant increases in prices and reductions in output, reflecting the significant use of fossil fuels in these industries. The reductions in output are accompanied by reductions in annual after-tax profits.

Table 3 shows the impact of the carbon tax on

Table 3. Equity Value Impacts and Overall Economic Costs of CO_2 Abatement Policies.

	Standard carbon tax	Carbon tax with industry-specific corporate tax cuts	Carbon permits, partial grandfathering	Carbon permits, 100% grandfathering
Equity values of firms in 2000				
(% change from reference case)				
Agriculture and noncoal mining	0.2	0.1	0.1	0.0
Coal mining	−27.8	0.0	0.0[a]	1,005.4
Oil and gas	−5.0	0.0	0.0[b]	29.2
Petroleum refining	−4.5	0.0	−4.5	−4.7
Electric utilities	−5.4	0.0	−5.4	−5.7
Natural gas utilities	−0.4	−0.3	−0.4	−0.8
Construction	1.5	1.8	1.5	1.0
Metals and machinery	−0.4	−0.6	−0.4	−0.5
Motor vehicles	0.2	0.2	0.2	0.1
Miscellaneous manufacturing	−0.1	−0.2	−0.1	−0.2
Services (except housing)	0.2	0.2	0.1	−0.1
Housing services	0.4	0.4	0.4	0.1
Total	−0.1	0.1	0.0	1.1
Efficiency cost ($)				
Per ton of CO_2 reduction	60.0	46.9	64.4	95.1
Per dollar of tax revenue	0.42	0.30	0.50	NA

[a] 4.3% of permits need to be grandfathered.

[b] 15% of permits need to be grandfathered.

Source: Author's model (see text).

equity values. In the model, equity values are the present value of after-tax profits (net of new share issues). Thus, the percent changes in equity values indicate the average percent changes in profit in the short, medium, and long term. The largest equity value impacts are in the coal industry, in which such values fall by about 28%. The reductions in equity values in the oil and gas, petroleum refining, and electric utilities industries are much smaller but still substantial, in the range of 4.5–5.4%. The impacts on equity values of other industries are relatively small.

The efficiency costs (which represent overall economic costs) listed in Table 3 are gross, rather than net, measures of economic impact because our model does not account for the benefits associated with the environmental improvement from reduced emissions. The standard carbon tax policy implies an efficiency cost of approximately $60/ton of emissions reduced, or $0.42 per dollar of discounted gross revenue from the carbon tax.

Policies that Address Distributional Concerns

Let us now consider policies in which distributional constraints are considered as the results for a $25/ton carbon tax accompanied by industry-specific corporate tax rate cuts sufficient to prevent equity values from falling in the coal, oil and gas, petroleum refining, and electric utilities industries (Tables 2 and 3). Any revenues from this policy remaining after financing the corporate tax cuts are used to finance cuts in the personal income tax, as under the previously considered policy. Table 2 shows that the impacts on prices and output under this policy are quite similar to those under the standard carbon tax. However, the profit or equity value impacts are very different—a consequence of the targeted cuts in corporate income tax rates.

What does this compensation cost? Model results suggest that this compensation does not add anything to the overall economic cost. Indeed, the overall economic cost is *lower* with this compensation than in the case of the standard carbon tax. As indicated in Table 3, the overall cost per ton is about $47 under this policy, whereas the cost would be $60/ton under the standard carbon tax. The reason is that, according to our analysis, the corporate income tax is

an especially distortionary tax, which involves greater efficiency costs than the personal income tax. The compensation schemes introduced under this policy lead to reductions in this especially distortionary tax. (Other efficiency considerations apply here, including the potential of industry-specific corporate tax cuts to produce inefficiencies in the way productive capital is allocated across industries.)

Results are also presented from policies that involve carbon permits. To make these policies comparable with the carbon tax policies, the number of permits issued is such as to generate a permit price of $25/ton. The amount of abatement under these policies is very close to the amount under the carbon tax policies already considered.

In the "partial grandfathering" case, just enough permits are freely allocated (or grandfathered) to ensure that equity values do not decline in the fossil fuel industries (oil and gas production and coal mining). The impacts on prices and output are very similar under this policy to the effects under the previous policies (Table 2). However, because this policy enables fossil fuel producers to supply a portion of their output without charge, profits do not fall. As shown in the table, only small percentages of the permits need to be grandfathered: 4.3% of the coal industry's permits and 15% of the oil and gas industry's permits. Together, these numbers imply that, overall, 10% of *all* of the permits issued (to domestic producers and to importers of fossil fuels) need to be grandfathered. Most of the permits can be auctioned, and thus the sacrifice in revenue is small. As a result, the added efficiency cost of neutralizing the impacts in these industries is fairly small as well. For example, the efficiency cost per ton of CO_2 reduction is $64.4, only 7% higher than the cost under the standard carbon tax.

The "100% grandfathering" permit policy involves free allocation of 100% of the CO_2 permits. This policy substantially overcompensates firms, causing equity values to rise substantially. In particular, the model predicts that coal industry equity values would rise by 1,200%! Because this policy sacrifices much more revenue, the efficiency cost is much higher: about $95/ton of CO_2 reduction, 58% higher than under the standard carbon tax.

In recent years, many policy discussions have focused on two tradable permit alternatives—auctioning all permits, and grandfathering all permits—but these cases are polar. Other, intermediate options might strike a better balance between the concern for cost-effectiveness and the desire to attend to distributional consequences and their implications for political feasibility. Under a program in which about 10% of the permits are grandfathered and the rest auctioned, for example, concerns about industry impacts (and political feasibility) can be met without greatly adding to the cost of CO_2 abatement.

Conclusions and Caveats

Overall, our results suggest that the potential adverse impacts on key energy industries can be avoided at fairly low cost. Under a standard carbon tax or carbon permit policy, key energy industries can shift a significant share of the burden of regulation to downstream industries and final consumers. The potential losses of profit are therefore small in relation to the potential revenues from CO_2 abatement policies. As a result, the amount of revenue sacrificed to avoid losses of profit in these industries is relatively small. Additional policy experiments support this idea. We obtain similar results when policies involve higher carbon taxes (up to $50/ton) and more extensive CO_2 abatement.

Some caveats are in order. First, our analysis focuses on the United States. The results might not transfer to nations with different characteristics of energy demand and supply or with different preexisting taxes or regulations.

Second, our research so far has considered only the fossil fuel, petroleum refining, and electric utilities industries. It is also important to consider impacts on other industries that might be especially vulnerable.

Third, the focus of this research is on preventing profit losses. It does not consider the expense of compensating labor for costs associated with job losses. This issue deserves close attention. Clearly, labor can have a significant political voice, and attending to employment impacts is highly relevant to political feasibility. A detailed analysis of this issue

has yet to be carried out. However, the potential cost of compensating workers is suggested by the following. Suppose that the average worker rendered unemployed under the standard carbon tax policy of Table 2 would require the equivalent of two years of salary to be compensated for lost earnings and other costs associated with unemployment. Under this assumption, the results from the numerical model suggest that the cost of labor compensation would amount to about $2.8 billion. In comparison, $15 billion would be needed to compensate capital owners for potential profit losses under the standard carbon tax. Thus, compensating workers would not raise the policy costs by a very large fraction. It should be kept in mind that these values are an initial, rough estimate. In addition, it is important to note that unemployment can cause serious local economic disruptions; the costs of such disruptions are over and above the costs of compensating unemployed workers.

Finally, it should be recognized that the determinants of political outcomes are complex and difficult to gauge. Political feasibility depends on factors beyond maintaining profits and offering compensation for induced unemployment. Although the modified CO_2 abatement policies described here are likely to enhance political feasibility, they surely do not guarantee it.

Notwithstanding these qualifications, these results come as good news for people who wish to improve the political feasibility of CO_2 abatement policies. The price tag for doing so may not be high.

Suggested Reading

Bovenberg, A. Lans, and Lawrence H. Goulder. 2000. Neutralizing the Adverse Industry Impacts of CO_2 Abatement Policies: What Does It Cost? In *Behavioral and Distributional Impacts of Environmental Policies*, edited by C. Carraro and G. Metcalf. Chicago, IL: University of Chicago Press.

Farrow, Scott. 1999. The Duality of Taxes and Tradeable Permits: A Survey with Applications in Central and Eastern Europe. *Environmental and Development Economics* 4: 519–35.

Fullerton, Don, and Gilbert Metcalf. 1998. Environmental Controls, Scarcity Rents, and Pre-Existing Distortions.

Working paper. Austin, TX: University of Texas at Austin.

Goulder, Lawrence H., Ian W. H. Parry, and Dallas Burtraw. 1997. Revenue-Raising vs. Other Approaches to Environmental Protection: The Critical Significance of Pre-Existing Tax Distortions. *RAND Journal of Economics* 28(Winter): 708–31.

Jorgenson, Dale, and Peter Wilcoxen. 1996. Reducing U.S. Carbon Emissions: An Econometric General Equilibrium Assessment. In *Reducing Global Carbon Dioxide Emissions: Costs and Policy Options*, edited by D. Gaskins and J. Weyant. Energy Modeling Forum. Stanford, CA: Stanford University.

Keohane, Nathaniel O., Richard L. Revesz, and Robert N. Stavins. 1998. The Choice of Regulatory Instruments in Environmental Policy. *The Harvard Environmental Law Review* 22(2): 313–67.

Olson, Mancur. 1965. *The Logic of Collective Action.* Cambridge, MA: Harvard University Press.

Parry, Ian W. H. 1997. Revenue Recycling and the Costs of Reducing Carbon Emissions. Issues Brief No. 2. Washington, DC: Resources for the Future.

Parry, Ian W. H., Roberton C. Williams III, and Lawrence H. Goulder. 1999. When Can CO₂ Abatement Policies Increase Welfare? The Fundamental Role of Pre-Existing Factor Market Distortions. *Journal of Environmental Economics and Management* 37(January): 52–84.

U.S. Department of Commerce. 1998. *Survey of Current Business*, November.

Williamson, Oliver E. 1996. The Politics and Economics of Redistribution and Efficiency. In *The Mechanisms of Governance*. Oxford, U.K.: Oxford University Press.

Carbon Sinks in the Post-Kyoto World

Roger A. Sedjo, Brent Sohngen, and Pamela Jagger

The Kyoto Protocol to the U.N. Framework Convention on Climate Change (UNFCCC) prepared in December 1997 contains several defining features. It provides for legally binding greenhouse gas (GHG) emissions targets for Annex I countries (as listed in the UNFCCC), but no targets are required for developing countries. Additionally, it recognizes carbon sequestration enhanced through human actions as a way of meeting emissions targets. The protocol specifically mentions emissions from sources and removals by sinks that result from direct changes in land use and forest-related activities—afforestation, reforestation, and deforestation—undertaken since 1990.

However, the protocol is silent on the role of other sinks in meeting national emission inventories. In Annex A of the protocol, for example, agricultural land is mentioned as a possible carbon source that must be included in a country's emission inventory, but no provisions are made for national credits for the buildup of the agricultural soil carbon sink. However, Article 3.4 appears to allow for expansion of recognized human-induced sink activities. Finally, the protocol is largely silent on how such credits would be calculated or verified. In 1998, the Intergovernmental Panel on Climate Change (IPCC) was asked by the Parties to the UNFCC to clarify some of the implications of various definitions. The Special Report on Land Use and Forests by the IPCC was released in 2000 (see Suggested Reading).

In this chapter, we first review some basic concepts related to the definition of carbon sinks, the function of forest ecosystems as sinks, forest-related activities that change carbon balances, and some estimates of the costs of using forests to sequester carbon. We then briefly examine trends in global forests that affect their carbon balances and review some of the questions that must be addressed in defining an effective system of forest carbon sequestration credits as part of a national strategy for meeting emission reduction targets. Next, we discuss the Kyoto Protocol and point out several ways in which the protocol is ambiguous as it applies to carbon sinks and sink credits. Finally, we discuss some broad concepts and practices for projects designed to provide carbon credits.

It should be noted that forests are not a panacea for addressing the buildup of carbon in the atmosphere. Only a limited amount of additional carbon can be sequestered into an expanded global forest system. Nevertheless, forests can "buy time" while the nations of the world devise a more fundamental way of addressing the problem of fossil fuel emissions. It is estimated that terrestrial ecosystems can sequester up to 100 gigatons (billions of tons) more carbon, much of that through forestry. At the current rate of GHG buildup in the atmosphere, the terrestrial system could offset about three decades of emissions.

Global Carbon Sinks: Basic Concepts

Global carbon is held in various stocks. Natural stocks include oceans, fossil fuel deposits, the terrestrial system, and the atmosphere. In the terrestrial system, carbon is sequestered in rocks and sediments; swamps, wetlands, and forests; and in the soils of forests, grasslands, and farms. About two-thirds of the globe's land-based carbon—exclusive of that sequestered in rocks and sediments—is sequestered in the standing forests, forest understory plants, leaf and forest debris, and forest soils. Carbon also is held in some nonnatural stocks. For example, long-lived wood products (such as buildings and furniture) and waste dumps constitute a separate human-created carbon stock. Given increased global timber harvests and manufactured wood products over the past several decades, these carbon stocks are likely increasing as the carbon sequestered in long-lived wood products and waste dumps is probably expanding.

A stock that takes up carbon is called a sink, and one that releases carbon is called a source. Shifts or flows of carbon from one stock to another over time (for example, from the atmosphere to the forest, or from fossil fuel to the atmosphere) are called carbon fluxes. Physical processes also gradually transfer some atmospheric carbon to the ocean. Biological growth involves the shifting of carbon from one stock to another. Plants fix atmospheric carbon in cell tissues as they grow, thereby transforming carbon from the atmosphere to the biotic system.

The amount of carbon stored in any stock may be large, even though if the changes in that stock (fluxes) are small or even zero (an old-growth forest that experiences little net growth would have this property). Also, the stock may be small and the fluxes significant (young, fast-growing forests tend to be of this type). The potential for agricultural crops and grasses to act as sinks and sequester carbon appears to be limited because of their short life and limited biomass accumulations. However, agricultural and grassland soils have substantial potential to sequester carbon. Their role for the human management of carbon could increase as we learn more about their characteristics and as new approaches such as conservation tillage are introduced.

How Forest Ecosystems Act As Sinks

Oceans, soils, and forests offer some potential to be managed as sinks, that is, to promote net carbon sequestration. The role of forests in carbon sequestration is probably best understood and appears to offer the greatest near-term potential for human management as a sink. Unlike many plants and most crops, which have short lives or release much of their carbon at the end of each season, forest biomass accumulates carbon over decades and centuries. Furthermore, the carbon accumulation potential of forests is large enough to offer the possibility of sequestering significant amounts of additional carbon in relatively short periods—decades. However, forest carbon also can be released fairly quickly, as in forest fires.

Fortuitously, forests managed for timber, wildlife, or recreation have carbon sequestration as a by-product. Forests may be managed strictly to sequester carbon. Such a focus on biomass accumulation could somewhat reduce the amount of other forest ecosystem uses, such as biodiversity. However, if forests managed for carbon sequestration are allowed to mature and remain unharvested, enhanced biodiversity may be one long-term effect.

The four components of carbon storage in a forest ecosystem are trees, plants growing on the forest floor (understory material), detritus such as leaf litter and other decaying matter on the forest floor, and soils. Carbon is sequestered in the process of plant growth as carbon is captured in plant cell formation and oxygen is released. As the forest biomass grows, the amount of carbon held captive in the forest stock increases. Simultaneously, plants grow on the forest floor and add to this carbon store. Over time, branches, leaves, and other materials fall to the forest floor and may store carbon until they decompose. Additionally, forest soils may sequester some of the decomposing plant litter through root–soil interactions.

Forest transitions from one ecological condition to another will produce substantial carbon flows;

forests can be carbon sources or sinks. It is important to carefully assess exactly what is happening to carbon as the forest changes to determine the forest's contribution (sink or source). The result may be a net release of carbon—thereby making the forest a source—in the case of biomass reductions from fire, tree decomposition, or logging. However, the forest may again become a carbon sink as it is restored through forest regrowth.

In much of the world, wood is burned as a source of energy, and wood burning releases carbon into the atmosphere. Where the fuelwood is taken from a forest and the forest grows back, no net carbon is emitted. Furthermore, to the extent that biofuels are produced sustainably and used as a substitute for fossil fuel energy, fossil fuel emissions are avoided and no new net carbon emissions are created, because biofuel regrowth offsets the initial biofuel emissions.

Costs of Carbon Sequestration

Numerous studies have been undertaken to estimate the costs of carbon sequestration using forests. They suggest that carbon-sequestrating forest activities may be one of the least expensive approaches to mitigating the buildup of atmospheric carbon. However, these studies are not without problems.

Many of the early studies looked at the costs of individual forestry projects—usually some form of forest establishment on lands previously not in forestry—and estimated the costs of establishing a planted forest and the amount of carbon that would be expected to be sequestered over some period of time. The average cost of sequestering a ton of carbon could then be estimated. Projects of this type often generated very modest cost estimates, often in the range of $1–$20 per ton. These studies have been criticized on several accounts. For example, in many cases, the opportunity costs of the land have been ignored, thereby underestimating the true costs. Additionally, these approaches typically assume small-scale projects in which it is assumed that other things, such as the cost structure and the behavior of other forestry activities, are not affected.

More recent studies have remedied some of these defects. They have provided estimates of cost functions, rather than simply point estimates of the costs of a specific project. These kinds of analysis recognize that the costs depend on the scale of activities and will rise, for example, as planting moves from land with low opportunity costs to land with high opportunity costs. These studies tend to generate sequestration costs in the range of $30–$80 per ton, especially where the volumes of carbon are substantial. However, most of these studies assume that other industrial forestry investments and activities (such as timber growth) will remain essentially unaffected by forest carbon activities.

Many recent studies recognize that large-scale forest sequestration activities probably will have secondary impacts on tree-planting decisions unrelated to carbon. For example, if large-scale tree-planting activities are undertaken for carbon sequestration purposes, they may also have an unintended impact on future timber availability and thus affect expectations of future timber prices. More tree plantings imply relatively lower future timber prices. In this situation, commercial wood producers may decrease their tree plantings, anticipating that the large numbers of trees being planted for carbon purposes will generate reductions in future timber prices. This effect has been called *leakage* because the overall effects of a particular project or policy are reduced when private timber growers respond to lower anticipated prices with reduced investments in commercial timber production. A recent study suggested that for a large-scale global project, leakage could be on the order of 50% (for example, for each ton of carbon captured by a forest carbon sequestration activity, half a ton is released elsewhere in the system) (see Sedjo and Sohngen in Suggested Reading). Nevertheless, forest-generated carbon sequestration is still cost-effective compared with various activities proposed in the energy sector.

Forest Management and Natural Disasters

Commercial timber harvests are typically followed by reforestation—whether natural or generated by human activity—but conversion of land use away from forests to other uses (primarily agriculture and pasture) is still common in the tropics. When commercial harvests are accompanied by reforestation, the land-clearing effects of the harvest on the forest carbon stocks are offset in the long term by carbon sequestration and the buildup of carbon stocks in the newly regenerated forest. The long-term change in carbon storage will depend directly on the kind and volume of forest harvested and regenerated. In some cases, second-growth forest will not sequester as much carbon as the original forest. For instance, when old-growth forest is harvested, the replacement forest is typically smaller, especially if it is being managed for timber harvests. However, when storage in long-lived wood products is considered, the net carbon of the managed replacement forest and its products will more closely approach that of the initial forest over a longer period.

Forest management can contribute to carbon sequestration by promoting forest growth and biomass accumulation. Additionally, it can be used to extend the harvest rotation, thereby increasing the average forest stock and the average carbon sequestration in a forest.

Finally, natural disasters can affect forest stocks and often result in forests becoming a carbon source, at least for a time. Large portions of the world's natural forests are subject to natural disturbances that occur periodically as part of natural cycles. Forests are subject to substantial carbon-releasing disturbances, particularly in the form of wildfires, which often occur after the forest is first disturbed by forces such as drought, disease, or pests. Natural disasters may release large amounts of carbon in a short time. However, where land is not converted to other uses, the forest typically reestablishes itself and again begins to sequester carbon. In many forests, natural disturbance regimes create a cyclical pattern of growth (sequestration), disturbance (emission), and regrowth (sequestration) over many hundreds of years.

World Forest: The Existing Situation

Although it is well-known that the world's tropical forests are getting smaller, it is less widely recognized that the world's temperate and boreal forests have been expanding, albeit modestly. Thus, while one carbon stock is shrinking, the other is growing. Some analysts believe that the failure to fully account for the sources of all of the carbon buildup in the atmosphere, the "missing carbon sink," is explained by the expansion of the forests of the northern hemisphere. However, because the Kyoto Protocol limits carbon credits to human-induced activities, it is unlikely that countries in the northern hemisphere will receive credit for this sink, even if forest expansion occurs within their national boundaries.

Nevertheless, the overall decline of the carbon stock in the global forest generates a net carbon source. Although it contributes to the buildup in atmospheric carbon, analysts widely agree that the primary cause of the buildup of atmospheric carbon is attributable not to changes in land use but largely to fossil fuel burning and its associated emissions.

The Kyoto Protocol and Sinks: Deforestation, Reforestation, and Afforestation

The Kyoto Protocol specifically mentions three forestry activities that may affect a country's carbon accounting balance: deforestation, reforestation, and afforestation. *Deforestation* occurs when forestland is cleared and not reforested; commonly, land clearing is associated with the permanent conversion of forestlands to other uses, such as croplands, pasture, or development. When forest is converted to some other use, there is a net loss of carbon in the terrestrial stock because most other land uses will sequester less carbon than the forest would. Under these circumstances, net carbon transfers occur. If the site is cleared and the vegetation burned, most of the carbon is released into the atmosphere. However, to the extent that the vegetation is converted into long-lived wood products or substituted for fossil fuel energy, only a portion of the carbon in the forest will be released into the atmosphere.

Although *reforestation* typically refers to the practice of reestablishing a forest on a site that was recently harvested, it also may refer to the reestablishment of forest on a site that was cleared some time ago. In either case, reforestation acts as a carbon sink because it builds up carbon stocks in the newly established biomass.

The creation of a forest on land never before forested or not forested for a long time is called *afforestation*. Often, the distinction between afforestation and reforestation blurs as the period during which the forest has been absent lengthens. Afforestation occurs when forests are established on grasslands, never previously forested. It also may be said to occur as lands that were once forested but have been in agriculture for long periods (as in parts of the southern United States) are converted to forests by either natural processes or tree planting. On afforested lands, the additional carbon stored in trees and other components of the forest ecosystem constitutes a net addition to the terrestrial forest stock.

Meanwhile, the protocol is silent on many other forestry activities, including management, conservation, and harvesting.

Some Unresolved Questions

Article 3 of the Kyoto Protocol focuses on the net change in GHG emissions by source and on removals by sinks resulting from direct human-induced changes in land use and forestry activities. The national commitments to emission targets for the developed countries are tied mainly to base-year 1990 emission levels. The sink changes that can be considered for meeting the commitments in the national emission inventories under the protocol are limited to deforestation, reforestation, and afforestation since 1990; they are measured as verifiable changes in carbon stocks in each commitment period. Initially, only sink accumulations during the five-year commitment period 2008–12 will be counted toward meeting net emission targets. However, subsequent commitment periods may be contiguous after 2012. Under Article 3.3, only a modest portion of the entire forest is eligible for carbon credits or debits, that is, the part of the forest that is being reforested or deforested and any afforested area. The forests affected by deforestation, reforestation, and afforestation are sometimes called the

Kyoto Forest in recognition of the fact that they are only subsets of a large managed forest and the still broader global forest.

How helpful the inclusion of sinks in the protocol will be in helping a country meet its emission targets will depend importantly on how the activities of deforestation, reforestation, and afforestation are ultimately defined for purposes of the protocol. The language of Article 3.3 has been described as confusing and complicated, resulting in different interpretations. Some analysts contend that the apparent omission of forest management, conservation, and protection implies that these activities would not qualify as emission reductions in the future. Others contend that forest management and conservation are encompassed within the terminology (deforestation, reforestation, and afforestation). Additionally, Article 3.4 leaves open the possibility that other management activities can be added at a later date. Depending on what additions might be made through the provisions of Article 3.4, forest management could be added to the list of recognized carbon sequestration activities. It could raise issues of what portion of the net growth experienced by the managed forest systems was eligible for carbon credits. Furthermore, Article 6, which mentions that projects can be developed in "any sector of the economy," could be interpreted as indicating that conservation and forest management could qualify for emission reduction credits. One problem with including forest conservation is that Article 3.3 states that the activity must be measured as a verifiable change in stock. It is not clear how much of what is considered conservation and protection would produce positive changes in verifiable stocks, because conservation usually protects only stocks that already exist. However, much of conservation's potential contribution to carbon sequestration would be through reducing deforestation (recognized in Article 3.3) and thereby reducing the accumulation of new debits in national emissions inventories.

The protocol provides for countries to gain credit toward their protocol requirements through afforestation and reforestation undertaken after 1990 on lands that were not forested before 1990. However, countries that are net carbon emitters from their forestry sector may lose credits, because under the protocol, countries must count the carbon activity (positive or negative) on land where the designated forestry activities have occurred since 1990.

The issue of how commercial timber harvests are to be treated is also unclear. One interpretation is that harvesting simply has not been included and should be ignored. In this view, because most commercial timber harvests involve logging followed by regeneration (often natural regeneration), one could argue that neither carbon released from commercial harvests nor the carbon sequestered from the reestablishment of a forest and growth after harvest is to be considered for meeting the levels called for in the protocol. A different interpretation is that the carbon accumulating during the compliance period (2008–12) that results from the reforestation after a post-1990 harvest can be treated as a credit. This issue may ultimately depend on definitions of "forests," "deforestation," and "reforestation" that are not yet pinned down. Additionally, Article 3.4 allows for the possible future inclusion of other categories of changes in land use and forestry activities. It may be an avenue for eventually including forest management, agricultural soils, and other managed sinks. At this time, all of these issues remain unclear.

Another potential role for sinks in the protocol is through the newly developed Clean Development Mechanism (CDM), which allows projects to be initiated by the developing countries themselves and also allows investments and participation by Annex B countries. It has not yet been determined whether Article 3.3—which limits the activities that can provide net changes to be included in Annex I national emission inventories to deforestation, reforestation, and afforestation—also applies to projects under the CDM.

Finally, the Kyoto Protocol is almost silent on the role of other sinks in meeting national emission inventories. Although agricultural land is mentioned in Annex A as a possible carbon source that must be included in a country's emission inventory, there is no provision for national credits for the buildup of the agricultural soil carbon sink. However, some provisions of the protocol (for example, Article 3.4) leave open the option that additional sink activities might

be determined eligible to provide carbon credits at some future time.

To sum up this discussion, at least four important points should be recognized:

- Forests are definitely included in the protocol.
- The protocol provides credit for some forest-based reduction of net emissions through human activities during the 2008–12 commitment period.
- Forest management, conservation, and agricultural soil sinks are not specifically mentioned; hence, their role in obtaining credits is currently subject to varying interpretations.
- It is unclear whether the emerging CDM will allow broader sink activities. How the CDM is ultimately treated depends on whether it is covered by the same limitation on sink activities as is called for in Article 3.3.

Designing Projects for Carbon Reduction Credits

In addition to the substantive issues associated with clarifying the language of the protocol so that forest-based emissions credits (and debits) can be efficiently incorporated into the portfolio of carbon reduction mechanisms, several practical issues need to be formalized. Current thinking indicates that carbon sequestration projects under the Kyoto Protocol will be managed on a project-by-project basis. What remains to be determined is how issues of baseline determination, "additionality" (the amount of "read" carbon reduction), leakage, verification, and the potential for perverse incentives to motivate forest management decisions will be addressed at the project level. Reconciliation of these issues tops the carbon sequestration agenda. In addition, the costs of afforestation and reforestation—the land use activities that support carbon sequestration—are being estimated with widely divergent results. The cost aspect of forest-based carbon sequestration as an offset mechanism is particularly important. It determines how carbon sequestration compares with other potential carbon offset mechanisms in the broader scheme of GHG reduction policies.

The determination of a baseline against which to assess carbon sequestration is critical because it provides the frame of reference for determining how carbon sequestration projects are contributing to the net carbon sink at the project level or the national level. Determining business-as-usual baseline measures of the total stock of carbon for a defined area has proved to be scientifically challenging thus far, particularly where heterogeneous forest ecosystems are the land use being examined. Evidence indicates that the measurement of total carbon is a complex process and may have high transaction costs because of the site-specific nature of forest ecosystems. Monoculture plantations appear to offer more straightforward measurement options, probably at much lower costs.

It may not be necessary to evaluate the total terrestrial stock of carbon in a defined area. Instead, one can focus on the carbon flows that result from changes in land use in this area over a specified period. This approach would allow carbon sequestration to be incorporated into an emissions trading policy in which changes in land use that gave rise to increases or decreases in long-term carbon sequestration would be covered as part of the national inventory of carbon flows relative to a national baseline. The alternative of allowing the voluntary opt-in of individual carbon sequestration projects provides a less clear-cut determination of overall changes in the carbon stock relative to a previously defined baseline.

A related issue is the unintended consequences associated with the development of a carbon sequestration system. Simply focusing on forests for carbon sequestration would probably lead to the almost exclusive establishment of single-species tree plantations. As noted earlier, old-growth forest has limited potential for absorbing additional atmospheric carbon. When ecosystem climax levels are reached, old-growth forests act primarily as fixed carbon sinks rather than continuing to sequester carbon. Without proper incentives, monoculture crops—which are known to sequester carbon rapidly and thus offer greater short-term carbon storage gains than the previously existing ecosystem—might replace biodiverse heterogeneous forest ecosystems. In fact, anecdotal evidence indicates that part of the rationale for converting some tropical forests into plantations

after 1990 in non-Annex B countries has been the interpretation that carbon credits might be obtained whereas no credits would be given for deforestation. Thus, it is important in the determination of baselines, through either regulatory enforcement or market-based mechanisms, that the incentives be appropriate to the entire set of social objectives to be met by the forest ecosystem.

Additionally, there is significant potential for local emissions reductions to simply shift deforestation and other carbon-emitting land use practices to other locations. Leakage is most likely if carbon sequestration is evaluated at the project level rather than within the more comprehensive framework of a national carbon budget. Again, a cap-and-trade policy—administered at the national level and requiring the monitoring of total additions and deletions in reference to a defined baseline—is one promising option for dealing with leakage issues. The problem of leakage is especially relevant to nations where forested land falls in regions with poorly defined property rights and where large populations rely on forest resources for subsistence needs, especially shifting cultivation.

Clearly, issues surrounding baseline determination, additionality, and leakage require serious attention. In many cases, after measurement issues are resolved and the rules that will address additionality and leakage concerns are agreed upon, verification probably will occur through a neutral third party. One of the probable institutional frameworks for the verification of forest-based carbon offsets will be third-party audits conducted in a fashion similar to those that are currently being undertaken to certify timber harvested from sustainably managed forests. If the scope of carbon sequestration develops to include agriculture and other land uses, then institutions established to verify carbon offsets will have to be adapted to accommodate the potentially infinite number of land use portfolios that will contribute to the global carbon stock.

Before carbon sequestration is widely accepted as a policy tool for reducing GHG emissions, the costs of carbon sequestration must be clarified. When analyzing carbon sequestration on the basis of the opportunity cost of land, proponents of carbon sequestration note that a great deal of forested land is located in remote regions, often characterized by high timber extraction costs because of rugged terrain and limited access. From a financial cost–benefit perspective, the opportunity cost of these forested lands is very low, which implies that if costs to establish tree cover are low, then the total cost of afforestation or reforestation to establish permanent tree cover probably will be lower than many other carbon offset mechanisms. In addition, these lands are easily viewed as prime candidates for conservation easement programs that will deter deforestation and promote conservation. However, the common convention of measuring the opportunity cost of land, in the form of forgone agricultural earnings because of carbon sequestration, often omits important considerations in decisionmaking about land use.

Research is currently under way to more clearly identify the costs of carbon sequestration, taking into account behavioral responses to alternative land use values. The inclusion of nonmarket benefits and the consequences of potential irreversibility of land conversion may significantly affect a landowner's willingness to agree to a major change in land use that will result in carbon sequestration. When these factors are considered, the costs of carbon sequestration are often higher than the majority of values generally observed in previous studies (approximately US$1–70/ton of carbon sequestered). However, many of the high cost estimates were obtained in the United States, where the opportunity costs of the land are often significant. Other parts of the world (for example, in regions of South America) appear to have the potential for low-cost carbon sequestration partly because of the inherently lower opportunity costs of land.

On the basis of current knowledge for the Annex I nations, afforestation on low-cost lands is among the less expensive GHG reduction policies, particularly in the near to medium term. The low-cost potential may even be greater in some developing countries. Given that reduction of atmospheric CO_2 is a long-term goal, carbon sequestration probably will be best used as one of a broader portfolio of carbon offset mechanisms. This carbon offset portfolio will differ from nation to nation, and developing countries may rely heavily on carbon sequestration offsets in the short to medium term.

Concluding Remarks

The Kyoto Protocol has provided a vehicle for considering the effects of carbon sinks and sources as well as addressing issues related to the emissions of fossil fuels. However, the approach of the Kyoto Protocol clearly is not comprehensive in its treatment of sinks; it is inherently restrictive in its focus. The protocol deals only with a small subset of the total carbon fluxes that are generated by selected sinks and sources, limiting its attention to human-induced carbon fluxes dealing with deforestation, reforestation, and afforestation undertaken after 1990. Additionally, the approach is further limited by its focus on changes in carbon stocks only during the 2008–12 commitment period. Thus, the protocol ignores carbon changes during other periods and from other sources, many of which are human-induced. For example, management and many human actions will generate far more carbon sequestration than credit received. The protocol also leaves unaddressed the issue of measurement and monitoring. However, in Article 3.4, the protocol provides for making the desired changes over time.

When the on-the-ground logistics of carbon sequestration projects are considered, several issues still must be clarified. Project details such as establishing baselines, controlling for leakage, and identifying the institutions that will support project verification are currently being debated. Carbon sequestration through forest activity has considerable potential to generate low-cost sequestration alternatives, especially in some developing countries. Nonetheless, care must be taken to recognize the true opportunity costs of alternative land uses and that, in many cases, social values other than carbon sequestration are involved and trade-offs are necessary.

Suggested Reading

General

Mendelsohn, Robert. 1998. Carbon Sinks: Management Tools or Bottomless Pit. Paper presented at the NBER

"Post Kyoto" Snowmass Meeting, Snowmass, CO, August.

Sedjo, Roger, and Brent Sohngen. 1998. *Impacts of Climate Change on Forests.* RFF Climate Issue Brief 98-09. Washington DC: Resources for the Future.

UNFCCC (U.N. Framework Convention on Climate Change). 1999. *The Kyoto Protocol to the Convention on Climate Change.* UNEP/IUC/99/10. Paris, France: Published by the Climate Change Secretariat with the Support of the U.N. Environment Programme's Information Unit for Conventions. http://www.unfccc.de (accessed March 14, 2000).

Technical

Marland, Greeg, and Bernhard Schlamadinger. 1999. The Kyoto Protocol Could Make a Difference for the Optimal Forest-Based CO_2 Mitigation Strategy: Some Results from GORCAM. *Environmental Science and Policy* 2(2): 111–24.

Plantinga, Andrew J., Thomas Mauldin, and D.J. Miller. 1999. An Econometric Analysis of the Costs of Sequestering Carbon in Forests. *American Journal of Agricultural Economics* 81(November): 812–24.

Sedjo, Roger A., and Brent Sohngen. 2000. Forestry Sequestration of CO_2 and Markets for Timber. RFF Discussion Paper 00-35. Washington, DC: Resources for the Future.

Sedjo, Roger A., R. Neil Sampson, and Joe Wisniewski (eds.). 1997. *Economics of Carbon Sequestration in Forestry.* New York: CRC Press.

Sedjo, Roger A., Joe Wisniewski, V. Alaric Sample, and John D. Kinsman. 1995. The Economics of Managing Carbon via Forestry: An Assessment of Existing Studies. *Environment and Resource Economics* 6(2): 139–65.

Sohngen, Brent, and Roger A. Sedjo. 2000. Potential Carbon Flux From Timber Harvests and Management on the Context of a Global Carbon Flux. *Climatic Change* 44: 151–72.

Sohngen, Brent, Roger A. Sedjo, and Robert Mendelsohn. 1996. Analyzing the Economic Impact of Climate Change on Global Timber Markets. RFF Discussion Paper 96-08. Washington, DC: Resources for the Future.

Stavins, Robert N. 1999. The Costs of Carbon Sequestration: A Revealed-Preference Approach. *American Economic Review* 89: 994–1009.

14 Environmentally and Economically Damaging Subsidies
Concepts and Illustrations

Carolyn Fischer and Michael A. Toman

A common adage in environmental policy discussions is that substantial opportunities for improving the environment and the economy exist in eliminating various subsidies that distort the decisions made by producers and consumers. Several publications in the past few years have addressed this claim, and it figured prominently in the recommendations of the President's Council on Sustainable Development in 1996. That body called for a systematic review of both expenditure and tax policies by a national commission to ferret out such dually unproductive elements.

Energy subsidy reduction has become an important component of the debate over the economic burden of reducing greenhouse gas (GHG) emissions and the policies that should be pursued to reduce these emissions. If substantial subsidies excessively encourage the use of fossil energy, then reduction of these subsidies would generate GHG savings at a negative cost; society as a whole would benefit in terms of improved allocation of scarce resources and in terms of reduced local environmental burdens.

If economists agree on one point, it is the elimination of subsidies that reduce economic efficiency. However, economists view both subsidies and economic efficiency from a perspective that is a little different from the common usage of these terms. For example, the failure to address environmental

spillovers in energy pricing is a source of inefficiency from an economic perspective; yet, it does not necessarily follow that elimination of this implicit subsidy would increase economic performance as conventionally defined. In other words, the elimination of economically inefficient subsidies may still involve trade-offs between environmental values (including climate change mitigation) and conventional goods in many cases. Identifying subsidies that harm both the environment and the economy also gives rise to several tricky methodological questions.

We address these issues in this chapter. First, we define "subsidy" in some detail. Then, we discuss some challenges that arise in measuring environmentally and economically damaging subsidies. We provide several illustrations of energy-related subsidies that clearly seem to be economically inefficient, but in several cases, the environmental benefits and GHG implications of their removal are unclear at best. We also discuss subsidies intended to promote environmentally desirable outcomes.

A Working Definition

A good working definition of *environmentally damaging subsidy* for our purposes must start with the general concept of economically inefficient subsidy, then progressively narrow the scope to identify more concrete meanings. Broadly speaking, an *inefficient subsidy* of a good or service occurs whenever its price does not correspond to the overall cost to society of producing and consuming a little more or less of the good or service. In such a case, society as a whole necessarily can be made better off in broad terms by correcting the subsidy (see the box below).

Several features of this general definition warrant emphasis. One key focus is on subsidies that change behavior by altering the relative prices of goods and services, as opposed to pure income transfers. As we discuss later, thinking about subsidies in terms of the effects of price changes on behavior can lead to conclusions that are very different from an accounting definition that simply divides a total payment or tax benefit by the amount of observed activity to de-termine an average. The relevant calculation from an economic perspective is a marginal one: how a change in price alters behavior.

In terms of scope, the general definition of sub-sidy is broad enough to encompass subsidies to pro-ducers that inflate prices as well as subsidies to con-sumers that lower prices. It also encompasses the provision of public goods that the market cannot offer well (such as urban infrastructure) and distor-tions in the pricing of private market goods and serv-ices (but see the discussion later in this chapter for important caveats on this point).

At this most general level, the concept of subsidy includes the failure to internalize environmental spillovers that result from unregulated market deci-

How Subsidies Work

Subsidies, in this context, are transfers to a particu-lar group that are based on certain characteristics or actions (for example, small oil producers or re-source depletion). Unless bestowed as an uncondi-tional lump sum, subsidies change the effective cost of a good. Some subsidies have a direct effect in the short term by changing the marginal cost of a good (for example, tax credits for ethanol). The re-sult is that producers are willing to accept a price lower than the actual cost of production, because the subsidy makes up the difference (or because consumers are willing to pay more because the dif-ference will be rebated). After markets adjust, some of the subsidy will be passed on in the form of lower prices to purchasers and an expansion of out-put (or some of the rebate will be passed back to producers as a higher price; in either case, produc-ers are receiving more and consumers paying less on the margin).

The incidence of a subsidy depends on the rela-tive elasticities of supply and demand: Even if the producer is the transfer recipient, if demand is rela-tively inelastic (unresponsive to price changes), then the subsidy will operate mostly to benefit consumers through lowered prices. On the other hand, con-sumers may receive a tax credit for purchasing, say, energy-saving durable goods; however, if supply is inelastic, most of that subsidy would be passed on to suppliers as consumers bid up prices.

Other subsidies lower the fixed costs to firms of producing a good; examples include subsidies to start-up costs such as oil exploration, research and development, or building a nuclear power plant. Once paid, these types of subsidies are "sunk" and generally do not affect short-run decisions involving the current marginal cost of production. However, if they continue to be offered in the long run, they can affect decisions to enter or exit the industry and af-fect long-run average costs. In the long run, supply tends to be elastic, and the subsidy primarily bene-fits purchasers.

Although certain groups benefit from subsidies, society as a whole loses from such policies unless they correct a preexisting market failure. Efficient re-source allocation requires prices to equal marginal costs, and minimum average cost in the long run. Subsidies distort relative prices and shift the alloca-tion of resources away from more productive sectors in the economy. Subsidies also can exacerbate preex-isting efficiency losses, such as when they are funded by government revenues raised through dis-torting labor income taxation.

sions as well as more direct subsidies with environmental side effects. De facto subsidies from the non-internalization of environmental externalities may be quite substantial, even though uncertainty and controversy continue to surround estimates of such subsidies. However, as already noted, such subsidies do not present the same set of "win-win" opportunities as more direct subsidies. The latter subsidies may make it possible to both reduce environmental damages and increase market efficiency in the narrower, more conventional sense.

We distinguish subsidies according to the deliberateness of the underlying policy. An *active subsidy* is created by the presence of government policy and causes financial resources to flow directly or indirectly from government to private actors or among private actors; a *passive subsidy* stems from the absence of a policy to correct externalities. An *inefficient active subsidy* is an active subsidy that reduces economic efficiency.

Active subsidies can flow through five possible channels. Direct payments, provision of in-kind services, and tax preferences all involve the government budget directly. Although the latter form may seem less direct, the forgone revenues from tax preferences represent real costs that must be made up with higher tax levies elsewhere in the economy. These three methods reduce the cost of the targeted good or services, benefiting producers and consumers but harming overall taxpayers; the latter two do not involve directly traceable government expenditures or tax benefits but nonetheless convey benefits of substantial value to a favored few. Trade preferences (such as the sugar import quotas) and regulatory mandates (requirements that certain products or technologies be used) raise revenues for suppliers by limiting market alternatives. Unlike the previous examples, consumers of the protected products do not share in the benefits of the subsidy; rather, they shoulder the burden through higher prices instead of the taxpayers paying higher taxes.

In this chapter, we consider environmentally damaging active subsidies—those economically inefficient active subsides that also have negative environmental side effects. One example in a nonenergy context would be government outlays in support of timber harvesting that are not justified on economic or environmental grounds, even when multiple-use benefits (for example, greater recreational access) are taken into account. The sugar import control program is another important example; this program roughly doubles the U.S. cost of sugar relative to the world price, putting an unwarranted floor under other sweetener sources as well (for example, sugar beets) and protecting ongoing sugar plantation output that damages the environmentally sensitive Florida Everglades. Where such subsidies have significant effects on economic behavior and the environment, neglecting them overstates the cost of environmental protection. However, the importance of subsidies in practice often is a complicated empirical question, as discussed below.

Although our main emphasis here is on active subsidies with negative environmental consequences, it is useful to subject to similar scrutiny active subsidies that attempt to provide environmental or other social benefits. A subsidy is not inherently inefficient; it may serve to correct some existing market failure. However, in practice, the existence of unrealized environmental or other social benefits does not offer a compelling argument for subsidization in and of itself. Subsidies usually are inefficient policy tools because they stimulate efforts to expand eligibility (rent-seeking behavior) and usually are not well targeted to address the problem (for example, tax breaks for open space preservation may not discriminate well in terms of importance of land area or economic circumstances of the landowners). They also can be quite costly compared with the behavior induced. Although some recipients will change their behavior to receive more of the subsidy, all qualifying actions that would have occurred anyway will also be rewarded.

Challenges in Identifying and Measuring Subsidies

In practice, several questions arise in evaluating environmentally damaging active subsidies and their consequences. Before the magnitude of subsidy effects can be determined, the portions of the subsidies that may be inducing environmentally damaging behavior must be identified. A prime example of a common error is seen in evaluations of how government support for the development of different technologies may adversely affect the environment. Average historical subsidies are easy to calculate—aggregate the cumulative budget outlays or tax expenditures for a new technology, and divide that sum by the cumulative output from the technology in question (say, a power generation technology)—but the resulting ratio in no way reveals a current, policy-relevant subsidy of the unwanted output or technology.

In fact, all past expenditures are irrelevant for current subsidy assessments, except insofar as they determine the baseline against which changes in current and future activity are measured. What's done is done; instead, the relevant questions are

- What public expenditures are occurring today? and
- How might these expenditures increase unwanted activities today and in the future?

In the case of government technology expenditures, the issue is how today's support for research and development (R&D) might be affecting the development and penetration of technology with unwanted environmental side effects by lowering its cost. Notice that this calculation inherently involves not only incrementalist reasoning (versus calculation of averages) but also the need for difficult counterfactual estimation.

This baseline problem is complicated by the introduction of other externalities that may be improved by the subsidy. Suppose, for example, the inability to patent the results of basic R&D means that the government supports research that also leads to the development of profitable but environmentally damaging technology. This support of R&D is a response to a nonenvironmental market failure and does not represent an environmentally damaging subsidy. But ongoing government support for the development and commercialization of the damaging technology arguably is an environmentally harmful subsidy (especially because private markets seem reasonably capable of undertaking profitable tech-

nology development and diffusion without government help).

Just as subsidies for other externalities can indirectly induce environmental damage, so can the subsidization of goods that are used in concert with environmentally damaging goods. With respect to the fees charged to road users, for example, there is debate about whether light-duty motor vehicles pay appropriate rates relative to those paid by heavy trucks as well as whether road fees are too low in general. In either case, the environment would be harmed not only by the subsidization of travel per se (which might, for example, increase noise and accidents) but also by policies that can encourage environmentally damaging energy use. Policies that affect energy use by affecting land use (for example, the deductibility of home mortgage interest that encourages larger dwellings and policies that encourage exurban development) may also stimulate energy use and put pressure on inadequate water infrastructure.

Important practical problems arise in detecting subsidies when looking at the pricing of public goods such as roads or the regulated pricing of "lumpy" services (including some utility services such as electricity transmission) where fixed costs are high relative to variable costs. In practice, the most economically efficient approach to pricing such services involves several departures from pure short-run marginal-cost pricing. Subsidies in the provision of lumpy services need to be calculated against this baseline—no mean feat in practice, given the difficulty in identifying so-called cross-subsidies (the portions of price premiums attributable to the costs of providing production capacity).

Additional problems in evaluating subsidies in the provision of public goods occur where there are many different types of beneficiaries. For example, the benefits of an urban road network generally fall to property owners, not only to drivers. If some road building and upkeep is paid through property taxes, then road use may not necessarily be subsidized. The question is whether users in each group are paying the right amount. Of course, efficiency would require drivers to pay for their marginal impact on road maintenance costs as well as congestion and other externalities; however, the amount of this complicated array of costs has little to do with general road construction costs and the accompanying benefits.

Thus, the magnitude of the relevant subsidy is sometimes as difficult to estimate as the magnitude of the environmental impact. For many of the same reasons already cited, the total expenditure on a subsidy is a poor indicator of the amount that induces a behavioral reaction and thereby the amount of harm. In other words, by determining the direction of the behavioral response, one can identify a win-win situation; however, the size of the win in terms of greater economic efficiency and less environmental degradation is harder to pin down. The bottom line is that the easiest environmentally harmful active subsidies to identify are those that involve more or less conventional market goods and services, without too large a fixed infrastructure component, and which benefit a well-defined subset of consumers or producers.

Whereas we have been addressing subsidies with environmental harms, we need to consider ostensibly good subsidies as well. In addition to the general problem of long-run inefficiencies induced by subsidies, costly and even perverse results can occur in the short run in cases where good behavior cannot be directly targeted. Consider, for example, the ubiquitous use of subsidies to promote better land management or agricultural practices with less damaging offsite effects in lieu of trying to deter bad management. In some cases, it might be argued that inducements are useful, or at least necessary, because of difficulties in directly monitoring bad behavior (for example, agricultural pollution from smaller non-point sources) or because part of the problem lies in polluters lacking adequate information, which the government can help supply. However, it may be equally hard to monitor good behavior and thereby decide whether the effects of subsidies intended to improve the environment are working or working cost-effectively. Well-intentioned but poorly targeted and supervised subsidies could even encourage environmental damage. For example, a land preservation or GHG policy could create an incentive to deforest land prematurely or excessively to gain eligibility for a reforestation credit.

Some Illustrations from Inefficient Energy Subsidies

Environmentally Damaging Energy Subsidies

All fossil fuels give rise to emissions of carbon dioxide, a GHG. Local environmental damage from energy use comes from two main sources: end-use combustion, primarily in the form of gaseous emissions, and by-products of production (such as effluents, leakages, accidents, spills, and reclamation costs). Any subsidies that encourage production increase the latter type of damage; however, only subsidies that lower the cost of energy use affect GHG emissions and local damages from energy end use. For example, subsidies to domestic oil production will increase oil-related wastes; however, because oil prices are largely determined on the world market, such policies are unlikely to lower domestic prices and increase end use. On the other hand, if coal prices are largely determined domestically, then coal subsidies could lead to lower prices and more domestic coal combustion. By the same token, policies that protect domestic high-cost energy supplies from foreign competition (as in Germany) will discourage the use of that kind of energy. Nor are production subsidies the only culprits. Subsidies to goods consumed in concert with energy can increase both combustion and, to the extent that producers respond to higher energy prices with increased supply, industrial by-products.

As described in more detail in the box on pages 148–149, one can identify in the United States several environmentally harmful active subsidies in the sense defined in the previous section: No economic justification for the subsidy is apparent, so eliminating it should enhance overall economic performance, and its removal probably will provide at least some environmental benefit. Several examples involve tax preferences for U.S. fossil fuel producers, such as extra tax write-offs for oil and gas depletion; preferential tax treatment of royalty income; exemptions from passive loss write-offs; and tax-preferenced financing for various governmental or quasi-governmental entities supplying electricity. Others involve in-kind transfers such as below-market mineral rights leasing and the potential subsidization of

More Detail about U.S. Energy Subsidies

Subsidies to Primary Energy Supply

Tax preferences are a common form of subsidy, although they are less transparent than direct forms. Such preferences distort the allocation of capital and raise the cost of public budget finance by requiring higher tax levies elsewhere in the economy. When a whole sector benefits from some tax credit not available to the rest of the economy, it makes sense to check for any justification in terms of offsetting other tax distortions and, if not, to label the preference a subsidy. Several examples are present in the energy sector.

The easiest tax expenditure subsidy to target is the percentage depletion provision ($840 million in 1999 for fuels). Rather than deducting cost depletion, producers can elect to deduct a certain percentage of gross income from resource production: 10% for coal and 15% for oil and gas. (However, the oil and gas provision is limited to small independent producers.) This option has no discernable economic justification, not even compliance simplicity, because taxpayers still have to calculate their cost depletion to figure out which one is greater. (Cost depletion, although not quite true economic depreciation, is at least comparable to the treatment of other capitalized assets in the tax code.)

Individual owners of resource properties have additional special tax breaks. Individual owners of coal leases have preferential tax rates ($50 million in 1999); their royalty income is taxed as capital gain. As these rates are being lowered, this subsidy will increase. (Owners who elect this option cannot take the percentage depletion provision. Both preferential treatments should be eliminated simultaneously to avoid shifting between subsidies.) Working interests in oil and gas properties are granted an exemption from the passive loss limitation ($50 million in 1999), meaning excess losses can be used to offset ordinary income rather than be carried over to offset future passive losses.

Some analyses argue that applied technology development expenditures and tax breaks constitute subsidies in that with a patent system, these technologies (unlike basic research) will be adequately rewarded in the marketplace. Conceptually, this argument seems correct. The problem is in estimating the subsidy. Total expenditures and some calculation of average expenditure per output are not accurate indicators of the level of behavioral impact. The appropriate question is, How much lower is the cost of energy supply today because of an unwarranted technology development expenditure in the past? And looking ahead, to what extent are today's technology development expenditures subsidizing future energy production? These questions are inherently hard to answer. Nevertheless, we can say that government support for technology development that the market could have handled is economically unwarranted and also may put pressure on the environment.

Some other less obvious production subsidies also exist. Below-market leasing of mineral rights on federal lands has no economic justification. The environmental harm that results from these leases is not easy to estimate, however. Firms still will have an incentive to plan the timing and intensity of resource development and extraction to maximize the market value of output. Of course, favorable terms could cause sites that have greater value in an undeveloped state to be developed. The economic harms are more recognizable. By depriving the government of revenue, the leases require greater revenue generation elsewhere in the economy, with the resulting economic distortions. Infrastructure investments, such as port and railway building and maintenance, may have other public good aspects but nonetheless lower transportation costs for fossil fuel producers. Unless beneficiaries of these investments shoulder the burden of financing them, the resulting lower transportation costs will stimulate fossil fuel production.

Subsidies to Primary Energy Demand

An important category of subsidies to primary energy demand—one that is declining in importance—involves electricity generation. Tax exemptions for interest on state and local bonds help
continued on next page

More Detail about U.S. Energy Subsidies, *continued*

finance large capital investments for public utilities. Cooperatives are exempt from income tax as non-profit organizations and are eligible for low-interest loans from the Rural Electrification Administration. Power marketing administrations are able to finance their debts at rates through the U.S. Treasury on favorable terms. All these policies lower the cost of electricity production, thereby stimulating electricity demand, which in turn leads to more purchases of primary energy sources to produce electricity. However, as wholesale electricity markets become more competitive, the effect of these subsidies on electricity prices will diminish in importance; they will simply reflect local windfalls. Moreover, direct subsidies for capital financing have been reduced over time.

Other potential demand-side subsidies involving transportation and land use are even more controversial. Automobile use (and thereby fossil fuel use) is encouraged by grants to states to develop and maintain roads and highways. Notwithstanding the public goods and safety aspects of highways, many see this form of financing as a subsidy. However, the magnitude of the subsidy is very difficult to estimate in practice, especially considering that the costs of road congestion and overbuilding are, to a consider-

able extent, borne by drivers (although environmental side effects are more broadly borne). A more straightforward subsidy is the tax preference employees can enjoy from the use of employer-provided parking, when benefits for offsetting the cost of mass transit use are more limited. Another controversial item is the tax preference for home mortgage interest deductions, which some have argued causes people to purchase larger, more energy-consuming dwellings that also may be farther from urban centers.

Another example is various tax expenditures for alcohol-based motor fuels, including an exemption from gasoline excise taxes and a tax credit for small ethanol producers. These tax breaks provide very limited local environmental benefits at best, because they chiefly displace cheaper substitutes in the market for fuel blenders. Corn-based ethanol also is a questionable means of reducing GHGs, after accounting for the fossil fuels used in cultivating and processing the feedstock. Meanwhile, the tax breaks reduce tax revenues by several hundred million dollars per year. (According to the *Budget of the United States Government*, Fiscal Year 1999, revenue losses in 1998 and 1999 are estimated to be $720 million and $750 million, respectively; projected future losses exceed $800 million/year beginning in 2001.)

goods and services used in concert with energy, such as roads, whose magnitudes and consequences are more controversial. Again, only some of these subsidies encourage increased energy end use and GHG emissions. Finally, whereas some analysts include past expenditures or tax benefits for R&D in current subsidy calculations, we argue that only current outlays that affect current and future energy decisions are relevant.

Moreover, it should be emphasized that the magnitude of environmental benefits that might result is uncertain and may be relatively small in many cases. Part of the reason is that subsidies to oil producers have relatively little effect on world oil prices, so eliminating them will do little to discourage energy use. The greater opportunity for environmental improvements may lie outside the scope of this chap-

ter, in restricting harmful environmental by-products of energy production and use to achieve environmental benefits even though real economic costs of doing so are likely as well.

Energy Subsidies for Environmental Goals

Several tax expenditure items in the United States are related to various renewable energy sources. The largest is the tax credit for ethanol fuels and exemption of these fuels from excise taxes. This item is a prime example of our earlier caution concerning the use of subsidy policy for environmental improvement. Numerous analyses have shown that this policy generates few if any environmental benefits: When the fossil fuels used in cultivation and processing are taken into account, the net GHG savings from corn-based ethanol are small at best, and the environmental advantages of ethanol in reformulated gasoline are controversial at best. The primary beneficiaries of this policy are ethanol producers who benefit from a subsidized demand for their output and, to some extent, Midwestern farmers who experience marginally firmer markets for their crops with the program.

Another example of a tax subsidy related to promotion of environmental goals lies in the 1997 Taxpayer Relief Act, which extended (though not to full parity) current tax preferences for employer-provided parking to other transit benefits. Doubtless this policy will encourage some shift toward more environmentally friendly travel. In this way, the subsidy differs from the corn ethanol example. But the larger question is why tax-preferred forms of employee compensation give rise to the need for such adjustments.

Beyond these more or less easily quantified tax preferences, numerous regulatory programs or proposals can or could affect energy use. The Energy Policy Act requires improved appliance efficiency and greater use by the government of certain kinds of products (such as alternative transportation fuels). Such regulatory mandates raise demand for the favored goods (energy-saving parts or alternative fuels) and create an implicit subsidy to producers. Several proposals are circulating that would require electricity producers in a more competitive, restructured marketplace to supply at least a minimum fraction of their power by using some kinds of renewable technologies. (More flexible versions of this mandate, called the generation portfolio standard, would allow some companies to fall below the threshold if they paid other companies with a comparative advantage in renewable resources use to go above the threshold.) Such a mandate would provide an implicit subsidy to renewable energy sources and an implicit tax on nonrenewable energy sources, because additional generation from a nonrenewable source must be accompanied by a certain additional amount from a renewable source.

What—if any—justification might there be for such policies, and are better alternatives available? The fact that renewable energy sources might be environmentally friendlier in some cases does not in it-

self justify a mandate. (Of course, in other cases, renewable energy sources may not be so friendly overall. For example, biomass burning generates particulate matter, and hydro projects create controversies over land inundation.) Mandates are poorly suited to distinguish differences within the targeted groups; all renewable resources are not equal in their actual environmental consequences, nor do all nonrenewable energy sources have identical emissions characteristics. The more efficient approach to pollution control would be to levy policy sanctions directly against pollution or GHGs, thus making clean technology choices more attractive.

Some analysts have argued that other market failures call for more intervention in energy markets. One argument is that renewable energy technologies need government protection for "market creation" to overcome market inertia or achieve scale economies. Another assertion is that the market's choices regarding energy efficiency are flawed (for example, tenants may not have adequate incentives to invest in energy efficiency and may not even pay for energy separately from the lease payment). Sorting out what are legitimate market failures in these cases is complicated, and it remains the subject of ongoing research and debate. However, in general, policies that target the problem most directly are most efficient. Mandating or protecting certain technologies is almost always less efficient than penalizing pollution or GHGs, because penalties will also create incentives for developing and using cleaner technologies and induce changes in institutional and contractual arrangements that impede energy efficiency. Any specific instances of market failure related to the development and diffusion of renewable or energy efficiency technologies are also best tackled directly (for example, through information and demonstration programs).

Environmentally Damaging Subsidies in Developing and Transitional Countries

Whereas the scale of environmentally damaging subsidies in the United States and other advanced industrialized countries is somewhat unclear, it has been almost an article of faith that massive subsidies exist in developing countries. The International Energy Agency estimates that in the eight largest countries outside the Organisation for Economic Co-operation and Development (OECD)—China, India, Indonesia, Iran, Kazakhstan, Russia, South Africa, and Venezuela—end-use energy prices are approximately 20% below their economic opportunity cost. This underpricing both spurs overconsumption and also leaves energy firms with inadequate financial resources for investing in productivity, capacity, or environmental improvements. The production inefficiencies associated with state control of the energy sector thus add to environmental woes as well as economic ones, from excess emissions due to overuse of fossil fuels to soil and water contamination from poorly maintained pipelines. In addition to this direct form, energy subsidies in developing countries also take indirect forms, such as trade barriers that limit the availability of more energy-efficient technologies. Estimates by the World Bank (see Suggested Reading) and other experts in the early 1990s put the magnitude of environmentally damaging subsidies (energy and otherwise) in the hundreds of billions of dollars. Although some of this amount reflected necessarily crude efforts to impute values of noninternalized environmental damages, the estimated magnitude of active subsidies still was substantial.

Although these subsidies still exist, they have been declining in recent years. According to recent World Bank figures, during the 1990s total fossil fuel subsidies in 14 developing countries declined by 45%, compared with OECD subsidy reductions of 21% during the same period. According to other figures, China in particular has engaged in a significant effort to reform its energy pricing (see Reid and Goldemberg in Suggested Reading). It cut coal subsidies from 37% in 1984 to 29% in 1995, and petroleum subsidies from 55% in 1990 to 2% in 1995. India, Mexico, South Africa, Saudi Arabia, and Brazil have also cut fuel subsidies significantly in recent years.

One important change lying behind this trend is the transition of many centrally planned or highly statist economies toward more market-oriented economic systems. A major part of this transition has been substantial increases in energy prices toward efficient levels. Although energy subsidies persist, especially for households, they are much smaller than in the past. This trend means fewer opportunities remain for win-win subsidy removal in this sector. A second force that has worked toward reduced subsidies is progress toward more open trade, which has lowered protections for inefficient and environmentally damaging domestic manufacturing sectors in some countries. (Outside the energy arena, subsidies also can include failure to effectively manage "commons" resources such as forests, making the effective cost of resource extraction well below the true opportunity cost. Inadequate regimes for management and protection of commons resources then often result in excessive exploitation and deforestation.)

Conclusions

We have attempted to illustrate both the potential importance and the controversies surrounding the assessment of environmentally damaging subsidies in the United States, particularly those related to energy. More research is needed to better understand the magnitudes of existing subsidies in terms of their effects on behavior. At the same time, a clearer public consensus on the effects of subsidies and what to do about them (including what to do with beneficiaries) is needed.

These needs suggest a two-pronged approach. At the national policy level, an institution could be impaneled to examine existing subsidies and recommend corrective actions for up-or-down congressional vote. To support the commission's work and increase public understanding generally, the relevant government agencies and concerned private foundations should maintain and increase analysis (in-house and with extramural funding) of environmentally damaging subsidies.

Acknowledgements

We are grateful to Marina Cazorla, Jennifer Lee, and Kelly See for assistance in preparing this chapter and to participants in the Wirth Chair dialogues at the University of Colorado, Denver, for useful comments about the issues addressed here. Responsibility for the content is ours alone.

Suggested Reading

Many of the figures presented in this document were taken from the *Budget of the United States Government*, Fiscal Year 1999, Analytical Perspectives, pp. 89–120.

Anderson, K., and W. McKibbin. 1997. Reducing Coal Subsidies and Trade Barriers: Their Contribution to Greenhouse Gas Abatement. CEPR Discussion Paper No. 1698. London, U.K.: Centre for Economic Policy Research.

Council of Economic Advisers. 1997. Economic Report of the President. February. Chapter 6.

Delucchi, Mark. 1997. A Revised Model of Emissions of Greenhouse Gases from the Use of Transportation Fuels and Electricity. Report UCD-ITS-RR-97-22. November. Davis, CA: University of California, Institute of Transportation Studies.

IEA (International Energy Agency). 1999. *World Energy Outlook, 1999 Insights, Looking at Energy Subsidies: Getting the Prices Right.* Paris, France: Organisation for Economic Co-operation and Development.

Koplow, Douglas. 1993. *Federal Energy Subsidies: Energy, Environmental, and Fiscal Impacts.* April. Washington, DC: Alliance to Save Energy.

Larsen, Bjorn, and Anwar Shah. 1995. Global Climate Change, Energy Subsidies, and National Carbon Taxes. In *Public Economics and the Environment in an Imperfect World*, edited by Lans Bovenberg and Sijbren Cnossen. Boston, MA: Kluwer.

OECD (Organisation for Economic Co-operation and Development). 1996. Subsidies and Environment: Exploring the Linkages. Paris, France: OECD.

Reid, Walter, and Jose Goldemberg. 1998. Developing Countries Are Combating Climate Change. *Energy Policy* 26(3): 233–37.

U.S. DOE/EIA (Department of Energy, Energy Information Administration). 1992. Federal Energy Subsidies: Direct and Indirect Interventions in Energy Markets. Report SR/EMEU/92-02. November. Washington, DC: U.S. Government Printing Office.

World Bank. 1992. *World Development Report 1992: Development and the Environment.* New York: Oxford University Press for the World Bank.

15

Electricity Restructuring
Shortcut or Detour on the Road to Achieving Greenhouse Gas Reductions?

Karen L. Palmer

For much of its 100-year history, the U.S. electric power industry has been organized largely as a collection of franchised monopolies that generate, transmit, distribute, and sell electricity at regulated prices to captive customers. As a result of changes in technology and regulatory policies, however, the generation and sales of electricity are being opened up to competition (and, to some extent, being separated from the transmission or "wires" side of the business). This transition from end-to-end regulation to greater competition is often referred to as *electricity restructuring*.

Concerns over carbon dioxide (CO_2) emissions as well as conventional pollution are playing an important role in debates over the course of restructuring. Several studies have looked at the effect of retail competition on emissions, but the issue is far from resolved.

The effect of electricity restructuring on CO_2 emissions is of particular interest to U.S. government officials and others charged with developing strategies for achieving the U.S. greenhouse gas emission reduction targets established at Kyoto, Japan. How restructuring will affect CO_2 emissions will depend largely on the behavior of participants in the newly competitive electricity marketplace, which in turn largely will be determined by the policies that initiate and govern the transition from regulation. Many of the basic questions regarding the effect of

restructuring cannot yet be definitively answered; nonetheless, we can draw important lessons from the current state of knowledge.

Background on Restructuring

Technology in electricity generation has evolved in ways that make competition more feasible. Particularly important has been the emergence of gas-fired generation technologies that can be efficiently operated on a small scale relative to the size of the plant needed to satisfy demand. Thus, it is technologically reasonable and economically efficient to have more than one company generating electricity in a market area.

Other competition-enhancing changes derive directly from public policy. Although nonutility generators have been making inroads into wholesale generation markets for nearly 20 years, competition in electricity generation was bolstered by the Federal Energy Regulatory Commission (FERC) decision in 1996 that ordered transmission-owning utilities to allow open access to their transmission lines to facilitate wholesale electricity transactions. The push for retail competition began with the passage of laws in California and New Hampshire in 1996. Under retail competition, electricity consumers are allowed to pick their electricity suppliers, but delivery to the consumers' premises will continue to be handled by the regulated local distribution utility. As of November 1, 1999, state utility regulators, state legislatures, or both in 24 states had made the decision to implement retail competition within five years or less. Several bills seeking to promote retail competition in electricity markets nationwide (including the Clinton administration's Comprehensive Electricity Competition Act) have been introduced in Congress, (including a 1999 version of the Clinton administration's restructuring bill).

Key Demand-Side Issues

Effects on Prices and Demand

Electricity restructuring is expected to lower the average prices for electricity as competition from new entrants and low-cost suppliers drives down the market price in markets that traditionally are priced higher. Lower prices are likely to generate higher demand from consumers which, holding the composition of the generating stock fixed, will produce higher CO_2 emissions from electricity generation.

How much electricity prices will fall as a result of restructuring depends on several factors. If the regulated utility is relatively inefficient and the new market is very competitive and provides options for all classes of customers, then restructuring could produce substantially lower electricity prices. On the other hand, if the local regulated utility supplies electricity at a lower cost than its neighbors, then prices in the local area could actually rise under competition, particularly if new entrants are unable to beat the incumbent's price.

The price-lowering effect of competition will be muted by the provisions found in most electricity restructuring laws and proposed legislation to allow at least partial and often substantial recovery of "stranded costs" in retail electricity prices. Stranded costs are costs previously incurred by a utility that it will be unable to recover at market prices. The larger the amount of stranded costs that must be recovered through an electricity surcharge, the smaller the price reductions arising from competition in the short run. Over time, the impact of stranded costs on retail electricity prices will diminish as the contribution of stranded cost recovery to electricity prices decreases and ultimately disappears.

Partially offsetting this effect is the guaranteed rate reduction provision of restructuring laws passed in many states. In these states, legislators, regulators, or both have guaranteed rate reductions either by imposing a retail rate cap or by establishing a "standard offer" rate below the existing regulated price. This standard offer rate is available to customers who elect not to choose a competitive supplier. Like the stranded cost recovery provisions, these "rate cap" or guaranteed rate provisions expire after a period, making it difficult to predict the long-run price effects of competition.

The effect of expected price declines on electricity demand and on CO_2 emissions from the electricity sector depends on how sensitive demand is to changes in electricity prices. Estimates of the price sensitivity of demand vary depending on region of the country, time of year, and type of customer. However, in general, they suggest that if prices fall by, say, 10–15% as a result of restructuring, demand could increase by anywhere from 1 to 6% or more in response to that price change.

Restructuring is also expected to produce more widespread use of time-differentiated pricing of electricity. With time-differentiated prices, customers face higher-than-average prices during periods of peak demand and lower-than-average prices during off-peak periods. This form of pricing will shift demand away from peak periods to off-peak periods, which also could contribute to higher CO_2 emissions, especially in those areas where off-peak, baseload generation tends to be coal-fired and peak generation tends to be gas-fired.

Fate of Management and Energy Conservation Efforts

As electricity markets become more competitive and utilities seek to eliminate excess costs, the number of utility-funded demand-side management (DSM) programs is already rapidly diminishing. Utilities argue that they cannot maintain these programs and effectively compete with independent power producers who are not required to fund such initiatives. In the absence of a new policy initiative, many program supporters argue that the CO_2 emissions savings attributable to past DSM and conservation efforts by utilities will be lost in a competitive market.

The flip side of that argument is that, even though greater use of time-differentiated pricing of electricity may merely shift some of the demand to off-peak periods, it also will result in stronger incentives for consumers to conserve electricity during peak periods. Moreover, restructuring may create opportunities for energy service companies to compete on the basis of holding down the customer's overall energy bill.

Many state restructuring laws and federal restructuring bills also explicitly include a mechanism for funding DSM initiatives that does not discriminate among electricity suppliers and could result in some energy and CO_2 emissions savings. These initiatives usually take the form of a surcharge (often called a systems benefits charge) imposed on all electricity customers, regardless of their energy supplier. Some portion of the revenues from the system benefit charge is used to fund DSM programs.

Key Supply-Side Issues

The Future of Nuclear Power

Nuclear power plants produce roughly 20% of the electricity sold in the United States. Nuclear generators emit no CO_2, so maintaining or increasing the contribution of nuclear power to U.S. electricity production could be a key component of a strategy to prevent or delay increases in CO_2 emissions from the electricity sector. How important nuclear generation will be in the competitive electricity market of the future depends on the relative contributions of two opposing effects.

Competition may result in the early retirement of a substantial portion of the existing nuclear capacity. In a regulated environment, most nuclear power plants would be expected to remain on line at least until the expiration of their current operating licenses. At market prices, many nuclear plants will be unable to cover the costs of fuel, operation and maintenance, and meeting safety requirements. Estimates of the annual amount of nuclear generation potentially subject to early retirement range from 40 billion kWh/year to more than 110 billion kWh/year,

or between 6.3 and 17.5% of current levels of nuclear generation.

Countering this effect, competition will likely improve efficiency at nuclear power plants. These improvements could take the form of fewer unplanned outages or shorter downtimes associated with refueling and therefore increased generation at existing plants over the course of the year. Efficiency gains could also result from reductions in operating and maintenance costs as nuclear operators actively seek to reduce their production costs and increase their operating returns. Higher returns will help to keep plants on line longer.

Role of Renewable Energy Sources

Like nuclear generation, all renewable generating technologies—including hydropower, solar thermal and photovoltaics, biomass, geothermal, and wind power—are zero-carbon emitters. Renewable energy sources account for only about 12% of electricity generation in the United States, and nonhydropower renewables account for only 2%. Nonhydropower renewables have been slow to penetrate electricity markets because of their high cost. Nonhydropower renewables also are generally nondispatchable (that is, the timing of their use cannot be controlled), which tends to diminish their value relative to other generating resources. If increased competition in electricity markets leads to lower electricity prices, as expected, then in the absence of cost-reducing technological developments, renewables will be less likely to penetrate the market.

Competition brings greater possibilities for the diversification of services in the market that could provide a boost to renewables. In states that have moved to competition, renewable generators and power marketers are developing "green power" service packages. Under these packages, customers contract for power that is, for example, 20, 50, or 100% renewables-based and generally pay a premium above the market price of conventional power. Some of these service packages are limited to nonhydropower renewables, but many are not.

Whether green power marketing increases the amount of electricity generated from renewable sources depends on whether the size of the green power market exceeds the contribution of the existing renewable generators and on the selectiveness of green power purchasers. Some green power packages specifically indicate that a certain percentage of the power will come from new renewable sources, presuming that technology penetration beyond current levels is something that customers care about. Penetration of new renewables also does not necessarily preclude the early retirement of existing renewables that find it difficult to compete in a more competitive marketplace.

In the traditional regulated environment, renewables have benefited from promotion policies such as the renewable purchase requirements under the federal Public Utilities Resource Policy Act (PURPA) and state public utility commission (PUC) orders that have often resulted in utilities offering or being required to build some new renewables-based generating capacity. With restructuring, PURPA is expected to become a thing of the past because it places requirements on utilities not faced by their nonutility competitors, and PUC oversight of generation capacity planning probably will be drastically reduced. In the absence of a new policy to promote renewables in restructured markets, renewables are likely to fare less well in a competitive environment than in a regulated environment.

Not surprisingly, new policies to promote renewables are part of almost all existing and proposed restructuring laws and regulations. Some jurisdictions, such as California, have adopted surcharge-funded subsidies to renewable generation as a way to bring renewables into the market. Other jurisdictions, including Maine and Massachusetts, rely on a renewables portfolio standard (RPS) approach. The RPS typically requires that a minimum percentage of all electricity generated (or sold) within a region must come from nonhydropower renewable sources (in Maine, hydropower sources are included under the RPS cap as well). Generally, this percentage exceeds the current contribution of renewables-based power by a substantial percentage. Under this policy, each megawatt-hour of renewables-based power generated creates a renewables generation credit. Generators (or retailers, depending on who is responsible for meeting the requirement) can meet their obliga-

tions through some combination of direct generation, purchase of renewables-based electricity generated under contract, and purchase of sufficient renewables generation credits to fulfill their obligation. This program is designed to allow the market to identify the least-cost set of generators using renewables-based technology to satisfy the renewables obligation.

The impact of a RPS policy on CO_2 emissions will depend on how high the standard is set. Recent federal bills contain proposed RPS levels for 2010 that range from 4 to 20% of total electricity generation or total retail sales.

Effects on Interregional Electricity Trade

Open transmission access makes it possible for low-cost utilities with excess generating capacity to sell their cheap electricity to retail electricity suppliers in higher-priced neighboring regions. If the low-cost suppliers are predominantly older coal-fired facilities and the generation being displaced comes from oil-fired or natural gas–fired facilities, then increased interregional electricity trade will result in higher CO_2 emissions. There is some evidence that interregional trading of electricity—particularly the shipment of electricity from the Midwest to the Northeast—has increased with the establishment of open transmission access for wholesale power transactions. The amount of interregional electricity transmission is expected to increase even more with the move to retail competition. Under competition, exporting generators will also have incentives to improve plant availability, which could further increase the amount of electricity generated for export and, therefore, the level of emissions. On the other hand, generators will also have an incentive to economize on fuel use (perhaps by actually improving their heat rates), which could reduce the CO_2 emission rate per kilowatt-hour.

The extent to which interregional power trade will increase under competition depends significantly on the amount of interregional transmission capacity that is available. Large differences in electricity prices between regions suggest at first glance that there will be greater incentives to expand transmission capacity under competition in an effort to exploit those price differences. In addition, the FERC open transmission access order requires transmission-owning utilities to expand transmission capacity if necessary to satisfy a demand for transmission service that cannot be met with existing capacity. But the incentives to expand transmission capacity will depend importantly on how transmission service is priced. If transmission is priced in a way that allows transmission owners to earn excess profits whenever lines are congested, then they will have incentives to delay expanding transmission capacity. Alternatively, if transmission users have rights to congestion revenues, then the incentive to delay investment in new capacity could be muted.

The potential for increased CO_2 emissions from greater electricity trading will diminish over time as older coal-fired generators are retired. Although in general, coal-fired plants have tended to outlive their original 30-year expected life, they will not last forever. Indeed, some older plants will require capital investment to extend their lives, and the costs of these investments may not be recoverable in a competitive market where new gas combined-cycle plants can cover their fuel, operating, and capital costs at prices under 3.5 cents/kWh. Increasing the output from older plants will also increase their maintenance costs, making them uneconomic after a time. New environmental regulations to limit emissions that contribute to the formation of fine particulates could also accelerate retirement of these plants. Finally, the economic lifetime of these older coal facilities will also depend on what happens to the relative prices of coal and natural gas in the future. If the difference between coal and gas prices grows faster than expected, then, all other things equal, generators may be reluctant to retire their older coal facilities. Exactly how all of these influences will come together to affect the remaining lifetimes of older coal plants is still highly uncertain.

Penetration of New Gas Combined-Cycle Units

When England simultaneously privatized and introduced competition into its power sector, it was phasing out price supports for the British coal industry. The result was a substantial penetration of new gas-fired generation plants owned and operated largely by new independent power producers. This switch from coal to gas reduced emissions of both CO_2 and—dramatically—sulfur dioxide (SO_2).

Many analysts suspect that a similar phenomenon will occur in the United States. Driven by low natural gas prices and the advantages of gas-fired combined-cycle turbines (which include high efficiency, low cost, modularity, and a short lead time for bringing the units on line), energy market entrepreneurs will see (and seize) opportunities to make money selling electricity with this technology. According to a recent article in *Public Utilities Fortnightly* (see Suggested Reading), plans to build more than 50,000 MW of new generating capacity were announced by project developers at of the end of 1998. The vast majority of this proposed capacity will be either natural gas–fired combined-cycle plants or simple gas turbines. Most of the proposed plants are concentrated in California and New England, the two regions that have progressed the farthest toward implementing retail competition. Also, in New England, existing transmission constraints limit the potential to import power from cheaper suppliers to the south and west, which suggests that these entrants would face limited competition from cheaper imports in the absence of growth in transmission capacity.

Despite these announcements, whether gas plants will penetrate U.S. electricity markets faster than they would have in the absence of competition remains an unanswered question. First, competitive markets are riskier for investors than regulated markets with more assured returns, and the cost of capital in a competitive market therefore will be higher than it would be with the continuation of the regulatory status quo. All else equal, a higher cost of capital will tend to yield lower levels of investment in new electricity-generating plants. Second, the siting of new power plants will continue to be a regulatory hurdle, even in a deregulated environment. Concerns over the effect of power plants on environmental quality will have to be addressed before regulators will allow these plants to operate. In addition, new generating plants will need to locate in areas that have access to gas pipelines and high-voltage transmission lines. The number of sites ideally situated for new gas plant development may be largely in the hands of existing generators, leaving a potentially limited number of sites for new entry by independent producers (at least in the short run). Lastly, the rate at which the industry can bring new gas-fired combined-cycle turbines on line may be limited by the capacity of the existing equipment manufacturers to deliver the equipment. If demand exceeds their capacity to produce, then it will bid up the cost of this equipment and could have a dampening effect on the rate of new entry. Increases in the price of natural gas relative to competitive fuels such as coal also could slow the rate of entry of new gas combined-cycle units.

An Overview of the Numbers

Several studies have attempted to quantify the likely effects of electricity restructuring on CO_2 emissions. Some of these studies focus on the effect of one of the factors identified in the previous section, whereas others focus more broadly on the emissions impacts of restructuring generally and do not attempt to sort out the contributions of the different supply- and demand-side factors identified earlier. Only one study separates out the effects of several of these factors on CO_2 emissions from the electricity sector (see U.S. DOE in Suggested Reading).

The U.S. Department of Energy (DOE) study finds that, on net, restructuring as envisioned in the Comprehensive Electricity Competition Act (CECA) will lead to between 40 and 60 million metric tons (MMT) less carbon emissions in 2010 than would occur in the absence of the restructuring policy. These numbers are a small fraction of the reduction of several hundred million tons relative to business as usual that the United States likely would face to meet its Kyoto Protocol target. The DOE result comes from adding the increases in CO_2 emissions associated with

- higher demand for electricity as a result of lower prices,
- incremental nuclear retirements, and
- increased availability of generating capacity

to the emission-reducing effects associated with

- improvements in heat rates at existing fossil-fueled generators,
- greater energy efficiency,
- increased reliance on renewable energy, and
- higher penetration of combined heat and power (CHP) and distributed generation.

The contributions to CO_2 reduction of the last three items are particularly worthy of discussion. The DOE study includes the presumed effects of a 7.5% RPS and continued DSM financed through a system benefits charge. It also presumes that even without these policy measures, energy service companies will find substantial opportunities for energy efficiency in a restructured market and green power markets will grow. The DOE analysis indicates that restructuring would lead to an increase of up to roughly 6 MMT in annual carbon emissions during the next decade relative to the regulated reference case if DSM, renewables, and CHP initiatives are factored out. Ongoing analysis at Resources for the Future suggests that, leaving aside efficiency and renewables initiatives, carbon emissions could increase by 8–16 MMT by 2003 from national retail restructuring. Earlier studies found even greater increases.

There are reasons to question the magnitudes of the DOE figures for carbon savings from energy efficiency and renewables. The DOE analysis uses a base case in which utility expenditures on DSM continue throughout the forecast period, leveling off at roughly 10% below 1995 levels by 2000. Given that much of the revenues from the system benefits charge are likely to be replacing utility-sponsored programs instead of adding to them, the emission reductions resulting from energy efficiency programs under restructuring are potentially overstated.

The DOE study also is more optimistic than many other studies about other effects of restructuring on CO_2 emissions beyond energy efficiency, renewables, and CHP impacts. Different assumptions could further raise carbon emissions by as much as 15 MMT. The differences are largely attributable to different assumptions regarding price responsiveness of demand, growth in transmission capability, and extent of incremental nuclear retirements.

The DOE analysis uses a significantly lower figure for the reduction in nuclear generation than in a recent study (see Rothwell in Suggested Reading). It also assumes that interregional transmission capability will not increase as a result of restructuring. Given that the FERC open transmission access rule requires firms to expand transmission capacity as necessary to satisfy demand for transmission service from a paying customer, the assumption of no growth in the presence of large disparities in regional electricity prices seems overly conservative. Expanding capacity through the construction of new lines can be very difficult and subject to long delays due to siting issues, but capacity also could be increased through efficiency improvements in the existing system. Allowing for transmission expansion increases generation from existing coal-fired facilities that might displace generation from cleaner facilities or delay investment in new facilities.

Lessons for Policymakers

Whether electricity restructuring is indeed a shortcut or a detour on the road to achieving the Kyoto Protocol targets will depend on how policymakers design a restructuring policy. This review of the current state of knowledge concerning the effects of electricity restructuring on CO_2 emissions from the electricity sector offers three important lessons for policymakers as they embark on this important task.

1. Restructuring will be climate-friendly only by design.

All of the analysis to date suggests that electricity restructuring is likely to lead to reductions in CO_2 emissions from the electricity sector only when the policy is designed to promote greater use of renewables, greater electricity conservation, and increased penetration of CHP. Although it might be possible for other measures (such as an explicit effort to retire old coal plants or to keep nuclear units on line) to augment or substitute for these policies, the analysis to date suggests that without some provision to promote conservation or greater use of renewables and CHP, restructuring will not yield lower CO_2 emissions—at least in the short run.

2. The economic benefits of restructuring are an important consideration in formulating a climate-friendly restructuring policy.

If policymakers are going to incorporate CO_2 emission reduction measures into a restructuring policy, it is important for them to be cognizant of the potential costs of these measures. These costs come in the form of reduced economic benefits of restructuring. Policies that are highly prescriptive—such as a requirement for renewables-based generation without tradable renewables generation credits, or a provision that mandates the retirement of older coal facilities—could severely limit the flexibility of the competitive market and the size of the efficiency gains from competition. More flexible policies, such as a tradable RPS or a small tax on CO_2 emissions from electricity generators, will place less of a burden on electricity suppliers and be more consistent with the goal of promoting efficiency in the electricity sector.

3. Policymakers should develop a policy regarding the early retirement of nuclear power plants before it's too late.

Nuclear power plants are already beginning to retire early because plant operators fear they will be unable to cover their costs at competitive electricity prices. More early retirements are expected. Replacing the generation from these plants with fossil-fueled generation will increase CO_2 emissions from this sector, moving the United States farther away from instead of closer to achieving the Kyoto Protocol emission reduction targets. U.S. policymakers need to evaluate the potential impacts of these nuclear retirements on the costs of achieving these targets and of achieving other environmental goals. A complete evaluation of the role of nuclear plants in a CO_2 reduction strategy must consider both the environmental benefits of early retirement (in the form of lower waste disposal burdens) and the environmental costs. Active consideration of this issue by U.S. energy policymakers should begin now, before many more nuclear power plants shut their doors forever and this important option for delaying increases in CO_2 emissions from the electricity sector is eliminated.

Suggested Reading

General

Energy Modeling Forum. 1998. A Competitive Electricity Industry. EMF Report 15 (Volume 1). April.

Palmer, K. 1997. Electricity Restructuring: Environmental Impacts. *Forum for Applied Research and Public Policy* 12(3): 28–33.

Thurston, C.W. 1999. Merchant Power: Promise or Reality? *Public Utilities Fortnightly* 137(1): 15–19.

Technical

Clemmer, S.L., A. Nogee, and M.C. Brower. 1999. *A Powerful Opportunity: Making Renewable Electricity the Standard.* January. Cambridge, MA: Union of Concerned Scientists.

Palmer, K., and D. Burtraw. 1997. Electricity Restructuring and Regional Air Pollution. *Resource and Energy Economics* 19: 139–74.

Rothwell, G. 1998. Air Pollution Fees and the Risk of Early Retirement at U.S. Nuclear Power Plants (mimeo).

October. Stanford, CA: Stanford University, Department of Economics.

U.S. DOE (Department of Energy). 1999. *Supporting Analysis for the Comprehensive Electricity Competition Act.* May. Washington, DC: U.S. DOE, Office of Economic, Electricity, and Natural Gas Analysis, Office of Policy and International Affairs.

U.S. EIA (Energy Information Administration). 1998. *Annual Energy Outlook 1999.* Report DOE/EIA-0383(99). January. Washington, DC: U.S. Government Printing Office.

U.S. FERC (Federal Energy Regulatory Commission). 1996. *Environmental Impact Statement, Promoting Wholesale Competition through Open Access Non-Discriminatory Transmission Service by Public Utilities and Recovery of Stranded Costs by Public Utilities and Transmitting Utilities.* FERC/EIS-0096. April. Washington, DC: FERC.

16

The Role of Renewable Resources in U.S. Electricity Generation
Experience and Prospects

Joel Darmstadter

The use of fossil energy is by far the largest source of human-induced CO_2 emissions in the United States and worldwide. This fact has stimulated interest in the development and deployment of lower-emitting renewable energy resources, or "renewables" (see the box on page 162). The United States has been developing and advancing renewable energy technologies for more than 20 years to ease problems—actual and anticipated—that arise from the use of fossil fuels. Nevertheless, the penetration of renewables remains minuscule, and many technological, economic, and regulatory uncertainties still persist in U.S. renewable energy markets.

In this chapter, I focus on nonhydro renewable energy resources for generating electric power. Unless otherwise indicated, by renewables I refer to wind, photovoltaic and thermal solar, biomass, and geothermal sources. A broader definition would also include agricultural residues, municipal and industrial wastes, and other combustible materials. As customarily used, the term "renewables"—which implies the possibility of replenishment of what is taken from the relevant resource stock—is more convenient than precise. Thus, geothermal resources are, strictly speaking, exhaustible because a given site may lose useful heat after a number of years of extraction.

I consider three aspects of renewables. First, I concentrate on the use of renewables in electricity generation, where an expanded role for renewables probably has greatest promise for CO_2 displacement over the medium term. Second, I largely consider developments in the United States, whose experience with and potential for renewables use in the electric power sector parallel the situation in numerous other industrial economies. Finally, and perhaps most importantly, my discussion is conditioned by the view that the role and prospects of renewable energy can be sensibly assessed only within an economic setting that considers a range of competing energy technologies and sources, both renewable and conventional. An electron is an electron, whether produced by a wind turbine or coal-based steam generator. What matter most are three issues: actual and expected market realities (that is, cost and price), the extent to which the market captures or masks imperfections brought on by environmental externalities and other distortions, and the role of public policy in promoting socially beneficial outcomes.

A Word on Renewable Energy Worldwide

Although my principal focus is on the United States, it may be helpful to put things into a global perspective. (Quantitative observations are based on references listed under Data Sources at the end of the chapter.) In terms of energy consumption in the ag-

CO_2 Emissions from Renewable Energy Sources

Subject to three provisos, one can intuitively, and for the most part legitimately, view CO_2 emissions from the use of renewable energy as inconsequential. First, it takes energy to produce energy-using capital stock—a coal-burning power plant or a photoelectric array. This aspect is unlikely to alter the balance of advantage of renewables over conventional energy from a CO_2 standpoint, but it needs to be recognized.

Second, the production of renewable energy inputs itself may involve fossil fuel emissions. For example, carbon emissions associated with a possible "hydrogen economy"—such as the use of hydrogen as the basis of automotive fuel cells—would be negligible only to the extent that fossil fuels play little role in the production of hydrogen. It would not be the case if fossil-based electricity were used in the electrolytic extraction of hydrogen from seawater.

The third and probably most important proviso has to do with the use of biomass as an energy source. Combustion of biomass, largely in the form of wood and wood wastes, has accounted for a bit more than 3% of total U.S. energy consumption in recent years. Such combustion releases CO_2 at a rate (carbon release relative to the heat content of the fuel) that is even greater—by around 15%—than that of coal. To the extent that such release is matched by new and equal biomass growth, then biomass fuel use is an effective CO_2-mitigating option.

As long as statistical treatment is internally consistent, it might not make much difference whether

CO_2 emissions from biomass combustion are shown as nil on the assumption of being netted out by equal photosynthetic uptake or whether the estimated releases are part of (gross) nationwide emissions, with the assumed or estimated uptake separately shown as a component of total sequestered CO_2. The least satisfactory—and unnecessarily confusing—way of handling the matter is that provided by the IPCC's Working Group II (see Data Sources). In one table, the use of biomass in a power plant is illustrated by indicating zero emissions (Table 19-2); elsewhere, an emission factor for wood is given as approximately 28 metric tons of carbon/1 billion Btu (Box B-2).

I prefer the second of the two measurement options, particularly given increasing interest in present and prospective interrelationships among deforestation, afforestation, reforestation, and productive use of biomass combustion worldwide. These interrelationships are especially important to consider in a context of potential competition between biomass products (crops and forest products) and fuels markets. But for now, on the assumption of reabsorption and no net change in the overall carbon budget, the statistical practice by the Energy Information Administration is to treat U.S. CO_2 emissions biomass fuels as zero (rather than, for example, adding their 75 million tons to total U.S. CO_2 emissions of approximately 1,500 million tons in 1996 to reflect gross emissions from combustion of all fuels).

gregate (not only resources going into electric power generation), the 7% of the worldwide total accounted for by renewables in 1995 was of some significance; it was almost identical to the nuclear energy share. In poorer regions of the world, the percentage was markedly higher. In Africa, for example, renewables (dominated by wood and other biomass) contributed 37% of total energy. But far from reflecting a productive and sustainable role for renewables in the contemporary and prospective energy scene, such numbers in fact signify something

quite different in desperately poor parts of the world: the need to gather energy, through foraging and other means (contributing to a loss of soil fertility in the process), to meet the most basic survival requirements of cooking and heating. The statistics constitute an artifice in a related respect, one which gives an altogether distorted picture of renewable energy use in the African example. As a metric, the British thermal units (Btus) contained in a lump of coal may be comparable to the Btus contained in a cubic meter of firewood. But because the latter is typically burned with incomparably worse efficiency, its effective importance recedes greatly.

In the electric power sector, the focus of my discussion, statistics are probably somewhat more meaningful, because the mere fact of renewables serving as an input into power generation implies a more sophisticated technological application than, say, their use in open fireplaces in rural households. In any case, it turns out that only about 1% of worldwide electric generation is based on nonhydro renewable resources. Around 80% of such generation occurs in North America and Western Europe. In developing economies, whose share is essentially nil, the use of renewables for electricity production seems to be limited to a few opportune circumstances, such as the exploitation of bagasse (agricultural waste) as a boiler fuel on sugar plantations. Probably for that reason, renewables account for approximately 3% of Brazil's electric generation. For now, therefore, the rational exploitation of renewable energy poses entirely different challenges for developing and developed countries.

The Prevailing Role of Renewables in U.S. Electricity Generation

Table 1 provides a broad perspective on how renewables fit into the present fuel and power picture in the United States. It is immediately apparent that the magnitude of renewable energy is negligible on the national level. Few nonhydro renewables figure in electric power generation. And outside the electric power sector, the balance of renewables use is concentrated in industrial biomass utilization—much of it undoubtedly in the form of wastes in wood processing and in pulp and paper mills. Nor is this picture likely to change appreciably over the next several decades, at least if the analysis of the U.S. Department of Energy's Energy Information Administration is to be believed. Specifically, conventional hydro power generation is projected to decline a bit (at a rate of 0.3%/year), whereas all other renewables in the aggregate are projected to grow by around 2%/year—a very small rate of increase, considering the low absolute values from which growth proceeds.

U.S. Policies toward Renewables

Over the past quarter of a century, several public policies have been introduced in support of renewable energy. Rather than providing an exhaustive account of these measures, I will mention and illustrate four principal ways in which the federal government has sought, or is seeking, to promote the development and use of renewables: various kinds of research and development (R&D) support, the role of the 1978 Public Utility Regulatory Policies Act, the use of other financial incentives, and the prospective role of a "renewable portfolio standard." Although federal policies have dominated, states have presented some significant initiatives as well. A 1998 report from the Energy Information

Table 1. Role of Renewable Resources in U.S. Energy Consumption and Production, 1997.

Electric generation (quads)	4.35
Conventional hydro	3.60
Geothermal	0.43
Municipal solid waste	0.23
Biomass	0.04
Solar (thermal and PV), wind	0.05
Nonelectric consuming sectors (quads)	2.57
Residential wood	0.60
Industrial biomass	1.84[a]
Industrial hydro	0.03
Ethanol in transportation	0.10
Total renewable resources (quads)	6.92[b]
Share of U.S. energy production	9.5%
Share of U.S. energy consumption	7.4%
Nonhydro renewable resources (quads)	3.32
Share of U.S. energy production	4.5%
Share of U.S. energy consumption	3.5%
Share of electricity generation	2.2%

Note: 1 quad = 1 quadrillion (10^{15}) Btu.

[a] About one-fifth of this figure can be attributed to on-site electric generation at wood-processing facilities.

[b] This figure excludes about 0.04 quads in nonmarket residential and commercial applications.

Source: U.S. DOE (Department of Energy). 1998. *Annual Energy Outlook 1999.* December. Washington, DC: U.S. DOE, Energy Information Administration, Tables A1, A18, and 8.14.

Administration (see Suggested Reading) provides additional information about renewables programs. In the discussion that follows, I will not try to assess—if, indeed, an approximate quantitative assessment is as yet possible—how these policies have shaped energy markets.

R&D Support

For various reasons—excessive risks, long time horizons, limits to capturing the returns from successful outcomes, nonmarketability of external benefits—industry is commonly believed to underinvest in basic science and technology. Therefore, a federal role to augment private efforts in advancing basic science and technology is widely accepted. In the case of renewable energy, that role largely involves R&D activities conducted at or supported by the U.S. Department of Energy (DOE) and its national laboratories, principally, the National Renewable Energy Laboratory (NREL) in Colorado.

The U.S. General Accounting Office (GAO) reported in 1999 that for the 20-year period 1978–98, $10.3 billion (in current prices) was thus disbursed (see Suggested Reading). Solar photovoltaics were the leading beneficiaries of this program. Over the 20-year period, photovoltaics received about $2 billion and wind power $1 billion. During fiscal year 1999, the respective funding was $72 million and $35 million. In both cases, the GAO sees program objectives having gradually shifted away from fundamental research to enhanced market opportunities, both domestic and international. As just one example of a recent wind power initiative, DOE's Turbine Verification Program has provided for cost-sharing with utilities to facilitate the development and deployment of wind turbines.

In critical comments on the GAO analysis (included in the GAO report), DOE questioned GAO's characterization of a programmatic shift emphasizing market potentials. Whether GAO or DOE is more on the mark in this dispute, a chastening point does perhaps emerge. Programs whose start-up rationale puts major stress on precommercialization challenges—basic science, research, and early developmental barriers—may, subtly or not, slide over into terrain dominated by sales prospects. Perhaps unfortunately, the labels "research" and "development" are broad enough to allow such slippage.

PURPA

The federal Public Utility Regulatory Policies Act (PURPA) of 1978 was one major instrument that encouraged a shift away from conventional energy to renewables. Under this statute, utilities were mandated to purchase power from nonutility producers at prices that were supposed to represent the "avoided cost" that utilities would otherwise have had to pay to produce power using conventional resources such as petroleum. Because numerous beneficiaries of this policy lacked technical expertise in alternative energy production (renewables and certain other innovative categories) and because the calculation of avoided cost was frequently quixotic, PURPA is widely judged to have fallen far short of its objectives. Contracts for utility purchases under PURPA will soon begin to run out, but transactions in 1995 still occurred at prices which, for renewables as a whole, were 150% above the national average electric generation cost of around 3.5 cents/kWh (see Figure 1).

Other Financial Assistance

Overlapping with PURPA, and continuing to the present, the federal government has provided significant direct financial benefits to renewable energy producers. Both solar photovoltaics and wind power benefit from investment tax credits, and under the Tax Reform Act of 1986, wind power was accorded a depreciation life of five years—much shorter than the depreciation life of conventional power supply investments. One provision of the Energy Policy Act of 1992 (extended in 1999) provides an inflation-adjusted 1.5 cents/kWh production tax credit for wind power plants. By 1999, this credit had increased to 1.7 cents/kWh.

Renewable Portfolio Standard

In the context of the deregulated electricity market that is presently emerging in the United States, a policy position developed by the Clinton administration during 1999 embodies provisions for a so-called renewable portfolio standard (RPS). Its goal is to en-

Figure 1. U.S. Electric Utility Average Price of Renewable Electric Power Purchased from Nonutility Facilities, by Energy Source, 1995.

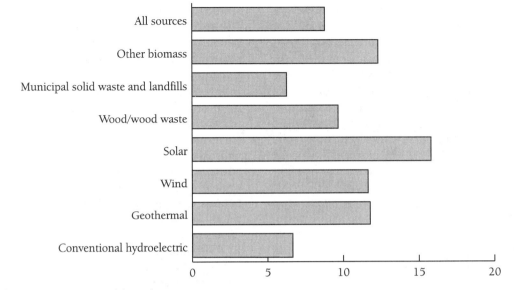

Note: Values are given in cents per kilowatt-hour.

Source: U.S. DOE (Department of Energy). 1999. *Renewable Energy: Issues and Trends 1998.* March. Washington, DC: U.S. DOE, Energy Information Administration, Fig. 11.

sure that some minimum percentage of generation originate with nonhydro renewable energy sources, regardless of whether or not it is justified by private market forces. An RPS target for 2010 calls for 7.5% of electricity sales to be based on renewable energy resources. (Separately, bills introduced in Congress call for RPS shares ranging from 4% to 20%.) If the RPS is implemented as presently conceived, the means envisaged for meeting the 7.5% target represent a much more economically efficient route to stimulating renewables-based electricity than PURPA does. That is because RPS incorporates a tradable permit system that encourages renewable power production to take place in the most cost-effective location. In addition, it would impose a ceiling on the increment to overall electric power costs that result from the mandate.

By means of various subsidies as well as surcharges on electric bills to consumers who are willing to pay a premium to ensure the presence of "green" power in their electricity mix, several states have introduced renewable minimums of their own. It is too early to judge the success of such efforts. One element of uncertainty is that even if these measures result in *new* investments in renewables generation, it is possible that *existing* facilities may be prematurely retired due to competitive pressures.

Why the Poor Showing for Renewables?

Despite the optimism regarding the emergence of renewables dating from the energy-market upheavals of the 1970s, and notwithstanding considerable policy support over the years, the reality, as noted, is sobering. It is evident from Table 1 that nearly 30 years later, renewable energy systems have not succeeded in emerging as a significant factor in the country's electricity infrastructure. Does this mean that renewable technologies have been such a great disappointment that continuing public policy support is misguided? To answer in the affirmative may be too casual a dismissal of an exceedingly complex matter. Evaluation of the available evidence indicates that renewable technologies have lived up to several significant expectations and public policy goals.

Several RFF colleagues and I recently analyzed what went right and what went wrong in the evolution of renewable energy inputs into U.S. electric power generation over the past quarter century (see the work of Burtraw and of McVeigh and others in Suggested Reading). We evaluated five technologies used to generate electricity: solar photovoltaics, solar thermal, geothermal, wind, and biomass. A principal aim of our study was to see how the actual performance of renewable energy technologies in the 1990s compared with specific goals of cost reduction and market expansion of earlier projections. Many groups (both analytically oriented and unabashedly proactive) that wrote in the 1970s and '80s had judged these goals to be attainable with the help of accommodating public policies.

In general, *market penetration* has been markedly lower than expected. However, the *cost* of renewable technologies has also been lower than projected—in several cases, significantly lower, even when compared with what seemed initially to be the optimistic forecasts of renewable energy advocates. Of course, with time, forecasts for the 1990s began to approach observed trends. Still, whereas 1980s wind power projections of generation costs a decade hence assumed roughly a 64% decline, to reach a level of 5.7 cents/kWh by 1995, costs actually declined by an estimated 67% to a level of approximately 5.2 cents/kWh. (Here and in the paragraphs that follow, costs are expressed in constant 1995 prices.) By contrast, although the volume of wind-generated electricity did show steadily rising absolute numbers in the course of the 1990s (from an almost zero level in the 1980s), it remained an inconsequential part of the nation's electricity system. Only at the end of the 1990s and in 2000 have we seen signs of some meaningful momentum in wind power capacity expansion.

One can argue about which of the two measures (market penetration or cost) has greater relevance in evaluating the performance of renewable energy re-source programs. To the extent that public-sector support was particularly driven by the need for and pursuit of cost reductions, the cost outcome seemed to us particularly important. Indeed, the cost outcome seems quite remarkable, because renewable technologies have not attracted large-scale investment and production that can contribute to technological development or economies of scale in production, as many people anticipated when forming their cost projections. Evidently, the characteristics of several renewable energy systems—high capital intensity, uncertainty about interconnections with the electric grid, variability in availability (the intermittency of wind, sunlight, and biomass wastes)—that have frequently been viewed as major barriers to economic viability have not precluded significant reductions in the reported cost of producing power.

The failure of renewables to emerge more prominently in the nation's energy portfolio is intimately linked to the concurrent decline in the cost of conventional generation. Consider that in 1984, the Energy Information Administration projected nationwide electric generation costs to rise from 6.1 cents/kWh in 1983 to 6.4 cents/kWh in 1995; in fact, they declined to 3.6 cents/kWh. That 41% decline, though less percentagewise than what was achieved by wind power, nonetheless preserved a sufficiently large margin of advantage—3.6 cents/kWh vs. 5.2 cents/kWh—for conventional over wind power as to foreclose more than a minute niche for the latter.

Several factors have contributed to keeping the cost of generation from conventional technologies low. They include developments in energy supply markets (notably, the emergence of a more competitive world oil market and productivity improvements in oil exploration and coal production); the successful deregulation of natural gas, oil pipelines, and railroads (the last a major factor in reducing the cost of coal shipping); technological progress in conventional generation itself (such as combined-cycle gas turbine systems); and the ongoing restructuring of the electricity industry. Although changes in the regulation, technology, and market structure of fossil fuels have thus been mostly beneficial for electricity consumers, they have hindered the development of technologies for renewable energy resources that have had to compete in this changing environment. Supporters of renewables have had to fix their sights on what has so far been a steadily receding target. As noted, nationwide electric generating costs of around 3.5 cents/kWh in the mid-1990s constituted a formidable target for even new renewables installations to meet, let alone for electricity based on renewable energy resources surviving from the distortions of the PURPA pricing regime. The inflated costs utilities found themselves having to pay as late as 1995 for such renewables-generated power are shown in Figure 1. All told, renewables have had to overcome something of a loser's image amid the favorable trends in conventional energy and electricity markets and the policy milieu that smoothed the path for those trends.

Is the persistence of the renewables–nonrenewables cost gap perhaps even understated when one considers the subsidies accorded the former? It is a fair question but not easily answered. Nonrenewables, after all, also receive a number of financial benefits. Nevertheless, a deeper probing of the extent to which certain tax and accounting benefits may distort cost comparisons between renewables-based and conventional generation would be a welcome contribution to the economic analysis of renewable energy.

Other countries have hardly fared much better than the United States in the extent of electricity market penetration by renewables (see IEA in Suggested Reading). A few heavily forested places (for example, Austria and the Nordic countries) have had some success exploiting fuelwood resources—aided, in some cases, by extremely favorable tax treatment and other subsidies. Denmark is developing a notable presence in wind energy. (It clearly helps when the wind resource and electricity load centers are close enough to each other to avoid costly transmission costs.) But, as in the United States, competition has not been kind to investment in renewables projects. And not surprisingly, competitive realities and policy dilemmas that face the United States are precisely those that arise when impediments to renewables are considered elsewhere.

Should Renewables Command a Premium Price?

Although the high avoided cost formulas under PURPA and some other financial inducements may have distorted the evolution of a more robust renewables sector, the notion that green power may deserve a price premium over conventional power rates is not thereby repudiated. Energy sources should trade in markets at prices that reflect both their private and social costs. There is as well a fairly common view that energy produced by various renewable systems imposes fewer of these social (external) costs than fossil-fired facilities—keeping in mind, however, that the latter have by now been compelled to internalize to a significant degree the cost of pollution abatement. The question for public policy intended to level the playing field is by how much various fossil sources should be penalized for their remaining externalities. This could be achieved through regulatory surcharges on fossil fuel use or, far less efficiently, through enhanced subsidies for renewables in recognition of their more benign environmental impact.

Quantifying external damages is complicated and controversial; the fact that many environmental impacts vary by location and the distorting nature of tax rates are just two of many complications. Some estimates from a study conducted several years ago by researchers at RFF, along with specialists in the European Community and the U.S. Department of Energy, are instructive. (See the report by Krupnick and Burtraw in Suggested Reading). The purpose was to monetize environmental damage throughout the entire fuel cycle, from resource extraction to final use. Coal was found to impose greater social costs than biomass (the only renewable resource covered in the analysis). The difference was reckoned at about 7 mills/kWh (that is, 0.7 cents/kWh) in the study. However, more than 90% of the differential (about 6.4 mills/kWh) is attributable to imputed values—however crude—of the impact of increased global warming from fossil fuel use. This imputed value is on the order of $18/ton of carbon emitted to the atmosphere, well within the range of plausible values derived from existing assessments of global warming risks. Nonetheless, these kinds of calculations are controversial.

Even if one accepts the estimated externality figures just discussed, the implied superiority of biomass over coal from an environmental costing perspective is far below the cost differential between these two fuels that prevails under current market conditions, as discussed earlier. Even a government subsidy equal to twice that difference, such as the 1.5 cents/kWh tax credit to wind power, does not bring renewables appreciably within the competitive range of conventional energy systems.

Whether a 1.5 cents/kWh tax credit or any other subsidy to renewables is a defensible estimate of externalities brought about by conventional energy systems should not blind us to the inherent defects of second-best ways of righting environmental wrongs. A system that held fossil fuel combustion fully accountable for its externalities would be more efficient in avoiding a proliferation of subsidies (hidden and explicit) and thus a waste of resources. It would also stem the political temptation to magnify exter-

nalities as a means of supporting one's favorite alternate energy system (see the box below).

Concluding Comments

All projections are conditional and inherently uncertain. One of the less uncertain ones, however, is that—in support of economic growth, particularly in developing parts of the world—the demand for electricity will increase substantially for many years to come. Several references included under Data Sources at the end of the chapter show a continued worldwide rate of increase ranging between 2% and 3% annually for much of the first half of the new century. (See, for example, the cited studies by the Electric Power Research Institute, International Energy Agency, and U.S. DOE *International Energy Outlook*.) Whatever else it may portend, this increase should be somewhat encouraging news for renewables; a growing electricity market, facilitation of scale economies, and movement up the learning curve are necessary, if insufficient, conditions for

The Ethanol Charade

Although my primary focus in this chapter is the role of renewable energy resources in electricity production, it is worth noting how a rationale for promotion and financial support of renewables—reminiscent of support for other energy resources in earlier times—can bring about an uneasy blend of policies and politics. A good example in the United States is a federal motor fuels sales tax exemption for producers of grain-based ethanol that works out to approximately $16 per barrel of oil equivalent. (The ethanol is designed to be blended with motor gasoline to produce "gasohol," the use of which is believed warranted seasonally in certain polluted areas.) No one who has observed this ultra-generous support program unfold and endure over the years has any illusions about its nature as anything but a political gift to grain processors and the U.S. agricultural constituency. I mention this as a reminder that noble sentiments on promoting clean energy may

mask motives that are neither clean nor economically justified by any stretch of the imagination.

President Bill Clinton's Executive Order of August 12, 1999, created a cabinet-level body charged with supporting a greatly expanded effort to promote biofuels. In official remarks made then, the President stated, "I am setting a goal of tripling America's use of bioenergy and bio-based products by 2010. That would generate as much as $20 billion a year in new income for farmers and rural communities while reducing greenhouse gas emissions by as much as 100 million tons a year—the equivalent of taking more than 70 million cars off the road." The initiative may reflect a genuine determination to pursue a sound and sustained research and development and demonstration program in renewable energy. But it is too early to say whether, once again, it is farm policy—or politics—masquerading as energy policy.

greater penetration of renewables. Depending on the extent of policies implemented to limit fossil fuel use out of concern over climate change and the possibility of rising costs of petroleum (a source of anxiety in some quarters), the attractiveness and viability of renewables may be strengthened.

But progress on the part of more traditional energy systems is sure to parallel further development of renewables, and there is no reason to expect that dynamic state of affairs to flag in the future. (On this point, see the views of Bradley under Suggested Reading.) Thus, for example, even as the size and technology of wind turbines improve and their costs decline, other systems aren't standing still. Efficient combined-cycle gas turbines seem to be rapidly becoming the configuration of choice in new utility plants. Fuel cells, other distributed systems of power supply, the emergence of advanced (and publicly acceptable) nuclear technology (even if presently unlikely), and across-the-board realizable improvements in energy efficiency are all possibilities to be reckoned with. Each could be a prospective competitor to power systems based on renewable energy sources.

What emerges from these final thoughts is an argument for retaining a reasonably wide range of options in our electricity and energy portfolio. The role of government is not only to help overcome market failures but also to ensure some degree of efficacy in its programmatic agenda, including injection of the broad public interest in its supportive activities. Policies should be sought that are more economically efficient and less politically influenced than the system of outright renewable subsidies that has prevailed in recent years. The nature of those policies is still being debated. For example, the introduction of an RPS into the nation's electricity mix would be more cost-effective than current and previous policies, but it would still be a forcing measure that may only loosely reflect the externality benefits of avoided fossil energy.

Prudently targeted programs in long-term R&D represent an important complementary strategy. Defending its proposed six-year (constant dollar) doubling of federal R&D support for renewable energy, the 1997 PCAST study (see Suggested Reading) indicated that such an increase

makes sense in light of the rapid rate of cost reduction achieved in recent years for a number of renewable energy technologies, the good prospects for further gains, and the substantial positive contributions these technologies could make to improving environmental quality, reducing the risk of climate change, controlling oil-import growth, and promoting sustainable economic development in Africa, Asia, and Latin America.

Opportunities exist for important advances in wind-electric systems, photovoltaics, solar-thermal energy systems, biomass-energy technologies for fuel and electricity, geothermal energy, and a range of hydrogen-producing and hydrogen-using technologies including fuel cells.... [T]he increased support for these renewable-energy technologies would focus on areas where the expected short-term returns to industry are insufficient to stimulate as much R&D as the public benefits warrant.

As of mid-2000, congressional deliberations pointed to a level of funding for renewable energy resources in fiscal year 2000–01 of around $440 million, nearly a 20% increase (in current dollars) over a year earlier. Whether consciously or fortuitously, the congressional path seems to embrace, and perhaps even leapfrog, the path recommended by PCAST. R&D funding levels apart, what deserves continued close attention is the extent to which environmental externalities and societal risks associated with energy production and use elude private-market transactions. Where they do, it would be surprising if the needed public sector initiatives seeking economically efficient correctives did not include a consequential role for renewables.

Suggested Reading

Bradley, Robert L. Jr. 1999. The Increasing Sustainability of Conventional Energy. Policy Analysis Paper No. 341. April 22. Washington, DC: Cato Institute.

Burtraw, Dallas. 1999. Testimony by Dallas Burtraw of RFF before the Senate Energy and Water Appropriations Subcommittee, U.S. Congress. September 14. *Congressional Record*, Daily Digest.

IEA (International Energy Agency). 1999. *The Evolving Renewable Energy Market.* June. IEA Renewable Energy Working Party. Sittard, the Netherlands: Novem BV.

Krupnick, A. J., and D. Burtraw. 1996. The Social Costs of Electricity: Do the Numbers Add Up? *Resource and Energy Economics* 18: 423–66.

McVeigh, J., D. Burtraw, J. Darmstadter, and K. Palmer. 1999. *Winner, Loser, or Innocent Victim: Has Renewable Energy Performed as Expected?* Research Report No. 7. March. Washington, DC: Renewable Energy Project.

Palmer, Karen. 1999. Electricity Restructuring: Shortcut or Detour on the Road to Achieving Greenhouse Gas Reductions? Climate Issues Brief No. 18. July. Washington, DC: Resources for the Future.

PCAST (President's Committee of Advisers on Science and Technology). 1997. *Federal Energy Research and Development for the Challenges of the Twenty-first Century.* Report of the Energy Research and Development Panel, Executive Office of the President. September, 10. Washington, DC: U.S. Government Printing Office.

U.S. DOE (Department of Energy). 1999. *Supporting Analysis for the Comprehensive Electricity Competition Act.* May. Washington, DC: U.S. DOE, Office of Policy. (See especially pp. 5, 22–24, 33, and 34 for details about the renewables portfolio standard.)

———. 1998. *Renewable Energy: Issues and Trends 1998.* March. Washington, DC: U.S. DOE, Energy Information Administration.

U.S. GAO (General Accounting Office). 1999. *Renewable Energy: DOE's Funding and Markets for Wind Energy and Solar Cell Technologies.* Report GAO/RCED-99-130. May. Washington, DC: U.S. GAO.

Data Sources

Electric Power Research Institute. 1999. Electricity in the Global Energy Future. *EPRI Journal* 24(3): 8–17.

IPCC (Intergovernmental Panel on Climate Change). 1996. *Climate Change 1995: Impacts, Adaptations, and Mitigation of Climate Change: Scientific-Technical Analysis.* Contribution of Working Group II to the Second Assessment Report of the Intergovernmental Panel on Climate Change. New York: Cambridge University Press, 80.

IEA (International Energy Agency). 1994. *World Energy Outlook*, 1994 Edition. Paris, France: IEA.

Nakićenović, N., and others (eds.). 1998. *Global Energy Perspectives.* New York: Cambridge University Press for International Institute of Systems Analysis/World Energy Council.

U.S. DOE (Department of Energy). 1999. *Annual Energy Review 1998.* July. Washington, DC: U.S. DOE, Energy Information Administration.

———. 1999. *International Energy Annual 1997.* April. Washington, DC: U.S. DOE, Energy Information Administration.

———. 1999. *International Energy Outlook 1999.* March. Washington, DC: U.S. DOE, Energy Information Administration.

———. 1997. *Emissions of Greenhouse Gases in the United States 1996.* October. Washington, DC: U.S. DOE, Energy Information Administration.

World Resources Institute/UNEP/UNDP/World Bank. 1998. *World Resources 1998–99.* New York, Oxford University Press.

17 Energy-Efficient Technologies and Climate Change Policies
Issues and Evidence

Adam B. Jaffe, Richard G. Newell, and Robert N. Stavins

Enhanced energy efficiency occupies a central role in evaluating the efficacy and cost of climate change policies. Ultimately, total greenhouse gas (GHG) emissions are the product of population, economic activity per capita, energy use per unit of economic activity, and the carbon intensity of energy used. Although GHG emissions can be limited by reducing economic activity, this option obviously has little appeal even to rich countries, let alone poor ones. As a result, much attention has been placed on the role that technological improvements can play in reducing carbon emissions and lowering the cost of those reductions. In addition, the influence of technological changes on the emission, concentration, and cost of reducing GHGs will tend to overwhelm other factors, especially in the long term. Therefore, understanding the process of technological change is of utmost importance. Nonetheless, the task of measuring, modeling, and ultimately influencing the path of technological development is fraught with complexity and uncertainty—as are the technologies themselves.

The carbon intensity of energy can be reduced by substituting renewable or nuclear sources for fossil fuels (and by substituting lower-carbon natural gas for coal) or by increasing energy efficiency. Recognizing this, recent policy proposals have included tax credits for residential and commercial purchasers of new energy-efficient homes and energy-efficient equipment such as electric and natural gas heat pumps, natural gas water heaters, advanced central air conditioners, and fuel cells as well as an investment tax credit for industrial combined heat and power systems. Extensions have also been proposed for existing tax credits for fuel-efficient vehicles powered by electricity, fuel cells, and hybrid power. In addition to tax incentives, other proposals include direct spending on research, development, and deployment of energy-efficient products.

Public–private partnerships have been created or proposed with the aim of developing and deploying energy-efficient technologies for houses (Partnership for Advancing Technology in Housing); appliances (Energy Star program, Golden Carrot Super Efficient Refrigerator Program); schools (Energy Smart Schools); commercial buildings (Energy Star Buildings, Green Lights); vehicles (Partnership for a New Generation of Vehicles); and industrial processes (Motor Challenge, Climate-Wise). Energy efficiency standards for many products have been established and in some cases revised since 1988 (Table 1). Many of these policies target technologies that embody a mix of improved energy efficiency and decreased carbon intensity (such as credits for natural gas heat pumps).

Although the importance of energy efficiency in limiting GHG emissions incites little debate, intense debate ensues regarding its cost-effectiveness and

Table 1. Effective Dates of Appliance Efficiency Standards, 1988–2001.

Technology	1988	1990	1992	1993	1994	1995	2000	2001
Clothes dryers	X				X			
Clothes washers	X				X			
Dishwashers	X				X			
Refrigerators and freezers		X		X				X
Kitchen ranges and ovens		X						
Room air conditioners		X					X	
Direct heating equipment		X						
Fluorescent lamp ballasts		X						
Water heaters		X						
Pool heaters		X						
Central air conditioning and heat pumps			X					
Furnaces								
Central and small			X					
Mobile home		X						
Boilers			X					
Fluorescent lamps								
8 ft					X			
2 ft, 4 ft						X		

Source: U.S. DOE/EIA (see Suggested Reading).

about the government policies that should be pursued to enhance energy efficiency. At the risk of excessive simplification, we can characterize technologists as believing that there are plentiful opportunities for low-cost or even "negative-cost" improvements in energy efficiency and that realizing these opportunities will require active intervention in markets for energy-using equipment to help overcome barriers to the use of more efficient technologies. These interventions would guide choices that purchasers would presumably welcome after the fact, although they have difficulty identifying these choices on their own. This view implies that with the appropriate technology and market creation policies, significant GHG reduction can be achieved at very low cost.

In contrast, most economists acknowledge the existence of market barriers to the penetration of various technologies that enhance energy efficiency and that only some of these barriers represent real market failures that reduce economic efficiency. This view emphasizes that there are trade-offs between economic efficiency and energy efficiency—it is possible to get more of the latter, but typically only at the cost of less of the former. The economic perspective suggests that GHG reduction is more costly than the technologists argue, and it puts relatively more emphasis on market-based GHG control policies such as carbon taxes or tradable carbon permit systems to encourage the least costly means of carbon efficiency (not necessarily energy efficiency) enhancement available to individual energy users.

In this chapter, we first examine what lies behind this dichotomy in perspectives. Ultimately, the veracity of different perspectives is an empirical question, and reliable empirical evidence on the issues identified above is surprisingly limited. We review the evidence that is available and find that although energy and technology markets certainly are not perfect (no markets are), the balance of evidence supports the view that there is not as much "free lunch" in energy efficiency as advocates would suggest. On the other hand, a case can be made for the existence of certain inefficiencies in energy technology markets, thus raising the possibility of some inexpensive GHG control through energy efficiency enhancement. We conclude with some reflections on the role of appropriate energy efficiency policy in climate change mitigation.

Understanding the Energy Efficiency Gap

Analysts have pointed out for years that there is an "energy efficiency gap" between the most energy-efficient technologies available at some point in time and those that are actually in use. On this basis, debate has raged about the extent to which there are low-cost or no-cost options for reducing fossil energy use through improved energy efficiency. It turns out that technologists and economists have very different views of this gap and of whether and to what degree it is the result of market failures that might be amenable to policy intervention or simply market barriers that would be surmountable only at relatively high cost. This debate is illustrated in the 1995 report from the Intergovernmental Panel on Climate Change (IPCC; see Hourcade and others in Suggested Reading). One part of this report states that energy efficiency improvements on the order of 10–30% might be possible at little cost or even with net benefits (ignoring climate benefits), whereas another part highlights the fact that most economic models indicate a significant cost for stabilizing or cutting OECD emissions below 1990 levels.

The basic dimensions of this debate are the subject of many studies. To understand the basic elements of the debate, it is helpful to first distinguish between energy efficiency and economic efficiency (Figure 1). The vertical axis measures increased energy efficiency (as decreased energy use per unit of economic activity), and the horizontal axis measures increased economic efficiency (as decreased overall economic cost per unit of economic activity, taking into account energy and other opportunity costs of economic goods and services). Different points in the diagram represent the possible energy-using technologies available to the economy as indicated by their energy and economic efficiency.

Consider two air conditioners that are identical except that one has higher energy efficiency and, as a re-sult, is more costly to manufacture—high-efficiency units require more cooling coils, a larger evaporator, and a larger condenser as well as a research and development effort. Whether it makes sense for an individual consumer to invest in greater energy efficiency depends on balancing the value of energy that will be saved against the increased purchase price, which depends on the value of the additional materials and labor that were spent to manufacture the high-efficiency unit. As we discuss below, the value to society of saving energy should also include the value of reducing any associated environmental externalities, but again, it must be weighed against the costs.

Adoption of more energy-efficient technology is represented in Figure 1 as an upward movement. But not all such movements will also enhance economic efficiency. In some cases, it is possible to simultaneously increase energy efficiency and economic efficiency. This will be the case if market failures impede the most efficient allocation of society's energy, capital, and knowledge resources in ways that also reduce energy efficiency. These are examples of what economists and others refer to as "win-win" or "no regrets" measures.

In Figure 1, the economist's notion of a "narrow" optimum is where failures in the market for energy efficient technologies have been corrected, increasing both economic efficiency and energy efficiency. This optimum is narrow in the sense that it focuses solely on energy technology markets and does not consider possible failures in energy supply markets (such as underpriced energy as a result of subsidies or regulated markets) or, more important, environmental externalities associated with energy use (such as global climate change). When analysts speak of no-cost climate policies based on energy efficiency enhancement, they often implicitly or explicitly assume the presence of market failures in energy efficiency.

Market failures in the choice of energy-efficient technologies could arise from various sources. Some of these are relatively uncontroversial, at least in principle, such as inadequate private-sector incentives for research and development and information shortages for purchasers regarding the benefits and costs of adopting technologies. Other potential mar-

Figure 1. Alternative Notions of the Energy Efficiency Gap.

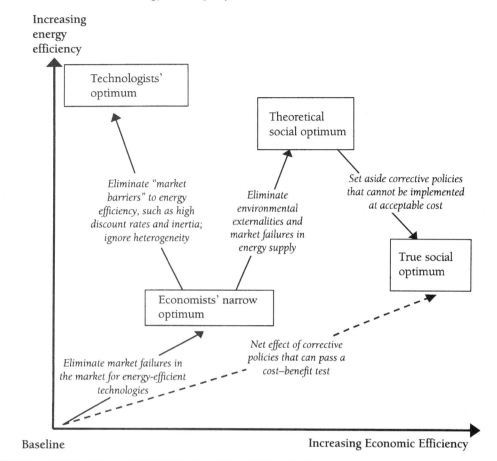

ket failures are more controversial: To what extent is small-scale investment in energy efficiency limited because of financing constraints (a failure of capital markets to efficiently allocate financial resources)? To what extent are there market failures because landlords (rather than tenants) pay utility bills, and landlords are not adequately rewarded in rental markets for providing energy-efficient dwellings (so-called principal agent problems)? To what extent are businesses not pursuing potentially rewarding energy efficiency investments because managers are not adequately rewarded (and capital markets do not adequately punish such inefficiency)? We discuss some evidence on these questions below.

Eliminating broader market failures takes us to what we call the *theoretical social optimum* in Figure 1, which represents both increased economic efficiency and energy efficiency compared with the economists' narrow optimum. But not all market failures can be eliminated at acceptable costs. In cases where implementation costs outweigh the gains from corrective government intervention, it will be more efficient not to attempt to overcome particular market failures; this level is what we refer to as the *true social optimum*. Market failures have been eliminated, but only those whose elimination can pass a reasonable benefit–cost test. The result is the highest possible level of economic efficiency, but a level of energy efficiency that is intermediate compared with what would be technologically possible.

In contrast to the economist's perspective, technologists have focused their interest on another notion of an optimum, which typically is based on a very simple engineering–economic model. The *technologists' optimal energy efficiency* is found by minimizing the total purchase and operating costs of an investment, where energy operating costs are discounted at a rate the analyst (not necessarily the purchaser) feels is appropriate.

The problem with this approach is that it does not accurately describe all the factors affecting investment decisions regarding energy efficiency. First, it typically does not account for changes over time in the savings that purchasers might enjoy from an extra investment in energy efficiency, which depends on trends and uncertainties in the prices of energy and conservation technologies. When making irreversible investments that can be delayed, the presence of this uncertainty can lead to an investment hurdle rate that is larger than the discount rate used by an analyst who ignores this uncertainty. The magnitude of this option-to-wait effect depends on project-specific factors, such as the degree of volatility in energy prices, the degree of uncertainty in the cost of the investment, and the rate of change in the prices of energy and conservation technologies over time. Under the conditions that characterize most energy conservation investments, this effect could raise the hurdle rate by up to 10 percentage points. The effect is magnified when energy and technology price uncertainty is increased, and when energy prices are rising and technology costs are falling more quickly. On the other hand, if there is no opportunity to wait, this effect can be ignored.

Second, the magnitude of important variables used in such engineering–economic analysis can vary considerably among purchasers—variables such as the purchaser's discount rate, the investment lifetime, the price of energy, the purchase price, and other costs. Heterogeneity in these and other factors leads to differences in the expected value that individual purchasers will attach to more energy-efficient or carbon-efficient products. As a result, only purchasers for whom it is especially valu-

able may purchase a product. For example, it may not make sense for someone who will only rarely use an air conditioner to spend significantly more purchasing an energy-efficient model—they simply may not have adequate opportunity to recoup their investment through energy savings. Analysis based on single estimates for the important factors listed above—unless they are all very conservative—will inevitably lead to an optimal level of energy efficiency that is too high for some portion of purchasers. The size of this group, and the magnitude of the resulting inefficiency should they be constrained to choose products that are not right for them, will of course depend on the extent of heterogeneity in the population and the assumptions made by the analyst.

Finally, evidence suggests that analysts have substantially overestimated the energy savings that higher efficiency levels will bring, partly because projections often are based on highly controlled studies that do not necessarily apply to actual savings realized in a particular situation. For example, studies have found that actual savings from utility-sponsored programs typically achieve 50–80% of predicted savings (see Sebold and Fox, as well as Hirst, in Suggested Reading). Metcalf and Hassett (see Suggested Reading) draw a similar conclusion based on an analysis of residential energy consumption data, in which they found that the actual internal rate of return to energy conservation investments in insulation was about 10%, which is substantially below typical engineering estimates that the returns were 50% or more.

This is not to say that profitable energy efficiency investments do not exist; rather, attempts to determine optimal or minimum energy efficiency levels for particular investments—as is done, for example, during the process of setting minimum energy efficiency standards—need to account for all costs, not overstate realizable benefits, and use appropriate discount rates.

An important implication of this perspective is that comparisons of an engineering ideal for a particular energy use with average practice for existing technology are inherently misleading, because the former does not incorporate all the real-world factors influencing energy technology decisionmaking. The overall economic costs of switching to more energy-efficient technology constitute what can be thought of as a market barrier to their use in that individual consumers and producers will not have incentives to use more costly technologies unless policy measures (such as technology standards or carbon taxes) compel or induce behavioral changes. Unlike market failures, however, market barriers cannot be lowered in a win-win fashion.

Constraining consumers to purchase appliances with a higher level of efficiency based on simplistic analysis will in effect impose extra costs on consumers. The result is higher energy efficiency but decreased economic efficiency, because consumers are forced to bear costs that they had otherwise avoided. Although it is possible that this effect may be justified by some larger societal goal to address certain environmental externalities associated with energy consumption, the problem should be approached from that broader perspective rather than from the narrow perspective of constraining energy efficiency decisions. Taking this broader perspective leads to a more direct focus on the real problem—climate change associated with CO_2 emissions—rather than constraining available technology options.

Technology Invention, Innovation, and Diffusion

To understand the potential for public policy to affect energy efficiency, we also need to understand the process through which technology evolves: invention, innovation, diffusion, and product use. Policies can affect each stage in specific and different ways. *Invention* involves the development of a new idea, process, or piece of equipment. This activity takes place inside the laboratory. The second stage is commercialization, or technology *innovation*, in which new processes or products are brought to market. The third stage is *diffusion*, the gradual adoption of new processes or products by firms and individuals who then decide how intensively to *use* new products or processes. From this perspective, we can now think of the energy efficiency gap discussed ear-

lier as a debate mainly about the gradual diffusion of energy-saving technologies that appear to be cost-effective.

Tying this all together, we could, for example, think of a fundamentally new kind of automobile engine being invented. It might be an alternative to the internal combustion engine, such as a system that depends on fuel cells. The innovation step would be the work carried out by automobile manufacturers or others to commercialize this new engine, that is, bring it to market and offer it for sale. The diffusion process would be the purchase by firms and individuals of automobiles with this new engine. Finally, the degree of use of these new automobiles will be of great significance to demand for particular types of energy. The reason it is so important to distinguish carefully among these different conceptual steps—invention, innovation, diffusion, and use—is that public policies can be designed to affect various stages and will have very specific and differential effects. Both economic incentives and conventional regulations can be targeted to any of these stages, but with greatly varying likelihood of success.

Diffusion

The S-shaped diffusion path has typically been used to describe the progress of new technologies making their way into the marketplace. Figure 2 portrays how a new technology is adopted at first gradually and then with increasing rapidity, until at some point, its saturation in the economy is reached. Some natural questions are what generates this typically observed gradual path of diffusion, how can public policy affect it, and how might public policy accelerate it?

The explanation for this typical path of diffusion that has most relevance for energy-conservation investments is related to differences in the characteristics of adopters and potential adopters. They include differences in the kind and vintage of their existing equipment, other elements of the cost structure (such as access to and cost of labor, material, and energy), and their access to technical information. Such heterogeneity leads to differences in the expected returns to adoption, and as a result, only potential adopters for whom it is especially profitable

Figure 2. The Gradual S-Shaped Path of Technology Diffusion.

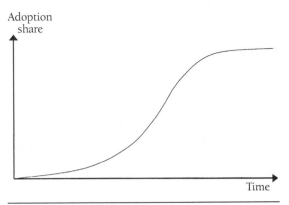

will adopt at first. Over time, however, more and more will find it profitable as the cost of the technology falls, its quality improves, information about the technology becomes more widely available, and existing equipment stocks depreciate.

Jaffe and Stavins (see Suggested Reading) investigated technology diffusion in the context of energy efficiency by carrying out econometric analyses of the factors affecting the adoption of thermal insulation technologies in new residential construction in the United States between 1979 and 1988. They examined the dynamic effects of energy prices and technology adoption costs on average residential energy efficiency technologies (that is, average R-values) in new home construction. The effects of energy prices can be interpreted as suggesting what the likely effects of taxes on energy use would be, and the effects of changes in adoption costs can be interpreted as indicating what the effects of technology adoption subsidies would be. The researchers found that the response of mean energy efficiency to energy price changes is positive and significant, both statistically and economically.

Interestingly, they also found that equivalent percentage cost subsidies would have been about three times as effective as taxes in encouraging adoption, although standard financial analysis would suggest that they ought to be about equal in percentage terms. However, this finding confirms the conven-

tional wisdom that technology adoption decisions are much more sensitive to up-front cost considerations than to longer-term operating expenses. In a study of residential conservation investment tax credits, Hassett and Metcalf (see Suggested Reading) also found that tax credits or deductions are many times more effective than "equivalent" changes in energy prices—about eight times as effective in their study. They speculate that one reason for this difference is that energy price movements may be perceived as temporary. One downside to efficiency subsidies, however, is that they do not provide incentives to reduce use, as energy price increases do. In addition, technology subsidies and tax credits can require large public expenditures per unit of effect, because consumers who would have purchased the product even in the absence of the subsidy will still receive it. In a time of fiscal constraints on public spending, this speculation raises questions about the feasibility of subsidies that would be sizable enough to have the desired effect.

Jaffe and Stavins also examined the effects of more conventional command-and-control regulations on technology diffusion, in the form of state building codes. However, they found no discernable effect. It is possible, of course, that stricter codes (that were more often binding relative to typical practice) might have an effect. However, proponents of conventional regulatory approaches should remember that although energy taxes, for example, will always have some effect, typical command-and-control approaches can have little actual effect if they are set below existing standards of practice.

Innovation and Invention

Now we can move back in the process of technological change from diffusion to innovation. In the area of energy efficiency, it is helpful to think of the innovation process as affecting improvements in the characteristics of products. In Figure 3, we represent this process as the shifting inward over time of a curve that represents the trade-offs between different product characteristics for the range of products available on the market. On one axis is the cost of the product, and on the other axis is the energy flow associated with a product—that is, its energy inten-

Figure 3. Innovation in Product Characteristics.

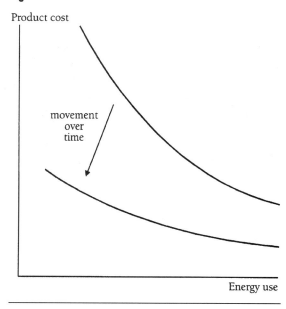

sity. The downward slope of the curves indicates the trade-off between equipment cost of energy efficiency. Innovation means an inward shift of the curve—greater energy efficiency at the same cost, or lower cost for a given energy efficiency.

Using data from 1960–90, Newell and others (see Suggested Reading) statistically estimated these characteristic transformation curves for a number of energy-consuming durables. By constructing a series of simulations, we can examine the effects of energy price changes and efficiency standards on average efficiency of the menu of products over time. As shown in Figure 4, which illustrates the findings for room air conditioners, a substantial amount of the improvement is what we would describe as autonomous (that is, associated with the passage of time); however, significant amounts are attributable to changes in energy prices and changes in energy efficiency standards. Energy price changes induced both the commercialization of new models and the elimination of old models. Regulation, however, works largely through dropping energy-inefficient models, because that is the intended effect of the energy efficiency standards (in other words, models

below a certain energy efficiency simply may not be offered for sale).

Invention is even farther back in the process of technological change. Popp (see Suggested Reading) analyzed U.S. patent application data from 19 energy-related technology groups from 1970 to 1994 and found that the rate of energy-related patent applications was significantly and positively associated with the price of energy.

All of these studies suggest that the response of innovation to energy price changes can be surprisingly swift, typically less than five years for much of the response in terms of patenting activity and introduction of new model offerings. Substantial diffusion can take significantly longer, depending on the rate of retirement of previously installed equipment. The longevity of much energy-using equipment reinforces the importance of taking a long-term view toward energy efficiency improvements—on the order of decades (see the box on page 180).

Energy, Technology, and Market Reform Policies

Aside from market influences, public policies also can affect the diffusion of more energy-efficient technologies. Policies that raise the cost of energy will induce the diffusion of extant energy-efficient technology as well as the development of new technology. Are additional nonprice policies needed to promote energy-efficient, climate-friendly technology advances and investment? Here, the debate mirrors that over the energy efficiency gap discussed above. Proponents of such policies argue that economic incentives are not adequate to change behavior. They advocate public education and demonstration programs; subsidies for the development and introduction of new technologies; institutional reforms, such as changes in building codes and utility regulations; and technology mandates, such as fuel economy standards for automobiles and the use of renewable energy sources for power generation.

No one doubts that such approaches might eventually increase energy efficiency and reduce GHG emissions. At issue is the cost-effectiveness of such programs. Advocates of technology mandates often

Figure 4. Historical Simulations of Energy Efficiency: Room Air Conditioners.

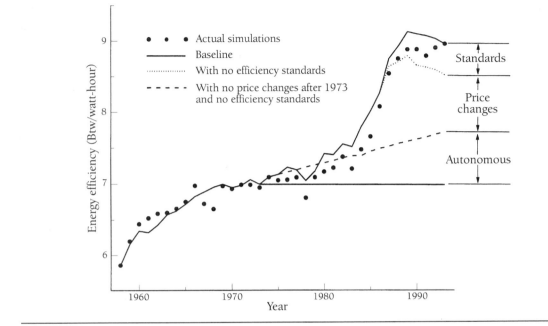

argue that the subsequent costs are negligible because the realized energy cost savings more than offset the initial investment costs. But as we noted earlier, this view ignores several factors that impinge on technology choices. Most economic analyses recognize that energy use suffers from inefficiencies but remain skeptical that large no-regret gains exist. They also acknowledge a role for government when consumers have inadequate access to information and if existing regulatory institutions are poorly designed. This role can include subsidies to basic research and development to compensate for an imperfect patent system, reform of energy sector regulation and reduction of subsidies that encourage uneconomic energy use, and provision of information about new technological opportunities.

Conclusions and Implications for Climate Policy

We have presented an overview of how to address the question of the appropriate role for government in energy conservation. In doing so, it is essential to decide first on the objective of government policy in this area: economic efficiency or energy efficiency. We find that market signals are effective for advancing the diffusion process, whereas minimum standards may not be unless they mandate certain technologies. We also find that market signals can have effects on the direction of innovation and invention, promoting increased energy efficiency when energy prices are rising. The bottom line is that technological studies that demonstrate the existence on the laboratory shelf of particular energy-efficient technologies are a useful first step. But such studies are not sufficient to address important policy questions. It is necessary to examine whether and how specific policies will affect the processes of invention, innovation, diffusion, and use intensity of products—and how much they will cost.

Although continued research is needed to pin down the precise magnitudes, it seems clear that economic motivations—operating directly through higher energy prices and indirectly through falling

Technology Diffusion and the Rate of Capital Stock Turnover

Technology diffusion is closely related to the concept of *capital stock turnover*, which describes the rate at which old equipment is replaced and augmented by new. New equipment can be purchased either to replace worn out and obsolete units, or as a first-time purchase. A primary driver of replacement purchases for durable energy-using goods is the product's useful lifetime. The rate of economic growth is also important, especially for first-time purchases of durable goods; the rate of home construction is particularly relevant for residential equipment.

The typical lifetimes for a range of energy-using assets are given here, illustrating that the appropriate time frame for thinking about the diffusion of many energy-intensive goods is on the order of decades.

Type of asset	Typical service life (years)
Household appliances	8–12
Automobiles	10–20
Industrial equipment/machinery	10–70
Aircraft	30–40
Electricity generators	50–70
Commercial/industrial buildings	40–80
Residential buildings	60–100

costs of technological alternatives due to innovation—are effective in promoting the expanded market penetration and use of more energy-efficient, GHG-reducing technologies. Some policies that support and enhance the effects of market signals, such as information provision and support for basic research and development, can be useful. In contrast, there are many more questions about the efficacy of conventional regulatory approaches, at least in developed market economies, where such policies are more likely to produce limited behavior changes or to incur excessive costs. There are good reasons to doubt the existence of a vast pool of cheap energy-reducing opportunities that offer a free lunch in reducing GHGs.

Although efficiency subsidies and tax credits may provide relatively strong incentives for the marginal purchaser, they also can require large overall public expenditure per unit of effect, because consumers who would have purchased the product even in the absence of the subsidy will still receive it. In a time of fiscal constraints on public spending, the large expenditure required raises questions about the feasibility of subsidies that would be sizable enough to have the desired effect. Energy efficiency improvements can certainly be relevant for climate policy; however, it is also important to remember that primary fuels differ substantially in terms of their GHG emissions per unit of energy consumed. Policies focused on energy use rather than GHG emissions run the risk of orienting incentives and efforts in a direction that is not cost-effective. In particular, policies focused on energy efficiency ignore the other important way in which GHG emissions can be reduced: namely, by reducing the carbon content of energy. Economists generally prefer to focus policy instruments directly at the source of a market failure. Policies focused on carbon emissions—such as tradable carbon permits or carbon fees—will provide incentives for conserving certain fuels in proportion to their GHG content. These policies would raise the price of oil by a higher percentage than the price of natural gas, for instance, thereby targeting incentives for energy efficiency improvements to oil-fired furnaces more than to gas furnaces. In addition, policies focused on GHGs rather than energy per se would also provide incentives for the purchase of gas-fired rather than oil-fired furnaces.

There may be market failures other than the environmental externality of global climate change associated with energy efficiency investments. If the magnitude of these nonenvironmental market failures is large enough and the cost of correcting them small enough to warrant policy intervention, then an argument can be made for attacking these other market failures directly. Any attendant reduction in GHGs can then be viewed as a bonus—a no regrets policy. In fact, this argument is often used by proponents of energy efficiency policy in the context of climate change policy discussions. Therefore, it becomes crucial to investigate the magnitude of these other market failures—in particular cases—and to assess which policies (if any) would be most cost-effective in addressing them. Policies that create clear incentives for changes in energy use and technology must be emphasized by raising the price of GHG emissions and targeting the institutional and other market failures that represent opportunities for cost-effective improvements in market performance.

Suggested Reading

General

Geller, Howard, and Steven Nadel. 1994. Market Transformation Strategies to Promote End-Use Efficiency. *Annual Review of Energy and the Environment* 19: 301–46.

Hassett, Kevin A., and Gilbert E. Metcalf. 1993. Energy Conservation Investment: Do Consumers Discount the Future Correctly? *Energy Policy* 21(6): 710–16.

Hourcade, J.-C., and others. 1996. A Review of Mitigation Cost Studies. In *Climate Change 1995: Economic and Social Dimensions of Climate Change*, edited by J. P. Bruce and others. Contribution of Working Group III to the Second Assessment Report of the Intergovernmental Panel on Climate Change. New York: Cambridge University Press, 263–366.

Jaffe, Adam B., and Robert N. Stavins. 1994a. The Energy-Efficiency Gap: What Does It Mean? *Energy Policy* 22(1): 804–10.

Levine, Mark D., Jonathon G. Koomey, James E. McMahon, and Alan H. Sanstad. 1995. Energy Efficiency Policy and Market Failures. *Annual Review of Energy and the Environment* 19: 535–55.

Metcalf, Gilbert E. 1994. Economics and Rational Conservation Policy. *Energy Policy* 22(10): 819–25.

Metcalf, Gilbert E., and Donald Rosenthal. 1995. The "New" View of Investment Decisions and Public Policy Analysis: An Application to Green Lights and Cold Refrigerators. *Journal of Policy Analysis and Management* 14(4): 517–31.

Sutherland, Ronald J. 1991. Market Barriers to Energy-Efficiency Investments. *The Energy Journal* 12(3): 15–34.

Technical

Hassett, Kevin A., and Gilbert E. Metcalf. 1995. Energy Tax Credits and Residential Conservation Investment: Evidence from Panel Data. *Journal of Public Economics* 57: 201–17.

Hassett, Kevin A., and Gilbert E. Metcalf. 1996. Can Irreversibility Explain the Slow Diffusion of Energy Saving Technologies? *Energy Policy* 24(1): 7–8.

Hirst, E. 1986. Actual Energy Savings after Retrofit: Electrically Heated Homes in the Pacific Northwest. *Energy* 11: 299–308.

Jaffe, Adam B., and Robert N. Stavins. 1994b. The Energy Paradox and the Diffusion of Conservation Technology. *Resource and Energy Economics* 16(2): 91–122.

Jaffe, Adam B., and Robert N. Stavins. 1994c. Energy-Efficiency Investments and Public Policy. *The Energy Journal* 15(2): 43–65.

Jaffe, Adam B., and Robert N. Stavins. 1995. Dynamic Incentives of Environmental Regulations: The Effects of Alternative Policy Instruments on Technology Diffusion. *Journal of Environmental Economics and Management* 29: S43–63.

Koomey, Jonathan G., Alan H. Sanstad, and Leslie J. Shown. 1996. Energy-Efficient Lighting: Market Data, Market Imperfections, and Policy Success. *Contemporary Economic Policy* 14(3): 98–111.

Metcalf, Gilbert E., and Kevin A. Hassett. 1997. Measuring the Energy Savings from Home Improvement Investments: Evidence from Monthly Billing Data. *The Review of Economics and Statistics* 81(3): 516–28.

Newell, Richard G. 1997. *Environmental Policy and Technological Change: The Effects of Economic Incentives and Direct Regulation on Energy-Saving Innovation*. Ph.D. thesis. Cambridge, MA: Harvard University.

Newell, Richard G., Adam B. Jaffe, and Robert N. Stavins. 1999. The Induced Innovation Hypothesis and Energy-Saving Technological Change. *Quarterly Journal of Economics* 114(August): 941–75.

Popp, David. 1999. Induced Innovation and Energy Prices. Working Paper. Lawrence, KS: University of Kansas.

Ruderman, Henry, Mark D. Levine, and James McMahon. 1987. The Behavior of the Market for Energy Efficiency in Residential Appliances Including Heating and Cooling Equipment. *The Energy Journal* 8(1): 101–24.

Sanstad, Alan H., Carl Blumstein, and Steven E. Stoft. 1995. How High Are Option Values in Energy-Efficiency Investments? *Energy Policy* 23(9): 739–43.

Sebold, Frederick D., and Eric W. Fox. 1985. Realized Savings from Residential Conservation Activity. *The Energy Journal* 6(2): 73–88.

U.S. DOE/EIA (Department of Energy/Energy Information Administration). 1999. *Analysis of the Climate Change Technology Initiative*. Washington, DC: EIA.

18 Climate Change Policy Choices and Technical Innovation

Carolyn Fischer

> Concern for man himself and his fate must always form the chief interest of all technical endeavors ... in order that the creations of our mind shall be a blessing and not a curse to mankind.
>
> —Albert Einstein, address at the California Institute of Technology, 1931

Climate change is a public policy issue because it may impose costs on society that include adverse human health impacts, productivity losses, and degradation of valued natural resources. On the other hand, policies to reduce greenhouse gases (GHGs) can have serious economic consequences, such as higher production costs for industry and increased energy expenses for households. This trade-off is the classic problem for policymakers trying to strike a balance between costs and benefits of environmental regulation.

The political balancing act would become much easier if policy could generate a "win-win" scenario with both environmental and economic benefits. Much attention has been given to the "Porter hypothesis," that environmental regulation can actually increase the profits of firms, chiefly by encouraging them to look for more efficient production technologies that ultimately lower their costs (see Porter and van der Linde in Suggested Reading). Economists traditionally doubt the concept of a free lunch: If

such gains in efficiency were worthwhile in the first place, why would firms not take advantage of them without regulation? What appears to be cost savings due to regulation often can be negated by proper accounting of management time and other human resource costs, for example. Although some theories have been developed to explain how win-win situations might arise, it is still unclear how widespread these opportunities would be.

Technology is increasingly being touted as the answer to the economy/environment trade-off. If technological progress is the engine of growth, then "green" technologies could be the engine of sustainable, climate-friendly growth. But beyond just looking for low-cost GHG abatement policies, some advocates propose that we can use clean technologies to take care of the environment without economic costs and without imposing environmental regulations. The search for such "win-win-win" scenarios is illustrated by a press release in which President Bill Clinton looked to promote biomass-based energy technologies to "help grow the economy, enhance U.S. energy security, and meet environmental challenges like global warming" (see The White House in Suggested Reading).

Can technology really allow us to have our free lunch and eat it too? Unfortunately, if it sounds too good to be true, it probably is. Good incentives for developing climate-friendly technologies depend critically on whether good climate policies are in place. Furthermore, the usefulness of public investment in green technology also depends on the incentives created by climate policies. Thus, whereas cases can be made for both GHG emissions regulation and promotion of innovation, pursuit of these goals will involve hard choices and real resource costs, not the least of which is the opportunity cost of research and development (R&D).

In this chapter, I present some of the economics of innovation as related to climate policy and explain how the two are fundamentally linked. I also discuss the appropriate role of the government in promoting R&D and GHG emissions abatement. These same arguments apply to environmental policy generally, but the focus here is on climate change policies.

Why Would Firms Want to Develop Cleaner Technologies Themselves?

Economists note that the driving force behind the development of new and improved technologies is the profit motive. Firms realize greater profits if they can produce more and better products for the same cost, or if they can make the same products with less. Thus, they are willing to spend resources on R&D if lower costs and higher productivity will result.

Even without any explicit climate policies, this drive for "more, better, and cheaper"—increased production, improved quality, and lower costs—can lead to lower-emitting techniques. Energy is costly, so firms want production processes that use less energy. For the same reason, consumers demand products that use energy more efficiently. However, market forces provide insufficient incentives to develop climate-friendly technologies if the market prices of energy inputs do not fully reflect their social cost (inclusive of environmental consequences). This incentive problem is magnified if energy prices do not even cover the costs of production (see Chapter 14, Environmentally and Economically Damaging Subsidies).

Even if energy prices reflect all production costs, without an explicit GHG policy, firms have no incentive to reduce their GHG emissions per se beyond the motivation to economize on energy costs. For example, a utility would happily find a way to generate the same amount of electricity with less fuel, but without a policy that makes carbon dioxide emissions costly, it would not care specifically about the carbon content of its fuel mix in choosing between, say, coal and natural gas. For firms to have the desire to innovate cheaper and better ways to reduce emissions (and not merely inputs), they must bear additional financial costs for emissions.

How Do GHG Emissions Abatement Policies Compare in Creating Incentives for Innovation?

Policies designed to abate GHG emissions can have significant impacts on incentives for innovation. However, the incentives can differ greatly across the range of possible measures.

Command-and-control policies that dictate by design which specific technologies or processes a firm must use give firms little leeway. As a result, the firm has little or no incentive to use a cost-effective approach to achieve the same emissions reduction, because it is not allowed to deviate from the specified technology. A policy mandating the "best available technology" might allow the firm to update or retrofit its production process with a more cost-effective technology; however, the firm would not want to develop a better technology that would achieve more abatement if that switch entailed higher total costs—even if the extra reductions in pollution would justify the added costs from a social standpoint.

The threat of future regulation can indeed induce innovation to develop the technology necessary for compliance; however, that threat must be credible. To the extent that governments express a second industrial policy goal—such as a preference for domestic technology—their credibility for implementing the environmental regulation is compromised, and the development of such a technology becomes less likely.

Performance standards, which dictate a certain level of abatement but leave the methods up to the firm, give the firm an incentive to develop a cheaper method of attaining the standard. However, one problem with performance standards is that firms have no incentive to perform beyond the standard, even if lower abatement costs make more abatement desirable from a societal perspective.

Market-based instruments, such as emissions taxes or permits, create a price for emissions. Firms then reduce their emissions as long as the cost of an additional amount of abatement is less than the tax (or permit) cost of additional emissions. For the remainder of their emissions, they pay the tax, buy permits, or use permits that they could otherwise sell. This system ensures that marginal abatement costs are equalized across firms, meaning that total abatement costs cannot be lowered by shifting some abatement from one firm to another. By developing a cost-effective way to reduce pollution, an innovating firm can then save in two ways: on its costs for achieving its current level of abatement and on its tax or emissions permit payments by performing additional abatement at the new, lower cost.

Thus far, the incentives discussed here are restricted to the firm's own gains from lower pollution abatement costs. Looking at this single-firm context, earlier analyses of environmental policy and innovation showed that emissions taxes and emissions permits generally provide more incentives for technological innovation than command-and-control policies such as technology mandates and performance standards. However, new technologies or processes may be useful for more than just the one firm performing the R&D. When innovations are more widely applicable, other incentives come into play, and additional differences arise among environmental policies.

What If One Firm's Research Is Also Applicable to Other Firms?

A technology developed to reduce one firm's costs of abating emissions may also help other firms do the same. As a result, the social gain from this technology consists of the cost reductions to all firms (and potentially gains from additional abatement after it becomes less costly to implement). Furthermore, if adopting the new technology helps the other firms' bottom lines, then they will be willing to pay to license it. Thus, with a well-functioning patent protection system, for example, the innovator

can sell the fruits of its R&D to the potential adopters and reap the full social gains from the invention.

However, reaping those gains may be difficult. New knowledge can create "spillover benefits" to other parties for which the innovator is not compensated. Not every advance may receive a patent; basic research may not have specific, patentable applications. Still, broadening the overall knowledge base can make future valuable applications possible. Even if a new technology or process receives a patent or copyright, it may be hard to enforce. For example, other firms also may have the opportunity to imitate the innovation despite the patent, for example, through reverse-engineering. Studies for commercial (or nonenvironmental) innovations suggest that appropriation rates vary considerably over different kinds of innovations, with an average rate of around 50% (see the papers by Griliches and Nadiri in Suggested Reading). The easier it is for other firms to imitate the innovation on their own, the less willing they will be to pay for a license, and the lower the gains the original innovator will be able to appropriate. As a result, the innovator will have less incentive to conduct R&D than would be best from a societal perspective.

On the other hand, the market for R&D also might encourage excessive research. For example, several innovating firms competing in a "patent race" for the rewards of a patent can collectively spend too much on redundant research. The consensus view, however, is that the private rate of return to R&D is well below the social rate of return. Thus, positive spillover effects prevail, and markets will tend to provide too little R&D.

How Do Environmental Policies Compare When Other Firms May Adopt a New Technology?

When we take into account interactions among innovating firms and other firms potentially adopting the new technology (at a price), the analysis of how different abatement policies affect innovation becomes more complex and ambiguous. Table 1 summarizes the various incentives for innovation and gives an idea of their relative importance.

Table 1. Incentives for Innovation Created by Environmental Policies.

Policy	Direct gains to innovating firm	Potential rents from adoption
Command-and-control	None	None
Best-available technology	Negative: New standard raises overall compliance costs	Positive: Tighter standard raises incentive to adopt
Performance standards	Positive, less than tax: Limited to existing abatement costs	Positive, less than tax: Limited to existing abatement costs
Emission tax	Positive: Lower abatement costs and taxed emissions	Positive: Lower abatement costs and taxed emissions
Auctioned emission permits	Positive, more than tax: Lower abatement costs and costs of all permits purchased	Positive, less than tax: Lower abatement costs, but buying permits becomes cheaper alternative
Grandfathered emission permits	Positive, less than tax: Limited to existing abatement costs	Positive, less than tax: Lower abatement costs, but buying permits becomes cheaper alternative
Tradable performance standards/output-allocated permits	Positive, may be less than tax: Initial abatement costs higher, but output subsidy also falls	Positive, less than tax: Initial abatement costs higher, but permits become cheaper

Source: Fischer and others (see Suggested Reading).

Command-and-control policies that specify a technology allow other adopters no more leeway than the innovator. Therefore, little or no incentive remains to implement cost-effective techniques for pollution reduction. However, a policy that mandates the best available technology offers the possibility that the innovator's new technology would be made the standard. Then, the other firms in the industry would have to pay to adopt it (if they could not come up with their own equivalent technology). Therefore, although the innovating firm's own pollution reduction costs could rise as a result of the new, stricter technology standard, the gains from other firms licensing the technology could be large.

Under performance standards, adopting firms receive the same gains from reducing abatement costs as the innovator. They are willing to pay for a cheaper way to abate their emissions up to the standard, but they will not want to push reductions beyond the standard.

With market-based mechanisms, the development and industry-wide adoption of an innovation can affect not only individual firms' costs but also the prices and quantities prevailing in the industry's market. Because different market-based mechanisms allow for different changes in prices and quantities of emissions, they have different implications for innovation. Taxes set the price of emissions, allowing total abatement to vary, whereas permits set the total amount of emissions, allowing the marginal cost of abatement to vary. Therefore, whereas one can choose a tax and an emissions cap that would generate the same outcome before innovation, product and technology markets will adjust differently after innovation, depending on the policy.

Under an emissions tax, as innovation makes abatement cheaper, the total amount of abatement will rise because it will be less expensive to increase emissions reductions than to pay taxes on all current emissions. On the other hand, with emissions permits, firms as a group will not abate more after the innovation, because that total amount is set by the cap. Therefore, the total abatement cost savings are less than under the tax, where more abatement was performed after innovation. Furthermore, as innovation lowers abatement costs, the price of emissions permits will fall.

From the point of view of the adopting firms, their individual decision of whether to license the technology does not affect the permit price; however, those firms will attempt to anticipate the fall in the permit price caused by collective adoption of the innovation. The lower permit price means that they now have a less costly option of foregoing the new technology and buying cheaper permits instead. Thus, adopting firms will not be willing to pay as much for the innovation under permits as under the tax. This "adoption price effect" lowers the maximum royalty the innovating firm can charge to license the technology.

However, the reduced price can benefit the innovator directly if permits are auctioned. Under this scenario, a lower permit price means the innovator will not have to pay as much for the rest of its emissions. This emissions payment effect raises the gains to innovation. On the other hand, if permits are freely distributed, then the reduced price lowers the value of the innovator's allocated permits, which counteracts the emissions payment effect. Thus, free permits give less incentive to innovate than auctioned permits or than taxes because of the smaller abatement cost savings and the adoption price effect.

It is hard to say whether auctioned permits or taxes generate more innovation, because the incentives depend critically on the adopting firms' ability to imitate the innovation. In general, with little imitation, unless the emissions payment effect is very large, taxes provide more innovation incentive. However, if imitation is substantial—preventing the innovator from recouping most gains, if any, from the adopters—auctioned permits provide more incentive.

With respect to climate policy, the more compartmentalized the emission permit programs, the larger the impact of a single innovation on permit prices. If a major fuel-saving invention is adopted by the trucking industry, it might have a small effect on permit prices in a broad-based cap-and-trade program, but it would have a large effect on a permit program limited to the trucking sector. Thus, in a broad-based carbon trading program, innovation in-

centives in a specific sector are going to more closely resemble those of a carbon tax.

Finally, it is worth mentioning another permit regime: output-allocated permits, which are very similar to tradable performance standards. Examples of tradable performance standards in other contexts include the phasedown of lead from gasoline and the Corporate Average Fuel Economy (CAFE) standards; related policies have been proposed for dealing with GHGs. These two kinds of market-based mechanisms require firms to hold permits to cover their emissions, and firms are allocated permits according to their output multiplied by an average emissions rate. Thus, above-average emitters must buy permits, whereas below-average emitters can sell them. This permit allocation method creates an implicit subsidy for output, because the more a firm produces, the more permits it receives.

This implicit subsidy affects innovation incentives in many ways that are not so obvious. First, it discourages firms from cutting back output, placing more of the abatement burden on lowering emission rates. As a result, for a given target level, total abatement costs are higher. This inefficiency then raises the potential gains to innovation, although in a costly way. Second, although it may seem that an innovator would want to push costs even lower to gain market share and more subsidies, widespread adoption eliminates any such market share advantage. Furthermore, the subsidy lowers output prices across the industry compared with the other permit or tax schemes, meaning that the extra subsidy gains from expanding market share are largely negated by lower revenue gains. Thus, the output subsidy seems more likely to reduce innovation incentives.

Of course, it is important to look beyond which environmental policy generates the most R&D and think about overall impacts on societal well-being. R&D spillovers and global warming are two separate problems. To correct for both ideally requires two separate policy tools. However, when R&D shortfalls cannot be ameliorated directly by available policies, such as research grants or tax credits, abatement policy is forced to balance its impacts on the two problems. Some environmental policies may help offset imperfections in the market for innovation. But the cost may be wasted expenses on abatement (such as with a high tax) or wasted opportunities for abatement once costs fall (such as with auctioned permits). As with any good thing, one can have too much innovation or abatement, because those resources could be used elsewhere. Therefore, an overly strict environmental policy to stimulate for shortcomings in R&D gives rise to its own costs; likewise, an overly generous R&D subsidy to make up for inadequately low emissions pricing can be wasteful.

How Do Innovation and Environmental Policy Interact over Time?

GHG emissions abatement policies determine not only the price signals (or regulatory requirements) for abatement but also the signals for innovation. However, just as the amount of innovation depends on those price signals, getting the right price signals depends on the amount of innovation.

For "stock" pollutants such as GHGs, which build up over time, current and future costs and benefits are dynamically linked. Future innovation can have a significant impact on the emissions rate, both now and in the future. To the extent that innovation reduces future abatement costs, it makes sense to postpone some emissions reductions until the future.

Ideally, policy would take into account future innovation and the evolving costs and benefits of regulation in setting current and future emissions taxes or quotas. Of course, in reality, the results of R&D investments are highly uncertain, and predicting such a path may be difficult. Furthermore, adjusting policies frequently may be difficult and involve significant administrative costs. In this case, policymakers need to weigh the benefits and costs of inducing more or less innovation against too much or too little abatement.

A fixed tax policy does not allow emission prices to adjust, thus creating a risk of too much abatement if costs fall (assuming the tax starts out reflecting the initial damage costs of emissions, which then decline as emissions fall). However, taxes do allow the amount of abatement to fluctuate according to cost

conditions. On the other hand, a permit policy does not allow an adjustment of the quantity of abatement, meaning that too little will be done if costs fall (and too much effort will be put into abatement if costs rise). In the case of GHG emissions, their potential damage is relatively insensitive to the rate of emissions at any particular time (although the total cost likely will rise as GHGs accumulate in the atmosphere). Thus, the presence of uncertainty in abatement costs due to unpredictable innovation favors a tax-based approach over a quantity-based permits approach. The reason is that under these circumstances, uncertainty about the volume of abatement imposes less of a burden on society than does abatement cost uncertainty.

Should Diffusion Be Aided?

The incentives to adopt a cleaner technology mirror those for creating it for oneself, namely, the cost savings. Higher energy prices raise the returns to adopting energy-efficient technologies and thus encourage their diffusion (see Chapter 17, Energy-Efficient Technologies and Climate Change Policies).

However, even when potential cost savings are substantial, a technological advance is not often adopted completely and immediately. Diffusion may lag because of adjustment costs (such as training personnel in a new production process that reduces energy use), information costs (such as educating consumers about energy efficiency), or irreversible investments (such as building a whole new plant with new generation technology), or because an even better technology might be on the way, making the wait worthwhile. Because these cases all represent real resource costs for the firm, avoiding or postponing adoption can be quite rational. Furthermore, in the absence of market failures or important returns to scale, delay or nonadoption also can be an appropriate allocation of resources from society's point of view.

On the other hand, the innovator may choose to limit diffusion. A patent is essentially a temporary monopoly, and unless the innovator can charge higher prices to firms that value the technology more (that is, price discriminate), royalties may be maximized by restricting licenses. A tension can then exist between the gains from spreading new technology today and the gains from promoting future innovation. Society would benefit most from the widest worthwhile diffusion of an existing innovation. For example, to convert as many vehicles on the road as possible, one might want to distribute freely the design of an engine modification that cost-effectively raises fuel efficiency. However, if automobile companies expected that to happen, they would have had little motivation to develop the modification in the first place. Therefore, when society values more immediate and complete distribution, such a policy must be done in a way that maintains the proper incentives for future innovation. In other words, the policy precedent should involve compensation (for example, buying the patent rights) rather than expropriation or forced distribution (see Chapter 24, The Economics of Technology Diffusion).

In many cases, however, government can aid diffusion without compromising R&D incentives. Support for high-spillover advances, such as in basic knowledge from fundamental research, offers such an opportunity. Reducing adjustment costs can aid diffusion and provide incentives for innovation. Public institutions, with their existing infrastructure of regulatory expertise and contacts, can take advantage of returns to scale in gathering and distributing information. Public policy toward the environment could include the dissemination of best business practices and the subsidization of training in environmental management, thus helping firms to take advantage of any win-win opportunities that might otherwise be ignored. However, the need for such subsidies depends on the extent to which firms fail to capitalize on such opportunities on their own, and the extent to which such problems can be distinguished from the pursuit of unnecessary subsidies by firms. These issues remain controversial. Similarly, investment subsidies can reduce the barriers created by capital market problems (such as impediments to borrowing), but they come at the cost of scarce public funds and offer windfalls to those who would adopt anyway.

A policy mandating the best available technology is a type of forced diffusion. How the policy is de-

signed greatly affects the payoffs to innovation. Requiring all firms to buy the new technology puts few limits on what the innovator can charge, possibly resulting in too much incentive to innovate. If costs are a consideration in imposing a new standard, some limit will be put on such opportunities. If other firms can use alternative means to achieve the new standard, then payoffs will be more in line with actual net benefits, unless imitation is an option. Finally, if the innovator is required to disseminate freely the advance, little incentive will exist for innovation. Thus, in situations where market-based instruments are not feasible and technology standards are used, policy design must also take innovation incentives into account.

What Is the Role of Government?

> Societies will, of course, wish to exercise prudence in deciding which technologies—that is, which applications of science—are to be pursued and which not. But without funding basic research, without supporting the acquisition of knowledge for its own sake, our options become dangerously limited.
>
> —Carl Sagan

Economists generally agree that the main role of the government is to stick to the basics: first, to create the proper private incentives for environmental protection and for innovation; second, to supplement the socially useful R&D that the private sector does not tend to provide sufficiently.

For the first part, the government can provide institutions such as a strong system of patents and copyrights to ensure inventors can reap the fruits of their efforts. To aid diffusion, the government can take cost-effective measures to lessen informational costs and other barriers that create transaction costs. If fast, low-cost distribution of new research and technologies is a priority, its implementation should remain consistent with preserving the long-run incentives for R&D, as discussed earlier.

With respect to climate change, the government should ensure that producers face financial costs for their emissions that bear some resemblance to the expected future environmental costs of those emissions, however inadequately these costs currently are measured. Market-based instruments not only promote cost-effective pollution reduction but also provide the most efficient incentives for investing in the development of environment-friendly technologies.

Even with strong patents and emissions pricing, private firms may not invest in the right kinds or the right amounts of R&D. Basic research is a good example: One can rarely foresee all the useful applications that may arise from the expansion of the general knowledge base. Therefore, government has a central role in funding high-spillover research. One way to achieve this goal in a decentralized manner is through a system of competitive grants.

With respect to R&D for specific applications (such as particular manufacturing technologies or electricity generation), governments are notoriously bad at picking winners. The selection of these projects is best left to private markets, whereas the government ensures that those markets face the socially correct price signals. One exception may be specific technologies that would aid in the implementation of more efficient GHG policies; for example, better emissions-monitoring devices could reduce compliance and enforcement costs and enable the transition to market-based mechanisms for emissions control. In this case, in a sense, the government is the relevant market for the innovation. But for most cases—to the extent that markets still tend to provide insufficient R&D—broad rather than project-specific R&D subsidies are in order.

Above all, policymakers concerned about climate change must recognize that incentives for abatement and innovation are inexorably linked. If failures in the market for R&D exist, and available policies cannot correct them, then the choice of climate policy will be affected. If innovation incentives are lacking, then the best climate policy response is likely to be looser regulation, reflecting the higher expected abatement costs, not stricter regulation to promote innovation. Similarly, if emissions are underpriced and too little abatement is performed, then less investment in R&D may be warranted, because the new climate-friendly technologies will not be sufficiently exploited. Because GHGs accumulate over time, incentives for their abatement and innovation are more intertwined, because current and future abatement can depend on each other as well as on innovation, which in turn depends on abatement.

Thus, a technology policy is no substitute for environmental policy. The primary gains to environmental protection always come from reducing environmental damages in the most cost-effective manner, regardless of innovation. Innovation then has the potential to increase those gains, but the benefits of emissions-reducing innovations are by definition limited to the potential environmental gains.

More than other policy instruments, market-based mechanisms offer incentives to both protect the environment efficiently and develop better methods of protection. Put simply, rather than look for a free lunch, perhaps we should make a nutritious lunch.

Suggested Reading

Bernstein, Jeffrey L., and Ishaq Nadiri. 1988. Rates of Return on Physical Capital and R&D Capital and Structure of Production Process: Cross Section and Time Series Evidence. Working Paper 2570. Cambridge, MA: National Bureau of Economic Research.

Cadot, Olivier, and Bernard Sinclair-Desgagné. 1996. Innovation under the Threat of Stricter Environmental Standards. In *Environmental Policy and Market Structure*, edited by C. Carraro, Y. Katsoulacos, and A. Xepapadeas. Amsterdam, the Netherlands: Kluwer Academic Publishers.

Cohen, Linda, and Roger Noll. 1991. *The Technology Pork Barrel*. Washington, DC: Brookings Institution.

DeCanio, Stephen J. 1994. Agency and Control Problems in U.S. Corporations: The Case of Energy-Efficient Investment Projects. *Journal of the Economics of Business* 1(1): 105–23.

DeCanio, Stephen J., and William E. Watkins. 1998. Investment in Energy Efficiency: Do the Characteristics of Firms Matter? *Review of Economics and Statistics* 80(1): 95–107.

Downing, Paul G., and Lawrence J. White. 1986. Innovation in Pollution Control. *Journal of Environmental Economics and Management* 13: 18–29.

Fischer, Carolyn. 1999. Rebating Environmental Policy Revenues: Output-Based Allocations and Tradable Performance Standards. Working paper. Washington, DC: Resources for the Future. http://www.rff.org/~fischer/papers/rebatepc.pdf (accessed October 18, 2000).

Fischer, Carolyn, Ian W.H. Parry, and William A. Pizer. 1999. Instrument Choice for Environmental Protection when Technological Innovation is Endogenous. Discussion paper 99-04. Washington, DC: Resources for the Future.

Griliches, Zvi. 1980. Returns to Research and Development Expenditures in the Private Sector. In *New Developments in Productivity Measurement*, edited by J. W. Kendrick and B. Vaccara. National Bureau of Economic Research (NBER) Studies in Income and Wealth No. 44. Chicago, IL: University of Chicago Press.

———. 1992. The Search for R&D Spillovers. *Scandinavian Journal of Economics* 94(Suppl.): S29–47.

Kneese, Alan, and Charles Schultz. 1975. *Pollution, Prices, and Public Policy*. Washington, DC: Brookings Institution.

Magat, Wesley A. 1978. Pollution Control and Technological Advance: A Dynamic Model of the Firm. *Journal of Environmental Economics and Management* 5: 1–25.

Mansfield, Edwin, John Rapoport, Anthony Romeo, Samuel Wagner, and George Beardsley. 1977. Social and Private Rates of Return from Industrial Innovations. *Quarterly Journal of Economics* 41: 221–40.

Nadiri, M. Ishaq. 1993. Innovations and Technological Spillovers. Working Paper No. 4423. Cambridge, MA: National Bureau of Economic Research.

The White House. 1999. Growing Clean Energy for the 21st Century. Press Release. August 12. Washington, DC: The White House, Office of the Press Secretary.

Palmer, K., W. Oates, and P. Portney. 1995. Tightening Environmental Standards: The Benefit–Cost or the No-Cost Paradigm? *Journal of Economic Perspectives* 9(4): 119–32.

Parry, Ian W.H., William A. Pizer, and Carolyn Fischer. 2000. How Important Is Technological Innovation in Protecting the Environment? RFF Discussion Paper 00-15. Washington, DC: Resources for the Future.

Porter, Michael E., and Claas van der Linde. 1995. Toward a New Conception of the Environment-Competitiveness Relationship. *Journal of Economic Perspectives* 9(4): 97–118.

Toman, Michael A., Richard D. Morgenstern, and John Anderson. 1999. The Economics of "When" Flexibility in the Design of Greenhouse Gas Abatement Policies. Discussion paper 99-38-REV. Washington, DC: Resources for the Future.

Sinclair-Desgagné, Bernard. 1999. Remarks on Environmental Regulation, Firm Behavior, and Innovation. Working paper 99s-20. Montreal, Canada: Centre interuniversitaire de recherche en analyse des organisations (CIRANO).

Weitzman, Martin, L. 1974. Prices vs. Quantities. *Review of Economic Studies* 41: 477–91.

Wright, Brian D. 1983. The Economics of Invention Incentives: Patents, Prizes, and Research Contracts. *American Economic Review* 73: 691–707.

Zerbe, Richard O. 1970. Theoretical Efficiency in Pollution Control. *Western Economic Journal* 8: 364–76.

19 Greenhouse Gas "Early Reduction" Programs
A Critical Appraisal

Ian W.H. Parry and Michael A. Toman

Several ideas have recently circulated in the United States for establishing some kind of national "early reduction" program to begin lowering greenhouse gas (GHG) emissions before 2008, when binding commitments under the (not yet ratified) Kyoto Protocol are scheduled to come into force. Although these proposals differ in important details as to what activities constitute allowable early reductions and how many early reductions could be earned, the critical element common to almost all the proposals is the voluntary creation of "early reduction credits."

These credits would be awarded to individual projects that reduce emissions below some business-as-usual level before 2008. Those holding credits in 2008 would be entitled to a greater share of the available national "emissions budget" during the 2008–12 Kyoto Protocol commitment period. If the United States were to implement an emissions control program during that period with tradable carbon allowances, then holders of early reduction credits would be allocated a share of the allowances, implying fewer allowances for others. (If the United States were to use a standards-based approach for individual GHG sources, then sources holding early reduction credits could offset some of their regulatory control requirements with these credits, and presumably, other sources would need to meet higher standards to satisfy the Kyoto Protocol budget. How-

ever, this is a less straightforward way to implement an early credits program.)

One proposal that sharply differs from these early reduction credit mechanisms is a proposal put forward by four researchers at Resources for the Future in 1999 (see Kopp and others in Suggested Reading). That proposal would establish a mandatory "cap-and-trade" system for GHG control in the United States starting about 2002. A national emissions budget would be allocated by auctioning tradable GHG allowances rather than tying credits to project-specific investments. This mechanism would operate "upstream," controlling GHG emissions by limiting fossil fuel supplies at various points where they enter the energy system.

In this chapter, we discuss the potential performance of these various early reduction proposals. First, we consider the economic efficiency of the different mechanisms. Proponents of early reductions claim that such programs can generate net benefits by increasing total emissions control over time, thus further ameliorating climate change. Another claimed efficiency benefit is the possibility of promoting technical progress in emissions control through learning by doing: It is asserted that by starting emissions control before 2008, individuals in the economy are stimulated to accelerate the use of lower-emissions technology, thus lessening unwelcome commitments to less-efficient technologies and lowering the cost of meeting the Kyoto Protocol targets.

Proponents also have advanced institutional and political arguments for early reduction programs. One argument is that by starting early to design and test the institutions of GHG emissions trading in the United States, valuable experience is gained that can be used to refine the regulatory approach. Proponents of voluntary early credit approaches also point to potential political benefits: If a broad cross section of business, environmental groups, and others could come together behind such a program, it would provide some political impetus for more ambitious goals, including eventual ratification of the Kyoto Protocol. Another component of this argument is the interest in providing a "safe harbor" for those businesses and other organizations that have made early investments in curbing GHG emissions and fear that they will be held to even higher standards by future regulations. Under early credit systems, these investments would be rewarded.

We find that the economic efficiency benefits of early reduction programs are ambiguous and very much connected to the design of the program as well as to the potential costs and benefits of GHG control. Under plausible assumptions, an early credit program may lead to costs from early reductions in excess of environmental benefits. On the other hand, early credit-earning activity may be minimal because of uncertainty surrounding the future of the Kyoto Protocol. A cap-and-trade approach has a greater economic potential than early credits because a clear regulatory target can be established for overall emissions control and because fewer problems arise in establishing and implementing baselines for the awarding of early credits. A cap-and-trade approach also avoids some adverse fiscal problems associated with early reduction credits.

The economic case for early reductions would be stronger if the Kyoto Protocol were to be amended to allow low-cost early emissions reductions to be banked for use during 2008–12. This approach could produce very substantial cost savings by allowing firms to smooth abatement over time, but it would not increase the overall amount of abatement.

Either early credits or an early cap-and-trade policy can generate additional benefits from technical advance through learning by doing. However, we expect that these benefits will be modest. Finally, the purported political and institutional advantages of early reduction programs probably are overstated.

Operation of Early Reduction Programs

To evaluate the potential economic benefits of early reductions, we must consider how early reduction programs would operate. We consider first an early reduction credit system. In this system, an eligible individual GHG emitter would invest in reducing emissions below some business-as-usual level to garner credit for the reductions. The credits would be an asset whose expected value is equal to the expected value of a GHG permit when and if the Kyoto Protocol is implemented. The incentive to invest in such an asset will depend on this expected value and on the cost of creating the asset through early reductions.

The cost of early reductions depends on the various factors that influence domestic GHG abatement possibilities: the degree of substitutability of lower-carbon and higher-carbon fuels, the degree of substitutability of energy and other inputs (for businesses and consumers, for example, in driving), and the degree to which energy markets are distorted (for example, inefficient subsidies lower the economic cost of GHG control). On the other hand, the volume of early reduction credits awarded depends on how the regulatory baselines for credits are defined and what rules for participation are established, as well as the expected future value of credits and the costs of GHG control. In this respect, early domestic credit programs are much like the Kyoto Protocol's clean development mechanism for generating GHG reduction credits through GHG-reducing project investments in developing countries. The volume of credits will be more limited if eligibility is limited (for example, to only a few sectors in the economy) or if stringent rules for defining GHG reductions relative to business as usual are used. The latter would be reflected in, for example, assumptions of substantial improvements in energy efficiency even without an early reduction program.

It should be noted that in this approach, the awarding of early reduction credits reduces the total emissions budget available to others during the Kyoto Protocol commitment period. Unless the economy as a whole can bank early reduction credits (not allowed now under the protocol), the economy's total emissions budget during the commitment period is fixed. Moreover, except for the possibility of changes in the capital stock, for example, additions to the knowledge stock through learning by doing (see later), the awarding of early credits will not alter the market price of a GHG permit during the commitment period. All that early credits do is alter the distribution of the economy's emissions budget in favor of those engaging in early reductions. This observation has important political implications.

A cap-and-trade program would operate quite differently from an early credit system. It would be a mandatory program that establishes an emissions target for the whole economy and then allocates responsibility for meeting that target to various individual actors. The upstream approach of Kopp and others (see Suggested Reading) would require permits based on the carbon content of fuels for domestic suppliers and importers. (In practice, the program would regulate oil inputs to refineries, sales by coal suppliers, and deliveries of natural gas into the pipeline system; GHGs other than CO_2 would be added on a case-by-case basis). This proposal also would auction GHG permits, raising revenue for the government that could be used to reduce other taxes in the economy and provide some transitional assistance to those parts of the economy most adversely affected by GHG controls. GHG cap-and-trade programs also could be established downstream, at the point of emissions (for example, a power plant), and they could involve permits issued at no charge rather than by auction, but these options are less efficient.

However the cap-and-trade program is implemented, the value of emissions reduction would depend on the incremental cost of reducing emissions in the early reduction period, not on the expected value of an emissions permit during the commitment period. (This statement assumes that early reductions cannot be banked for use during the commitment period, consistent with the Kyoto Protocol.) This incremental cost depends in turn on the stringency of the (mandatory) economy-wide early reductions target. It makes the implications for economic efficiency of cap-and-trade quite different from those of an early credit system. Moreover, the cap-and-trade approach does not require the development of rules for project-specific assignment of credits.

Economic Efficiency Benefits of Early Reductions

We recently examined the potential efficiency of early reduction programs under different assumptions about abatement costs and environmental benefits (see Parry and Toman in Suggested Reading). The results of that analysis are summarized here. The values are not the result of a detailed economic cost–benefit model; rather, they derive from a set of simple numerical simulations that provide plausible orders of magnitudes for the costs of early reduction and the resulting benefits. First, we consider the relationship of abatement costs to environmental benefits from early reductions. We then add in the possible benefits from learning by doing. Finally, we consider the potential value of early abatement if credits could be banked and used during the commitment period for the Kyoto Protocol.

Abatement Costs and Environmental Benefits

Environmental benefits are taken to be the present value of future damages avoided as a consequence of slowing down global warming. These benefits are highly uncertain; accordingly, we considered a range of values from $5/ton of carbon emissions reduction to $100/ton. These figures do not include the potential ancillary benefits of early GHG reductions as a result of reductions in conventional pollutants, which also are uncertain. The ancillary health-related benefits from air pollution reduction in the United States could be on the order of $5–20/ton of carbon emission avoided, depending on how the GHG policy is structured (see Chapter 8, "Ancillary Benefits" of Greenhouse Gas Mitigation Policies). Additional benefits could result from, for example, reduced nitrogen deposition in water bodies.

The incremental cost of GHG abatement and the price of a GHG permit during the Kyoto Protocol commitment period also are uncertain. We considered two possible figures: $50/ton and $150/ton. These figures are the incremental costs of the last ton of carbon abated per year during the commitment period. We further assumed that the incremental cost of the first ton of carbon abated is zero (meaning that the economy contains some "low fruit" in terms of GHG abatement). The $50/ton figure can be interpreted as reflecting either an optimistic scenario for domestic abatement cost or extensive reliance on international emissions trading through the "Kyoto mechanisms" (in particular, use of low-cost emissions credits obtained from clean development mechanism investments in developing countries and/or acquisition of low-cost surplus emission allowances from Russia). The $150/ton figure is correspondingly less optimistic regarding these factors.

Using standard figures for future U.S. business-as-usual emissions from the U.S. Department of Energy/Energy Information Administration, we pegged the total emission reduction required under the Kyoto Protocol as about 30% below business-as-usual emissions. To simplify the analysis, we assumed that the incremental cost of GHG control before the commitment period was proportional to the incremental cost during the commitment period. Under this assumption, the total volume of early reductions under an early credit program is about 23% of business-as-usual assumptions under either the lower or higher abatement cost scenario.

Table 1 shows the net benefits of early reduction relative to early reduction costs (costs during the precommitment period). We express the results in this form because, although the absolute magnitudes of benefits and costs depend on the particular assumptions used to generate the scenarios, the relative magnitudes are probably more robust. The key message from Table 1 is that if the cost of GHG control is relatively low (as would be the case if the permit price is $50), then reductions stimulated by an

early credit program are likely to yield net benefits unless the environmental gains are quite low. On the other hand, if the cost of GHG control is larger (that is, the permit price is $150), then early reductions yield abatement costs in excess of environmental benefits unless the environmental value is very high. In both cases, the magnitudes of cost or benefit are likely run to several billion dollars per year, a significant amount.

Our own view is that the higher-cost scenario probably is more representative of what costs will be under the Kyoto Protocol. Lower costs would be possible with a longer adjustment period, but the protocol requires a relatively substantial reduction in GHG emissions in a relatively short period of time. The Kyoto Protocol mechanisms could serve to hold down domestic abatement costs; however, that result is not assured, because debate continues over how the mechanisms will be structured, and the mechanisms inevitably will not function as well as theory indicates they might in an ideal world. Many economists believe that the environmental benefits of early reductions are $25/ton or less, although there is certainly the possibility of much higher values. Given the data in Table 1, it seems that an early credits program could generate costs substantially in excess of environmental benefits. The reason is that emissions abatement under the Kyoto Protocol is likely to be excessive from a cost–benefit perspec-

Table 1. Net Benefits Gained from Unlimited Early Reduction Credits Relative to Early Reduction Costs (%).

Permit price in commitment period	Marginal environmental benefit from early emission reduction			
	$5/ton	$25/ton	$50/ton	$100/ton
$50/ton	−0.74	0.29	1.56	4.12
$150/ton	−0.91	−0.59	−0.15	0.71

Notes: Benefits from early emission reduction are the value of avoided long-term climate change damages. Negative entries indicate early reduction environmental benefits less than abatement costs; positive entries indicate early reduction environmental benefits greater than abatement costs.

Source: Author calculations (see text for description).

tive. The high permit price would generate too much early abatement before the commitment period (in the absence of limits on total available credits).

A cap-and-trade program offers significant advantages over an early credits system. The early reductions cap can be set at whatever level policymakers think is best, as opposed to having incentives tied to the uncertain implementation of the protocol. The proposal by Kopp and others accomplishes this by setting a ceiling price, or a safety valve for the permit price, which the government would implement by making additional permits available any time the price went above the ceiling. It provides the opportunity to avoid excessive early abatement if environmental benefits are seen to be low relative to incremental costs. At the same time, a cap-and-trade system provides a degree of regulatory certainty in creating incentives for early abatement that is absent in early credit systems, and it avoids inefficiencies that result from the need to undertake project-by-project assessment of early emissions abatement.

The other important advantage of a cap-and-trade policy is more indirect but no less important: It is possible to raise revenue by auctioning permits. This revenue can be used to offset other tax burdens that reduce the overall performance of the economy. In contrast, early credits not only do not raise revenue before the commitment period but also reduce the ability to undertake a tax shift during the commitment period, because some permits have already been given away to the holders of early credits. Moreover, early reduction credits can "crowd out" government revenues in the event that future permits are auctioned. This difference in the fiscal implications of the two policy options greatly magnifies the advantages of a cap-and-trade policy. Although many distributional and other political factors enter into the decision to auction carbon permits or distribute them free of charge, the cost difference identified here is a major consideration in the relative efficiency of the policy options for early reductions.

Learning by Doing

We now turn to the possible advantages of early reductions in stimulating future abatement cost reductions through learning by doing. We considered two

scenarios: In one, early reduction brings down the incremental cost of later emissions control by 5%, and in the other, the cost reduction is a fairly optimistic 30% (see Parry and Toman in Suggested Reading). We also considered different assumptions for how extensively the private sector pursues early reductions out of self-interest. (Some of the benefits of any one emitter's learning by doing may spill over to other emitters, but these spillover benefits will be hard to capture in many cases.)

We concluded that only a very modest level of early abatement (well under 10% of business-as-usual emissions) is justified by learning-by-doing benefits alone. An early credit program will generate substantially more early abatement. Learning-by-doing benefits offset only about 10–50% of the costs of early abatement, depending on how much the cost of later abatement is reduced. Thus, learning by doing on its own is not a powerful rationale for early abatement.

Nationally Bankable Early Credits

Finally, we consider the potential cost savings of banking low-cost early reductions at the economy-wide level and using them to offset the most expensive GHG reductions required to meet the Kyoto Protocol target during the commitment period. Here, total emissions abatement over time does not increase (unlike the case without banking), but abatement is more evenly distributed across time rather than being concentrated during the commitment period. If banking were allowed, then early reductions would be undertaken until the marginal cost of early abatement equaled the expected present value of a future emissions permit (because the latter also reflects expected future marginal abatement cost). When this "arbitrage condition" is satisfied, cost savings from banking early reductions are at a maximum, and no one has any incentive to undertake further early reduction.

The potential cost savings from nationally bankable early credits are on the order of 40% of the (discounted) cost of meeting the Kyoto Protocol target during the commitment period without early reductions (see Parry and Toman in Suggested Reading). These savings could amount to several billions, even tens of billions of dollars per year depending on the cost of abatement. The amount of the potential cost savings underscores an inherent tension in the current design of the protocol, which allows banking of only early credits generated internationally through the Clean Development Mechanism. The postponement of commitments under the protocol was intended to provide Annex I countries (the industrialized countries that would cap their total emissions under the protocol) with some transition time to implement their targets. However, given the potential incremental cost of abatement during the commitment period relative to the cost of early reduction, the postponement also creates incentives to seek the early banking of domestic as well as international credits.

Institutional and Political Issues

The values in Table 1 assume that there are no regulatory limits on early reduction credits and no uncertainty about the implementation of the Kyoto Protocol. The introduction of limits on the total volume of early credits and the recognition of the uncertainty surrounding implementation of the protocol both imply less investment in early credits (the latter because uncertainty implies a lower expected value of a credit). In light of the discussion in the previous paragraph, such limits on early reduction activity actually could be beneficial under various plausible scenarios.

However, regulatory limits on credits also will create added incentives for different sectors to engage in costly lobbying (known as rent-seeking in the economics literature) to ensure favorable treatment under the program, raising the total cost of the program to society. Such incentives exist even without credit limits, because different sectors can profit from a favorable definition of business-as-usual baselines in the awarding of credits. In addition to lobbying costs themselves, politicking over the rules for early credits could reduce efficiency by favoring costly abatement technologies (seen by proponents as needing a boost) rather than allowing early reductions to be undertaken in the most cost-effective fashion possible. These problems are avoided with a cap-and-trade approach to early reductions. For this reason, the argument that early credits provide benefits from testing out mechanisms for full-fledged emissions trading is not persuasive.

Another potential problem of early reduction credits is the difficulty of judging a firm's business-as-usual emissions (with which actual emissions should be compared) to award the appropriate amount of credits. This problem arises in connection with "anyway reductions," abatement that occurs as a result of actions that would be taken anyway to reduce production costs. For example, it might be in a firm's own interest to adopt a new energy-saving technology. Such asymmetric information can generate a modest amount of inefficiency, because different firms do not end up with the same incremental cost of abatement. This problem is avoided under an early cap-and-trade program, because permits are auctioned or awarded on the basis of emissions before the precommitment period rather than on the basis of unobservable business-as-usual emissions during the commitment period.

Finally, we return to the potential political benefits of early credits in developing a broader base of support for ultimately implementing the Kyoto Protocol. This could be true in enhancing support among businesses seeking (for perfectly legitimate reasons) both economic advantages and a "greener" image. However, in the absence of economy-wide banking of early reductions, early credit programs essentially are a zero-sum game for emitters. While early actors enjoy the fruits of low-cost abatement that they can bank for use in 2008, other regulated entities must incur higher compliance costs during the commitment period because the total remaining pool of allowed emissions for these actors is smaller. The transfer of no-charge permits to holders of early credits also represents in part a transfer of income from households to these credit holders. Given these distributional implications, the ultimate political advantages of early credits may be open to question. Such an inefficient transfer mechanism is not necessary to provide a safe harbor to those who have, for whatever reason, chosen to pursue early action on their own. These actions are not penalized in a cap-and-trade system with auctioned permits or one with no-charge allocation, provided permits are allocated based on a historical base year that precedes most early action.

Concluding Comments

Given the political uncertainties surrounding future climate policy in the United States, the fate of any early reduction program also is difficult to fathom. If such a program is to be undertaken, strong economic reasons suggest pursuing a modest but mandatory cap-and-trade policy rather than a voluntary but less well targeted and less transparent early credit program. The case for early action is especially strong if permits can be banked at a national level, which is not possible under the Kyoto Protocol. The economic virtue of the mandatory cap-and-trade approach may also be its political weakness, because such an approach makes the cost of GHG control transparent and applies that cost widely throughout the economy. Given recent aversion in the United States to *any* increase in energy costs, the prospects for such a program in the near term may be doubtful.

On the other hand, the political virtue of early crediting also is open to question. Moreover, a real commitment to GHG control in the United States will necessarily require an acceptance of the need for some increases in energy prices; in our view, the alternative of a completely win-win scenario of GHG reductions with no costs (or even economic benefits) is not plausible. If this basic acceptance develops, then it is easier to politically rationalize a modestly sized early cap-and-trade program and to put in

place more effective regulatory institutions from the start.

Suggested Reading

General

Hahn, Robert, and Robert Stavins. 1999. What Has Kyoto Wrought? The Real Architecture of International Tradable Permit Markets. RFF Discussion Paper 99-30. March. Washington, DC: Resources for the Future.

Kopp, Raymond, Richard Morgenstern, William Pizer, and Michael Toman. 1999. A Proposal for Credible Early Action in U.S. Climate Policy. http://www.weathervane.rff.org.

Repetto, Robert. 1998. *Designing an Early Emission Reduction Credit System: Efficiency Aspects.* November. Unpublished report prepared for the U.S. Environmental Protection Agency.

U.S. GAO (Government Accounting Office). 1998. *Climate Change: Basic Issues in Considering a Credit for Early Action Program.* GAO/RCED-99-23. November. Washington, DC: U.S. General Accounting Office.

Technical

Parry, Ian, and Michael Toman. 2000. Early Emission Reduction Programs: An Application to CO_2 Policy. RFF Discussion Paper 00-26. Washington, DC: Resources for the Future.

Roughgarden, Tim, and Stephen Schneider. 1999. Climate Change Policy: Quantifying Uncertainties for Damages and Optimal Carbon Taxes. *Energy Policy* 27: 415–29.

Weyant, John P., and Jennifer N. Hill. 1999. Introduction and Overview. *The Energy Journal*, Special Issue (The Costs of the Kyoto Protocol: A Multi-Model Evaluation): vi–xiiv.

Appendix
Climate Policy and the Economics of Technical Advance
Drawing on Inventive Activity

Raymond J. Kopp

As the United States and other nations debate the pros and cons of binding targets for emissions of greenhouse gases (GHGs), interest in the possibility of reducing GHG emissions at low cost through the use of advanced technology is growing. In considering such a possibility, however, it is useful to understand the various ways that innovation can affect GHG emissions. It also is important to consider carefully the costs as well as the benefits of innovation aimed at improved GHG limitation. As with other uses of society's scarce resources, there is no free lunch. Technical innovation focused on reducing GHG emissions likely will lower the cost of abatement below the minimum cost possible with today's technologies. However, it may also have important effects on productivity growth elsewhere in the economy by crowding out other investments in technical improvement.

Background: Innovation and GHG Reductions

Technical advance that eventually reduces the cost of GHG policies proceeds through three stages: invention, innovation, and adoption. Invention is a creative process in which the end result is a new product, concept, process, or idea. Innovation is the adaptation (or evolution) of an invention into a us-

able, useful, and desired commercial product or process. Adoption is the process by which households and firms accept a new product or process into their economic activities.

Technical advance (or technical change, or progress) means doing things better than before as a result of enhanced knowledge. Changes in information processing brought about by enhanced knowledge of chip manufacture and CPU (central processing unit) design have resulted in computers that grow faster and more powerful each year and therefore represent a technical advance. Similarly, growing knowledge of solar panel manufacture gained through repeated experience with the process has led to technical advances.

Technical advances can lower the cost of GHG abatement policies by directly affecting the products produced by the economy and the processes used to produce them. Within each of these two categories, at least three effects of inventive activity can be identified. Product innovation brings

- more energy-efficient existing products,
- more energy-efficient new products, and
- advances in exotic technologies such as CO_2 scrubbing and advanced forms of carbon sequestration.

Process innovation, on the other hand, results in

- improvements in the energy efficiency of any manufacturing process,
- cost reductions in the manufacture of low-emission energy conversion technologies (for example, solar panels), and
- improvements in fossil energy conversion technologies that lower GHG emissions.

Technical advance is driven by two somewhat overlapping processes: learning by doing (LBD), which is accumulating knowledge through routine experience with an activity (such as solar panel manufacture) and then making conscious investments in research and product development (R&D) (exemplified by investments in new CPU design). Accumulation of technical knowledge leading to a fall in the cost of GHG abatement is facilitated by LBD; technical knowledge that results in an entirely new process, activity, or products is driven by R&D investments. R&D is naturally associated with inventive activity but LBD can, and often does, lead to invention in the form of new production processes and techniques.

The two paths leading to technical advance operate differently, and the differences between them influence the ways in which GHG abatement affects technical advance and the ways in which technical advance can affect the cost of GHG abatement. Technical knowledge gained through LBD resides with the people (and firms) that acquire the knowledge, manifesting itself in changes in processes or activities that reduce costs. On the other hand, knowledge gained through R&D investments finds its way beyond the confines of the firms where it originates and into the market as the new processes, activities, or products are sold. Thus, technical knowledge begins to take on the attributes of a public good.

The Costs of Technical Advances for GHG Mitigation

Although much attention has been paid to the direct costs of various policies designed to reduce GHG emissions, it is important to recognize the costs of making technical advances in response to a GHG-reduction policy. These costs reflect the value of economic resources that could have been allocated to other uses. This category of cost is not well accounted for by current climate policy models and usually is overlooked by policymakers. Ideally, the costs of making technical advances—along with the benefits—are subsumed within any analysis of the net cost of climate policy, not simply ignored.

The cost of an advance depends in part on its nature. For example, as we build solar panels, we gain knowledge about how to build them more cheaply. We gain the knowledge as a by-product of the act of producing the panels, without the investment of additional resources. (However, to be sure, firms are cognizant of the LBD process, actively seek to learn from their activities, and therefore consciously put forth effort and incur costs to gain that knowledge.) On the other hand, technical advance through investments in R&D can involve large direct resource costs. Accelerating the pace of R&D-driven technical advance means accelerating the pace at which resources are expended on R&D. If a GHG abatement policy is intended to bring forth a noncarbon "backstop" technology, for example, a full analysis of the policy's cost includes the additional resources devoted to the required R&D. A complete analysis links the stimulus of the policy to the behavior of economic agents (most importantly, firms), accounts for the additional resources devoted to R&D as a result of that stimulus, and estimates the results of these resource expenditures.

The redirection of R&D activity to incentives created by GHG abatement policy can be expected to reduce the rate of technical advance in other activities and sectors. This result, known as "crowding out," can occur within and across firms and is an important factor in accounting for the full cost of achieving technical innovation in response to GHG policy. Goulder and Schneider find that ignoring the crowding-out effect could lead to seriously understated GHG abatement costs (see Suggested Reading).

In practice, of course, forgone opportunities for advances elsewhere are hard to identify and evaluate, but this difficulty does not make the cost any less real. If the private markets for R&D investment are assumed to function reasonably well, then on average, the rates of return to R&D in one sector of the economy (or one area of a single firm) are likely to equal rates of return in other sectors. Under such an assumption, the market value of the lost technical advance in the sector where R&D has been crowded out will be roughly offset by the gain derived from the improved energy efficiency. (This result will tend to be true for modest reallocations of R&D activity but might not hold for very large reallocations.)

Once again in practice, however, the net overall gains to society from the reallocation of R&D are more difficult to assess. Even with patent and other intellectual property rights, a privately produced technical advance gives rise to benefits that extend beyond the originating firm to the public at large. Indeed, recognition of these potentially substantial external benefits is what undergirds the economic argument for government-sponsored R&D. (This argument applies to R&D generally, not only to R&D for GHG reduction.) These external benefits will change in ways that are hard to evaluate when R&D is redirected. For example, society can be made better off if the GHG abatement benefits from the redirection are large enough to offset the lost social benefits from diverted R&D resources invested in areas such as information technology and health care, or if other substantial ancillary benefits accrue. However, the reverse also could be true.

Given the current state of knowledge, we are relatively ignorant about the size of these different trade-offs. Policy modeling approaches that rely on exogenous characterizations of technical advance but do not properly account for the economic effort needed to improve energy efficiency are inherently inadequate and ultimately should be abandoned. One important example of this approach is the representation of trends in energy efficiency improvement in the economy through an autonomous energy efficiency index.

Technical Advance and Policies for GHG Abatement

If we can make existing products and processes more energy efficient and produce more efficient products to replace old products (at similar costs to purchasers), then we increase our capacity to lower our consumption of energy and GHG emissions to desired levels without substantial adverse effects on economic growth. However, although we can identify the possible paths by which technical advance can reduce costs, it is difficult to quantify the likely cost savings. Quantifying the savings is difficult because many factors influence the pace, scale, and direction of technical advance. Important factors include the timing of GHG abatement activities and the policies that induce inventive activity as well as the nature of these policies.

Timing of Abatement

The size of the cost reduction depends on when we abate. Because technical advance takes time, the longer we wait to accumulate technical knowledge and experience, the larger will be the store of cost-reducing technical advance to be drawn upon when abatement is undertaken. This assertion does not imply that the best approach is simply to defer abatement until technological possibilities have improved. The appropriate policy might be a mix of options that includes abatement and pursuit of technical advance (as well as adaptation and continued assessment of the risks of climate change).

Of course, we cannot gain technical knowledge from LBD unless we are engaged in abatement activities. Moreover, if an abatement policy is the sole incentive for R&D investments, then no technical advance will occur otherwise. Conceivably, however, policies might induce R&D expenditures without necessarily undertaking abatement activity. Such policies might include R&D-related tax incentives, or announcements of carbon taxes to be levied at a specific future date—as long as such announcements are credible in the private sector.

A longer time horizon for abatement conveys other cost-reducing possibilities, such as greater opportunities for avoiding premature turnover of highly productive capital; it also raises political questions about the credibility of commitments to reduce emissions. These issues are well beyond the scope of this paper. I simply note that we can attempt to accelerate the pace at which we accumulate knowledge through the use of strong policy-induced incentives.

Generally speaking, it will be costlier to abate and acquire knowledge sooner rather than later.

Inducing Inventive Activity

Some advocates who believe that technical advance will lower the cost of GHG abatement to acceptable levels are nonetheless skeptical about the extent to which technical advances are likely to come about from appropriately corrected market signals. These advocates argue that, to attain the necessary technical advances, firms will have to be induced more directly to put additional R&D resources into activities such as energy efficiency research, and the use of technologies will have to be mandated to stimulate cost-reducing LBD (for example, with renewable energy technologies).

However, if the inventive process responds to economic signals, as I believe it does, then creating appropriate incentives through GHG control policies will be the most cost-effective way to encourage invention. Such policies include carbon taxes and tradable permit systems that put a price on carbon emissions. Mandates and subsidies inherently direct inventive activity to specified technologies that may not be the most cost-effective opportunities for GHG reduction. For example, an automobile fuel economy standard does not include incentives for reducing distances traveled, which will increase after fuel economy is increased. And subsidies (or mandates) for the use of renewable energy resources by utilities may bypass lower-cost GHG opportunities through upgrades of existing generation capacity. Technology mandates also limit the choice of products available in terms of cost, convenience, reliability, familiarity, and other attributes. Limiting choice carries with it subtle but real costs in terms of consumer well-being.

Beyond policies to expand inventive activity are policies to increase demand for new products and technologies with lower GHG emissions. Consumer demonstration programs and low-cost financing for new energy-using capital are two examples. The adoption stage of technical advance, as opposed to invention, is beyond the scope of this chapter. However, similar controversies about the capacity to change behavior with economic signals arise in this context.

Concluding Remarks

If GHG policy costs are to be reduced to politically acceptable levels, some inducement to undertake more inventive activity seems to be required, and this requirement no doubt will lead to all sorts of novel approaches. In the face of such novelty, it is important to proceed with care.

Economic growth—with appropriate correction for environmental spillovers where needed—is the major source of improved material and social conditions worldwide, and technical advance is the force that ultimately drives increases in per capita living standards. Therefore, any tinkering with the economic processes that bring forth technical advance risks disturbing this all-important source of growth. The importance of technical advance to economic growth suggests caution when governments consider GHG policies with the goal of redirecting the forces of technical advance. The need for caution is further underscored by our primitive understanding of the economic processes underlying technical advance and the law of unintended consequences that so often exerts itself when policies are formulated in haste and on the basis of less-than-adequate understanding.

Suggested Reading

Council of Economic Advisers. 1998. *The Kyoto Protocol and the President's Policies to Address Climate Change: Administration Economic Analysis.* July. Washington, DC: Executive Office of the President.

Goulder, Lawrence H., and Koshi Mathai. 2000. Optimal CO_2 Abatement in the Presence of Induced Technological Change. *Journal of Environmental Economics and Management* 39(1): 1–38.

Goulder, Lawrence H., and Stephen H. Schneider. 1999. Induced Technological Change and the Attractiveness of CO_2 Abatement Policies. *Resource and Energy Economics* 21(3–4): 211–53.

Grubb, Michael. 1997. Technologies, Energy Systems and the Timing of CO_2 Emissions Abatement. *Energy Policy* 25(2): 159–77.

Ha-Duong, M., Michael J. Grubb, and Jean-Charles Hourcade. 1997. Influence of Socioeconomic Inertia and Uncertainty on Optimal CO_2 Emission Abatement. *Nature* 390: 270–73.

Interlaboratory Working Group (IWG). 1997. *Scenarios of U.S. Carbon Reductions: Potential Impacts of Energy Technologies by 2010 and Beyond*. Report LBNL-40533 and ORNL-444. September. Berkeley, CA, and Oak Ridge, TN: Lawrence Berkeley National Laboratory and Oak Ridge National Laboratory.

Jaffe, Adam B., and Robert N. Stavins. 1994. The Energy Paradox and the Diffusion of Conservation Technology. *Resource and Energy Economics* 16: 91–122.

Jaffe, Adam B., Richard G. Newell, and Robert N. Stavins. 1999. Energy-Efficient Technologies and Climate Change Policies: Issues and Evidence. RFF Climate Issue Brief 19. December. Washington, DC: Resources for the Future.

Romer, Paul M. 1990. Endogenous Technical Change. *Journal of Political Economy* 98(5): S71–102.

Toman, Michael A., Richard D. Morgenstern, and John Anderson. 1999. The Economics of "When" Flexibility in the Design of Greenhouse Gas Abatement Policies. *Annual Review of Energy and the Environment* 24: 431–60.

Part 4

International Considerations

20 Policy Design for International Greenhouse Gas Control

Jonathan Baert Wiener

Making international climate policy is difficult. The issues are complex, the impacts of policies are large and unevenly distributed, and the costs of erring in any direction are high. International climate negotiators must address, among many tough topics, the costs and benefits of reducing greenhouse gas (GHG) emissions; the distribution of these costs and benefits across countries; the incentives for countries to participate in reducing global emissions; and the innovation and diffusion of low-emissions technology, especially for use in developing countries.

How do we get from here to there? Whatever overall goal we set for climate policy, our success in achieving that goal effectively and efficiently will depend on the policy instrument we choose. No single instrument is universally best for all environmental problems; all policy instruments have strengths and weaknesses, and the choice among them is a pragmatic and contextual matter. For global climate policy, five major options for policy instruments have been suggested:

- technology standards, such energy efficiency standards for vehicles and appliances or fuel types for electric power generators;
- taxes on GHG emissions;
- subsidies and other rewards for GHG emissions abatement;

- quantitative limits on each nation's GHG emissions; and
- quantitative national limits, with a market in tradable GHG emissions allowances.

A growing number of countries (including the United States, through both the Bush and Clinton administrations) and experts have championed the last of these options—the creation of an international market in tradable GHG allowances. The U.S. government enacted this approach in 1990 to curb domestic sulfur dioxide (SO_2) emissions that yield acid rain and suggested the concept in the same year as a potential remedy for global GHG emissions. But at that time, the White House opposed quantitative "targets and timetables" on GHG emissions. Without such aggregate quantitative constraints or "caps," a formal system of tradable allowances (a cap-and-trade policy) could not function. The 1992 U.N. Framework Convention on Climate Change (UNFCCC) adopted at the Earth Summit in Rio de Janeiro, Brazil, imposed no quantitative target on aggregate emissions and asked countries only to undertake "policies and measures" to reduce emissions—with the proviso that such abatement actions may be "implemented jointly" by countries. This idea of joint implementation represented a window for "informal" emissions trading through project-by-project collaborations but not a formal market in tradable allowances.

Negotiations soon began on a protocol to implement the UNFCCC. The European Union had long advocated fixed national quantitative emissions limits and expressed doubts about international emissions trading. In 1993, President Bill Clinton announced that the United States would also endorse quantitative targets and timetables on GHG emissions; and the Clinton–Gore climate change action plan advocated international allowance trading as a way to meet that goal. In 1997, more than 2,000 economists—including several Nobel laureates—signed a statement endorsing formal allowance trading to control global GHG emissions. After intense negotiations, the Kyoto Protocol to the UNFCCC was signed in December 1997 (however, as of 2000, it had not yet been ratified by many countries, including the United States). The Kyoto Protocol included differentiated quantitative targets on GHG emissions of industrialized countries, with aggregate emissions by these countries to be cut to about 5% below 1990 levels by 2008–12. No quantitative emissions limit was applied to developing countries. And the Kyoto Protocol authorized three different versions of emissions trading to help accomplish this goal: joint implementation of abatement projects among industrialized countries (Article 6); a system for developing countries to sell emissions reduction credits to buyers in industrialized countries, called the Clean Development Mechanism (CDM) (Article 12); and a formal system of tradable allowances among industrialized countries (Article 17).

If the Kyoto Protocol enters into force, the stage will be set for international markets in GHG abatement. Both economic theory and actual experience suggest that creating such GHG emissions markets offers great promise for lowering the cost of GHG control, enhancing the spread of lower-emissions technology, and broadening participation in the international agreement to limit GHG emissions.

Designing such a market is not simple. Concerns have frequently been expressed that international emissions trading could be difficult to initiate, monitor, and manage. There are concerns about negotiating allowance allocations, ensuring participation, de-

terring free riding, reducing cross-national "leakage," measuring emissions, and enforcing compliance. These concerns are important, but they pertain to any GHG emissions control policy, using any policy instrument. They are generic concerns about global climate policy. Indeed, a market-based approach could actually ease these challenges. A second set of concerns, less frequently discussed, relates uniquely to a market-based instrument: concerns about transaction costs, the behavior of national governments, and market power. These specific concerns will need to be confronted in the design details of a GHG trading system.

The Case for International Emissions Trading

An international climate agreement could, in theory, use any of the five policy instruments outlined above. The advantages of emissions trading can be seen most clearly by comparing national caps with and without emissions trading; the actual treaty negotiations have focused on these two options. Without trading, a treaty would simply require every participating country, acting on its own, to limit its emissions to a certain level (its target or cap) by a certain date. Different countries could have different caps that imply different degrees of stringency; for example, the caps on industrialized countries could require absolute reductions in projected emissions, whereas developing countries in the future could be afforded substantial "headroom"—flexibility to increase their emissions as they develop economically. By contrast, under a market-based emissions trading approach, countries would achieve the same aggregate cap but with the flexibility to trade (that is, reallocate by mutual agreement) their emissions abatement efforts across countries. Again, different countries could initially receive allowances equivalent to national caps of different degrees of stringency.

Two basic kinds of international markets for GHG emissions abatement can be envisioned. One is a formal emissions trading market—a system of tradable allowances often called cap and trade. In this market, the international agreement sets a cap on aggregate emissions for some period, and the agreement also allocates GHG emissions allowances (often also called emissions permits) among the participating countries for that period. The national governments then allocate these allowances to businesses within their countries. Emitters must hold allowances to cover every unit they emit; they can control emissions, buy additional allowances if their abatement costs are high, and sell allowances if their abatement costs are low. (In practice, carbon dioxide emissions could be regulated upstream by associating allowances with the carbon content of fossil fuels.) Organized exchanges are established to facilitate trading. To ensure compliance with the allowance cap, at the end of each period, each country's report of its actual emissions (subject to monitoring and verification) is compared with the allowances held for that period by its emitters; if national emissions exceed total allowances held, then the country is out of compliance with the treaty and subject to whatever penalties for noncompliance the international agreement stipulates. (The penalties under the Kyoto Protocol were left for future negotiation.)

The allowances in this system could be fully interchangeable, or fungible—each one representing a unit of GHG emission, without reference to the country from which the allowance originated—so that buyers could rely on the value of the allowance without investigating the seller's behavior. In that case, seller compliance would be ensured in the same ways as under a national cap without trading. Or, allowances could be denominated by the seller country, with the provision that a selling country's violation of its national abatement commitments would devalue the seller's allowances, thereby giving buyers an incentive to purchase from the sellers most likely to comply with the treaty (and sellers an incentive to ensure compliance if they hope to sell additional allowances).

In an informal emissions trading market, on the other hand, the international agreement does not allocate formal allowances. Instead, each country may meet its abatement commitment through contracts for project-by-project abatement services (GHG reductions or sequestration) obtained both within and outside its territory. Thus, emitters seeking to invest in abatement services may do so at home, and they may also purchase credits for emissions reductions generated in other countries, including countries not subject to an overall emissions cap. These are the project-based trading systems envisioned by Article 6 (joint implementation among industrialized countries with emissions caps) and Article 12 (CDM credits from developing countries without emissions caps) of the Kyoto Protocol. The value of such credits will depend on the actual abatement undertaken at the specific site; the investor or a certifying entity must closely monitor the ongoing performance of the project. Abatement is measured against a baseline forecast of emissions from that project in the absence of the joint implementation or CDM investment. Investor countries' compliance with emissions caps is determined by comparing the country's cap with the country's actual national emissions, minus the abatement achieved at the specific overseas project sites credited to the country's investors. This approach is similar to the system of pollution offsets and emissions reductions credits adopted by the United States prior to the acid rain program to control other air pollutants. It is also essentially the system of joint implementation launched in the UNFCCC signed in 1992 at the Earth Summit. That pilot phase of joint implementation, however, did not allow official quantified credits to be earned by joint implementation activities, and investment in joint implementation was predictably muted. Under Articles 6 and 12 of the Kyoto Protocol, official credits could be earned by joint implementation and CDM investments, presumably encouraging a more vigorous market in such project-based trading than had been the case under the UNFCCC pilot phase.

In both of these approaches to emissions trading, all transactions would be voluntary. The motivation for trading is the desire to find more advantageous ways to comply with GHG limits. The operative principle is mutual benefit: Sellers and buyers will enter into transactions only when the terms of the deal, including financial and nonfinancial rewards, make them better off.

One of the major advantages of a market-based approach to global climate policy is its cost-effectiveness. Because the cost of GHG emissions abatement varies significantly from place to place, and because the global environmental benefit of GHG reduction is independent of where the emissions are reduced, allowing any agreed level of total GHG abatement to be undertaken where it is least costly will minimize the overall cost of achieving the policy goal. Emissions trading allows that least-cost strategy to be identified and pursued; fixed national caps (or worse, technology standards) do not.

Numerous studies indicate that allowing global flexibility in the location of GHG emissions abatement through emissions trading would cut the estimated total cost of emissions controls considerably. For example, the models used in a recent study of Kyoto Protocol compliance costs suggest that the cost savings from emissions trading within the Annex I countries (the industrialized countries that agreed to cap their total emissions, including Russia and Ukraine, with their likely surplus of emission allowances) are on the order of 30–50%, compared with a similar emissions reduction treaty with efficient domestic policies but without international trading (see Weyant and Hill in Suggested Reading). Even greater cost savings could be reaped from global trading—on the order of 65–85%. The cost savings also would be higher when compared with technology standards, because both tradable and nontradable national quotas allow important flexibility in the choice of abatement method. And these figures may understate the cost savings, because many of the models already assume some degree of cost-minimizing coordination of abatement among members of the European Union.

To put these figures in context, consider the Energy Modeling Forum results reported by Weyant and Hill on the costs of meeting the Kyoto Protocol targets (measured in lost gross domestic product [GDP]) without flexible international policy design. Figures are reported for the United States, Japan, the European Union, Canada, Australia, and New Zealand. The annual costs summed over all these countries in 2010 range from just under $100 billion to almost $500 billion. The cost of maintaining or further reducing emissions beyond the initial commitment period would be larger still. Thus, the 50% cost savings offered by a flexible market-based policy would mean considerable savings in absolute terms.

Of course, real-world cost savings may be different from these model estimates. The degree of cost savings associated with flexibility will depend on the stringency of the emissions target, the cost of meeting that target without flexibility, the specific countries that participate in the treaty, and other factors. Real-world cost savings might be greater if emissions trading induces innovation that further lowers abatement costs, if trading is allowed over time as well as across countries, or if controlling GHG emissions in industrialized countries turns out to be even more expensive (relative to developing countries) than predicted. But real-world cost savings from flexibility might be lower if arranging allowance transactions proves costly, or if it turns out that industrialized countries can control GHG emissions at home at costs much lower (relative to developing countries) than predicted.

Allowance trading markets have demonstrated substantial cost savings in practice when applied to several national pollution problems. The United States has used market-based approaches to phase out lead in gasoline, to cut SO_2 emissions as a means to reduce acid rain, and to control urban air pollutants in Los Angeles. The cost savings in the lead and SO_2 cases were substantial—50% or more compared with a control policy in which no trades were allowed. (GHGs seem to be an even better prospect for trading than lead and SO_2, because the variation in global abatement costs is even wider and because there are no local "hot spot" problems for most major GHGs.) The SO_2 trading policy also stimulated energy efficiency investments and the use of new abatement technologies. The SO_2 experience suggests that the more cost-effective market-based policy enabled the U.S. Congress to "buy" more pollution control than it would have if control were more expensive. Similarly, reducing the cost of GHG abatement could well lead countries to undertake more abatement than they otherwise would.

GHG allowance trading could also mobilize substantial resource flows to developing countries, assuming that abatement costs are lower in developing countries and that the treaty allocates GHG emissions control obligations to constrain industrialized countries while giving developing countries some headroom to grow. These resource flows would help poorer countries shift to a development path that is more prosperous but with lower emissions, invest in local health and environmental needs, and pursue other social priorities. Resource flows to poorer countries under a climate treaty employing emissions trading could grow to exceed all official international development assistance. Some studies suggest that, compared with no treaty at all, developing countries would be net losers under a no-trading policy in which industrialized countries cut their own emissions (and, in the process, cut their product imports from developing countries) and net winners under a global allowance trading system in which the cost of abatement is reduced and the developing countries can profit from allowance sales. The prospect of such gains from allowance sales could, in turn, attract developing countries to participate in the GHG abatement regime. (One caveat to these conclusions is that large resource inflows to developing countries could create adverse terms of trade that would crimp the growth of the developing countries' regular exports, as distinct from their exports of allowances.)

Participation is a fundamental issue in an international climate agreement. Under international law, no country is bound by a treaty unless it consents to participate. (This is quite different from national law, under which majority rule or perhaps fiat [a government decree] can impose constraints on emitters without the emitters' consent.) Thus, under international law, countries must find treaty participation to be in their interest. Without the participation of developing countries in particular, industrialized countries' GHG emission controls could be futile; the emissions of developing economies are increasing rapidly and soon will account for more than half of global emissions. Controls on emissions imposed only in industrialized countries could wind up missing the larger share of global emissions and, worse, could induce emissions-related activities to shift to unconstrained developing countries (as world fuel prices fall and emitting industries relocate), thereby exacerbating the growth in developing country emissions. Such leakage of emissions from constrained to unconstrained countries could offset much of the climate protection sought by a treaty. Moreover, legislators in industrialized countries will be loath to vote to ratify less-than-global emissions constraints that drive emissions-related jobs out of their electoral districts; this situation appeared to be part of the predicament with ratification of the Kyoto Protocol in the United States after it was negotiated. Thus, some way of getting developing countries to participate in the global GHG treaty on terms they find attractive seems crucial to ensure environmental effectiveness, reduce global abatement costs, direct needed resource flows to poorer countries, and encourage industrialized countries to ratify the Kyoto Protocol. Major developing countries such as China may see global climate change as a low priority (or even a benefit to their farmers) and hence may resist joining an emissions abatement treaty. Emissions trading offers a built-in feature—the resource flows to developing countries in return for their abatement efforts and associated allowance sales—that could accomplish broad participation. It also implies that assignments of "headroom" allowances (such as those to Russia in the Kyoto Protocol) are not a mistake but rather constitute an essential mechanism for securing participation by otherwise reluctant countries.

GHG taxes could, in principle, generate gains in cost-effectiveness equal to those of an allowance trading market. Theoretically, a tax imposed to achieve a given level of emissions would exactly equal the market price for the allowances issued to achieve that same quantity of emissions. A tax would offer more certainty about overall costs, because the tax would be fixed in advance, whereas the price of emissions allowances could vary. But a tax would offer less certainty about environmental results, because the tax by its nature does not constrain the quantity of emissions (and tradable allowances do). Depending on the relative importance of cost escalations versus emissions escalations, this factor could be important. Moreover, a tax would raise revenue that could be used to offset other distortionary taxes; tradable allowances could raise revenue only if they were initially sold or auctioned, not if they were issued free of charge. Meanwhile, a tax would not involve the transaction costs of structuring and operating allowance trading but would involve the administrative costs of tax collection.

More problematic is that an internationally agreed emissions tax (whether administered globally or nationally) could be circumvented by national subsidies and other tax code changes targeted to buffer high-emitting industries. This problem of "fiscal cushioning" is very difficult to monitor from outside, is worsened by national tax administration, and undermines the ability of the tax to constrain actual emissions.

Furthermore, a GHG tax would not create an automatic mechanism for resource transfers to developing countries, which are crucial to getting developing countries engaged. Politically difficult side payments to developing countries would have to accompany the tax to secure these countries' participation in the tax regime. Such payments could claim much of the tax revenue that might have been put to offsetting other distortionary taxes. Moreover, if poorly designed, these side payments could undermine the incentive effect of the tax by reducing the net price paid for emitting. Such will be the case if the revenue redistributions are in proportion to economic activity or emissions abatement, rather than being decoupled from emissions. And to attract countries' participation, the side payments would have to be in proportion to the actual economic burden imposed by the emissions tax (net of environmental benefits to that country), or countries will not agree to the tax. Thus it seems hard to escape the prospect that participation-attracting side payments would undermine the incentive effect of an emissions tax. Side payments delivered through developing countries' sales of tradable allowances, in contrast, would not undermine the incentive effect of the emissions trading system, because the quantity constraint on aggregate emissions would still be binding.

In developing countries, subsidies for abatement are often proposed in domestic programs or in the guise of official development assistance targeted at emissions-abating activities. Developing countries tend to resist such conditionality, unless it is truly additional aid money. Industrialized countries sometimes resist making larger taxpayer-financed budget outlays for foreign aid. From an environmental point of view, a serious concern with subsidies for abatement is their potential to yield perverse consequences. Although the subsidy for abatement (such as a tax on emissions, or tradable allowances) can reduce emissions at the margin by each subsidy-earning firm, the subsidy also could reduce the cost of doing business in the emitting industry if it is poorly designed—particularly if it is a subsidy per unit of output or emissions versus a lump-sum transfer. A subsidy for abatement can attract entry or investment into the emitting industry and perversely increase net emissions. Side payments delivered through developing countries' sales of tradable allowances, by contrast, would not yield perverse impacts on emissions, because the quantity constraint on aggregate emissions would still be binding.

Generic Concerns about Emissions Policy

Emissions trading therefore has several salient advantages over the other options for international GHG policy, such as technology standards, taxes, subsidies, and fixed national caps. But despite the apparent advantages of an emissions trading system, many countries and commentators continue to express concerns about using this approach to control GHG emissions—concerns about negotiating allowance allocations, ensuring participation, deterring free riding, enforcing compliance, reducing cross-national leakage, and measuring emissions. These concerns, however, apply generically to any emissions control regime, and allowing trading could actually *ease* these concerns.

Critics worry that it will be difficult to negotiate initial allowance allocations among countries. But the problem of allocating control and cost responsibilities is unavoidable in any climate agreement, with or without trading, and trading could ease this problem in two ways. First, trading makes the cost allocation transparent. Technology standards and national caps implicitly impose cost burdens that vary across countries. The use of formal emissions trading makes this allocation explicit rather than disguised, and this transparency can facilitate a bargain. Second, allowance trading enables flexibility after the initial allocation. Countries facing potentially high cost burdens will know that they can purchase abatement from countries with lower costs, and countries with low-cost abatement opportunities will know that they can earn substantial resource flows by selling allowances to those with higher costs. This flexibility would significantly relax the pressure on negotiators to devise ideal, once-and-for all allocations in the treaty itself. As Nobel Prize–winning economist Ronald Coase pointed out, the lower the obstacles to reallocation of entitlements, the less the initial assignment matters. Thus, without the opportunity for trading, initial GHG policy targets would be more difficult to negotiate.

Some argue that a cap-and-trade system is "a political nonstarter" because developing countries will refuse to accept caps and industrialized countries will refuse to make large resource transfers. But any control regime with binding targets will have to face this problem because under international law, countries must be attracted to participate. A cap-and-trade system has important advantages. The opportunity to earn significant resource transfers via allowance sales can make accepting phased-in caps (with headroom for future growth) more attractive to developing countries. And wealthy countries would undertake these transfers because they yield significant cost savings compared with a treaty without such trading. Moreover, the myriad private transactions involved in formal emissions trading would not raise the political specter of increasing official foreign aid financed by taxpayers. The other possibility, a set of domestic GHG taxes imposed separately by individual countries, probably would founder on the lack of transparency of fiscal systems and the apparent unwillingness of developing countries to substantially increase their own energy tax burdens.

Fundamental problems for any international treaty are engaging participation and deterring free riding (that is, benefiting from the group's efforts without participating). Treaty restrictions can be imposed only on consenting countries, making any kind of international collective action difficult. How-

ever, formal emissions trading has signal advantages over other policy options in securing participation while limiting the perverse consequences of side payments. Reluctant countries can be assigned headroom allowances (an in-kind side payment) that they can then sell, and the aggregate cap maintains the environmental effectiveness of the regime. If free riding is not deterred, the entire collective regime may unravel. Using allowance trading in a GHG treaty can reduce free riding: It dramatically lowers the cost of participation to industrialized countries and raises the profits from participation for developing countries. Because noncompliance is just a form of free riding, these incentives for participation in an emissions trading regime can also enhance compliance.

Emissions leakage—that is, emissions abatement achieved in one location being offset by increased emissions in unregulated locations—will afflict any subglobal treaty, whether it uses trading or not. Leakage can arise in the short term as emissions abaters reduce energy demand or timber supply, influencing world prices for these commodities and increasing the quantity emitted elsewhere. It also can arise in the long term as industries relocate to avoid controls. Both informal and formal allowance trading would reduce such leakage by reducing the abatement cost (thereby reducing the incentive for industry to relocate) and by expanding participation and inhibiting free riding (thereby enlarging the group of countries agreeing to constrain emissions).

An informal market in joint implementation or CDM credits from abatement projects in uncapped host countries may raise a special concern about local (within-country) leakage from these projects to other emissions sources in the same host country. However, the question is whether such local leakage within the project host country would be greater than the leakage from capped to uncapped countries that would have occurred if the same amount of abatement had been undertaken only within the project investor country. Meanwhile, local leakage would not be a problem in a formal allowance trading market with national caps, because all emissions of all participating countries would be counted in the national inventories used to assess compliance.

Another generic concern regards the ability to measure the magnitude of abatement efforts. Any treaty, with or without trading, requires forecasts of baseline emissions and subsequent monitoring to evaluate the likely cost of limitation options, the effectiveness of abatement efforts, and the extent of emissions leakage. A formal cap-and-trade system would not complicate these tasks; it would measure results using the same national inventories as under an agreement without trading. The CDM may add some uncertainty about what the emissions would otherwise have been in the uncapped project host country. But prohibiting CDM credits because of such uncertainty would forfeit both the opportunity to engage countries without national emissions caps in early GHG control efforts, and the opportunity to obtain low-cost abatement services in those countries. A better approach might be to allow both the cap-and-trade system and the CDM but to exercise caution by adjusting the credit for all abatement efforts (not only the CDM) in proportion to the projects' measurement credibility, and to invite investors to augment the credit calculation by showing more reliable emissions accounting—thus providing incentives for investors to improve measurement capabilities.

Specific Concerns about Emissions Trading

Other concerns apply with special force to international market-based emissions trading regimes. Yet these specific concerns often have tended to receive less attention in the debate over different international GHG control strategies.

Transaction costs include the costs of searching for trading partners, negotiating deals, securing regulatory approval, monitoring and enforcing deals, and insuring against the risk of failure. High transaction costs in the emissions abatement market would impede trades and raise total costs. Evidence from previous U.S. environmental markets such as the lead phasedown, the Los Angeles smog control program, and an experiment with water pollution trading on the Fox River in Wisconsin suggests that transaction costs can determine the success or failure of the trading system.

The transaction costs of joint implementation and the CDM, as currently structured, appear to be very high. Partners are hard to identify, each negotiation is novel, each project must be approved by the host and investor governments (and potentially by the CDM governance system), and each investor must monitor its own projects. Moreover, if joint implementation and the CDM require investors to support entire projects, each investor bears a large risk of project failure. The transaction costs of joint implementation and the CDM could be reduced through brokers (many of which are emerging), information exchanges, streamlined approval processes, accredited monitoring agents (including environmental nongovernmental organizations), mutual funds and other means of risk diversification, and official credit.

The transaction costs of a formal market for allowance trading would be much lower, especially if fungible allowances are traded on organized exchanges. Indeed, reducing transaction costs would be a central goal of such a formal system. Lower transaction costs would improve cost-effectiveness and also would promise easier opportunities to reallocate control burdens, thereby helping to facilitate the initial allocation negotiations. (A GHG tax would avoid interfirm transaction costs but would entail significant administrative costs in collecting the tax.)

A second concern is the role of national governments in the market. National governments might try to influence the market to their advantage, obstruct allowance trades, or otherwise depart from the conditions of well-functioning abatement markets assumed in the estimates of cost savings. Governments might pursue such strategies to favor domestic interests against foreign rivals, to redistribute wealth within a country, or for other purposes that in practice conflict with the operation of the international climate treaty regime. In the U.S. SO_2 trading system, for example, several states have attempted to intervene in the national market (for example, by trying to prevent electric power companies from switching to lower-sulfur out-of-state fuels, and by trying to prevent sales of allowances to upwind sources), but these efforts have so far been blunted by the limits on states' power to restrict interstate trade under the U.S. Constitution. National interpositions in a global GHG market could be limited by international trade law, but this depends on untested legal questions about whether and how General Agreement on Tariffs and Trade (GATT) and World Trade Organization (WTO) law applies to trade in GHG allowances.

Moreover, for trading to be fully cost-effective, national governments must let the trading be conducted by the entities actually responsible for GHG abatement and sequestration. Assigning allowances and credits to these entities will mobilize decentralized competition, creativity, and flexibility. It seems likely in the United States but might not occur so well (or at all) in countries where the state is a more active supervisor or owner of industry. (A GHG tax would also be vulnerable to national government manipulation, chiefly via subsidies and changes in the domestic tax code to buffer the impact of the GHG tax on domestic industries. Such fiscal cushioning would be exceedingly difficult for outsiders to monitor and penalize.)

A third specific concern is international market power. Concentrated power over allowance or credit prices could arise on the sellers' side (for example, a GHG "OPEC") or on the buyers' side (for example, a central sole purchasing agent for industrialized countries). Although there should be a plethora of competitors in an international GHG allowance market, market power could be enhanced by large state-run energy or forestry companies and by countries' efforts to prevent additional countries from entering the market. Many trading models show Russia as the main seller of allowances in an Annex I (industrialized country) trading system, and China as another main seller in a global system. And unlike domestic antitrust (competition) law, international law has no basic framework to combat market power; even if it did, enforcing such rules against nations (or cartels of nations) could be quite difficult. A successful international emissions trading system could require the evolution of new international antitrust remedies, either in general or specific to the climate treaty. "Thickening" the market is probably the best tonic for fears of market power, again underscoring the importance of broadening participation in the treaty noted above.

Furthermore, centralized purchasing or sales agents advocated to reduce search and approval costs (for example, the CDM governing board, or proposals to route project investments through a few multilateral institutions such as the World Bank) could also invite market power. A better way to reduce transaction costs is through organized exchanges and fungible allowances.

The International Political Economy of Policy Design

Given the advantages of international emissions trading—lower cost, valuable resource flows, greater participation—what is one to make of its opposition? Several speculative hypotheses are worth exploring.

Opponents may misunderstand or genuinely doubt the advantages of international emissions trading. For example, developing countries' fears of "carbon colonialism" may reflect a view that the market power and trading savvy of wealthy investors would depress allowance and credit prices, leading poorer countries to sell out their future at a loss. This concern may be legitimate, and it warrants efforts to combat market power and build the capacity of developing countries to bargain effectively in an allowance trading market.

Opponents may also have nonclimate agendas. Some may favor a high-cost regime because their objective is moral condemnation or broader social change in industrialized countries, not cost-effective climate protection. Some may worry that fairness requires industrialized countries to take the lead—but allowance trading does require wealthy countries to foot the bill for global abatement efforts; and as noted above, global allowance trading makes developing countries net winners instead of net losers, which seems far more fair.

Strategic behavior may also be at work. First, some government decisionmakers, in both developing and industrialized countries, may prefer official government aid to private market transactions because they believe they can control the former more effectively. Indeed some developing country government officials may oppose trading because they see the market sector, which would gain from allowance trading, as a domestic political rival—a replay of similar struggles in the transition from feudal and state-run societies to market capitalism. Second, some industrialized countries may prefer a less flexible control regime because it limits access by their trade rivals to lower-cost abatement opportunities. This may explain the European Union's preference for an E.U. "bubble" (trading region) under the climate treaty, but its opposition to unrestricted global allowance trading and its embrace of cumbersome "supplementarity" restrictions on trading volumes.

Opposition might also be a move to gain leverage over the goal (target or cap). Advocates of aggressive climate protection may withhold support for trading until it is paired with a more stringent cap—risking a costly treaty, or no agreement at all. Meanwhile, skeptics of aggressive climate policy may fear that cost-effective policy tools are an all-too-enticing "fast train to the wrong station," inducing premature adoption of an overly stringent cap. Of course, the goal of climate policy should be chosen with great care. Yet the skeptics' gambit of urging a higher-cost "slow train" (in the hopes that it will derail any GHG limitations agreement) may only invite "Murder on the Orient Express"—an unholy alliance behind a treaty that both costs more and is less environmentally effective—a "lose-lose" luxury train to the wrong station.

Toward Successful International Emissions Trading

Efficient policy design, such as emissions trading, has enormous advantages to offer global climate policy—in climate protection, cost savings, innovation, and resource flows to developing countries. GHGs offer an even more attractive case for application of emissions trading than have many local and regional pollution problems already well handled with emissions trading.

But designing efficient global climate policy also presents several novel challenges. One of these is the challenge of translating the accumulated national experience with flexible market-based policy designs (such as allowance trading) to the international con-

text. The U.S. acid rain trading system cannot just be transplanted to global climate policy. The legal and institutional terrain is different at the international level, and international environmental markets need to be designed with that terrain in mind. In particular, the requirement of consent to international treaty law puts extra emphasis on securing participation, which in turn makes emissions trading even more attractive at the international level than it is at the national level, but also requires the use of headroom allowance allocations to engage otherwise reluctant emitting countries. To be environmentally effective and economically efficient, the Kyoto Protocol needs to be expanded to engage developing countries in a global emissions trading regime.

Another challenge is bridging from normatively desirable policy design to actually adopted policy design. The economics of climate policy may not match the politics of climate policy. Even if smart policy designers address the generic and specific concerns surrounding international emissions trading and thereby deliver an efficient policy design to global climate negotiators, such a policy design will still confront the political marketplace in which efficiency may be viewed with indifference or even antipathy. Empirical research is warranted to explore and reveal the positive politics of international climate policy design. Without such inquiry, designing the most efficient global climate policy may be for naught.

Suggested Reading

Baumol, William J., and Wallace E. Oates. 1988. *The Theory of Environmental Policy* (2nd ed.). New York: Cambridge University Press.

Bohm, Peter. 1997. Joint Implementation as Emission Quota Trade: An Experiment among Four Nordic Countries. Report Nord 1997:4. Copenhagen, Denmark: Nordic Council of Ministers.

Cooper, Richard. 1998. Toward a Real Global Warming Treaty. *Foreign Affairs* 77(2): 66–79.

Dudek, Daniel J., and Jonathan Baert Wiener. 1996. *Joint Implementation, Transaction Costs and Climate Change.* OCDE/GD (96)173. Paris, France: Organisation for Economic Co-operation and Development.

Hahn, Robert W., and Robert N. Stavins. 1999. What Has Kyoto Wrought? The Real Architecture of International

Tradable Permit Markets. RFF Discussion Paper 99-30. March. Washington, DC: Resources for the Future.

IPCC (Intergovernmental Panel on Climate Change). 1996. *Climate Change 1995: Economic and Social Dimensions of Climate Change*. Contribution of Working Group III to the Second Assessment Report of the Intergovernmental Panel on Climate Change. New York: Cambridge University Press.

Jacoby, Henry, Ronald Prinn, and Richard Schmalensee. 1998. Kyoto's Unfinished Business. *Foreign Affairs* 774(4): 54–66.

Jepma, Cristinus J., and others. 1996. A Generic Assessment of Response Options. In *Climate Change 1995: Economic and Social Dimensions of Climate Change*, edited by James P. Bruce, Horsang Lee, and Erik F. Haites. Contribution of Working Group III to the Second Assessment Report of the Intergovernmental Panel on Climate Change. New York: University of Cambridge Press, 225–62.

Manne, Alan, and Richard Richels. 1996. The Berlin Mandate: The Costs of Meeting Post-2000 Targets and Timetables. *Energy Policy* 24: 205–10.

Stavins, Robert N. 1997. Policy Instruments for Climate Change: How Can National Governments Address a Global Problem? *University of Chicago Legal Forum* 1997: 293–329.

Stewart, Richard B., Jonathan B. Wiener, and Phillippe Sands. 1996. *Legal Issues Presented by a Pilot International Greenhouse Gas Trading System*. Geneva, Switzerland: U.N. Conference on Trade and Development.

Tietenberg, Thomas. 2000. *Environmental and Natural Resource Economics* (5th ed.). Reading, MA: Addison-Wesley, Chapters 15–19.

Tietenberg, Thomas, Michael Grubb, Axel Michaelowa, Byron Swift, and Zhongxiang Zhang. 1999. *International Rules for Greenhouse Gas Emissions Trading*. UNCTAD/GDS/GFSB/Misc.6. New York: U.N. Conference on Trade and Development.

U.N. Conference on Trade and Development. 1992. *Combating Global Warming: Study on a Global System of Tradeable Carbon Emission Entitlements*. Geneva, Switzerland: U.N. Conference on Trade and Development.

Weyant, John P., and Jennifer N. Hill. 1999. Introduction and Overview. *The Energy Journal*, Special Issue (The Costs of the Kyoto Protocol: A Multi-Model Evaluation): vi–xiiv.

Wiener, Jonathan B. 1999. Global Environmental Regulation: Instrument Choice in Legal Context. *Yale Law Journal* 108: 677–800.

Wiener, Jonathan Baert. 1999. On the Political Economy of Global Environmental Regulation. *Georgetown Law Journal* 87: 749–94.

Wiener, Jonathan Baert. 1995. Protecting the Global Environment. In *Risk vs. Risk: Tradeoffs in Protecting Health and the Environment*, edited by John D. Graham and Jonathan B. Wiener. Cambridge, MA: Harvard University Press, Chapter 10.

21 Establishing and Operating the Clean Development Mechanism

Michael A. Toman

The negotiation of the December 1997 Kyoto Protocol to the U.N. Framework Convention on Climate Change (UNFCCC) represented a significant development in international environmental policy. Not only had a large group of industrialized countries (known as the Annex I countries) agreed in principle to quantitative limits on their net greenhouse gas (GHG) emissions; all the participants also had agreed in principle to the development and implementation of novel "flexibility mechanisms" for international compliance with the numerical targets (see Chapter 20, Policy Design for International Greenhouse Gas Control). In addition to mechanisms for emissions trading among Annex I countries, a Clean Development Mechanism (CDM) was created in Article 12 of the protocol. The CDM would allow for the creation within non-Annex I developing countries of units of certified emission reductions (CERs) that could be traded and used within Annex I for complying with Kyoto Protocol emissions limits.

Operationalizing the CDM requires carrying out six distinct steps: identifying a project; assessing a project's net GHG emissions reduction; assessing a project's economic and social effects; financing a project; creating and certifying CERs through project implementation, with monitoring and independent verification of project performance; and distributing development benefits as well as CERs from the proj-

ect. Decisions made with regard to each of these elements will influence the CDM's capacity to deliver on the multiple aims of Article 12 and thus to satisfy the concerns of both Annex I and non-Annex I countries. Fundamentally, these choices will determine the extent to which the CDM is oriented more toward

- a market-based mechanism, in which nongovernmental actors take the lead in CER creation subject to certain basic and general rules (including respect for host country sovereignty) to ensure that the CDM is cost-effective, environmentally responsible, and fair or
- a multilateral governmental institution with substantial regulatory involvement in the operational details of project and credit creation.

In my view, a market-based mechanism is strongly preferable for meeting the needs of both Annex I and non-Annex I countries, as I explain in this chapter.

In the next section, I introduce the CDM's purposes, mandate, and institutional structure as authorized in the Kyoto Protocol. Then, I focus on some of the primary technical and administrative issues that will arise as the CDM is designed and implemented. The subsequent section is a discussion of some of the main issues that arise in considering how the CDM can contribute to the development goals of non-Annex I countries. Finally, I briefly consider the relationship of the CDM to broader questions about international climate agreements.

Purpose and Organization of the CDM

Like joint implementation investments in GHG control among Annex I countries, the CDM is a project-based mechanism. Under the CDM, emission reductions can be generated in non-Annex I countries and then transferred to Annex I countries for use in meeting Annex I Kyoto Protocol targets. CERs can be generated through joint ventures between entities in the host developing country and partners from developed countries, activities undertaken within developing countries that lead to credits that can be sold on an open market, or the intermediation of multilateral institutions such as the World Bank. And unlike the Annex I mechanisms, the protocol states that CDM activities and credits could be initiated as early as 2000 (versus 2008); however, there is controversy about which "early" projects (those coming online prior to formal implementation of the CDM) would be allowed to generate credits.

Three purposes of the CDM, which are given equal weight, are identified in the protocol text (Article 12, paragraph 2). They are, in order of appearance,

- to assist non-Annex I countries in achieving sustainable development;
- to assist non-Annex I countries in contributing to the ultimate objective of the UNFCCC (described in Article 2 of the UNFCCC as stabilization of GHG concentrations in the atmosphere at a level that would prevent dangerous anthropogenic interference with the climate system within a time frame sufficient to allow ecosystems to adapt naturally to climate change, to ensure that food production is not threatened, and to enable economic development to proceed in a sustainable manner); and
- to assist Annex I countries in achieving compliance with their quantified emission limitation and reduction commitments under Article 3 of the protocol.

In practice, the third objective emphasizes the provision of cost-effective emission control opportunities.

Meeting the aims of Article 12 explicitly requires *environmental additionality* (see paragraph 5 of the Article), which refers to generating GHG reductions that are additional to any that would occur in the absence of the project and that provide real, measurable, and long-term benefits related to the mitigation of climate change. There is, however, enormous controversy in practice about what constitutes environmental additionality.

Notions of financial and development additionality are implicit in the aims of Article 12. Broadly, *financial and development additionality* refers to encouraging long-term emission paths in non-Annex I countries that are compatible with the long-term objective in the UNFCCC of limiting global damages from climate change while not unacceptably restricting the "development space" of these countries. Specifically, many countries take financial additionality to imply that financial resources provided by the investor country should be additional to official development aid and funding through the Global Environment Facility (GEF). They also argue that for the investment to be additional, it should not have occurred under business as usual, in the absence of an international GHG control regime. Yet another set of proposed constraints involves the quality of the technologies transferred to developing countries as a consequence of CDM projects. As with environmental additionality, enormous controversy surrounds the practical determination of financial and development additionality.

Article 12 establishes three bodies to oversee the CDM: the representatives of the Conference of Parties (COP), an executive board established by the COP, and independent auditors to verify project activities. However, the protocol provides almost no guidance on what exactly the CDM would do or how it would operate. Instead, the structure and authority of supervisory bodies and the CDM are left for future negotiation.

The principles that govern this institutional development are subject to debate. Nevertheless, I suspect that the following general points would garner support in many quarters.

- **The institutions developed should be effective at promoting shared opportunity.** As in any trade dispute, it is always possible to imagine some constraint that generates a short-term advantage for one set of parties (nations or sectors) over others. The challenge is to ensure that the mechanisms do not degenerate into rent-seeking without real material advantages in facilitating environmental protection and sustainable development.

- **The institutions should be credible, particularly with respect to environmental performance.** If environmental goals are subverted by a poorly designed or operated flexibility mechanism, neither economic nor environmental interests will be served. Attending to this concern raises technical and legal issues as well as economic issues.

- **The institutions should be internally efficient and adaptable.** This point puts a high premium on linking the CDM (and the other Kyoto Protocol mechanisms) as closely as possible to existing markets and other institutions while recognizing the legitimate role of governments in ensuring the credibility of the mechanisms and in evaluating where their own sovereign interests lie with respect to participation. This last point links in turn to the first item: Countries should be free to decide the degree to which they wish to participate in the Kyoto Protocol mechanisms, but the structure of the mechanisms should not unduly cater to special interests.

CDM Design Issues

Implementing the CDM requires a basic structure that identifies responsibilities for operating the mechanism as well as rules for defining, measuring, and certifying emission reductions. Also required is the establishment of general compliance mechanisms within the Kyoto Protocol that would induce Annex I countries to demand CERs to meet emission reduction targets as well as sanctions for the misrepresentation or misuse of CERs. To make the mechanism acceptable to developing countries, issues related to the distribution of benefits and their authority over the mechanism must be clear.

Basic Structure of the CDM

The CDM is a novel international legal and institutional structure that will involve both governmental and nongovernmental actors. Several specific issues related to the basic legal and institutional structure of the mechanism remain to be addressed:

- the function of the CDM Executive Board and its relationship to member countries and to the operational entities that carry out the daily mission of the CDM,
- the way in which developing countries can expand or restrict their participation in the mechanism (from initiating projects to blocking access to certain potential investments),
- the criteria for project eligibility and certification, and
- the role of the CDM itself in collecting and disbursing funds.

How these issues are resolved will substantially influence the performance of the CDM in meeting the goals sketched above. One model would involve substantial ongoing involvement by governments and operational entities in the functioning of the CDM. In this approach, the entities would be charged with not only auditing project performance and calculating CERs (see later) but also project organization and financing. Governments would retain substantial involvement and discretion in reviewing the findings of operational entities and would exercise their sovereign self-interest with respect to the hosting of CDM projects. In addition, certain kinds of projects or technologies would be favored or rejected as a matter of principle, quite separate from the details of the project's ability to generate development benefits and CERs. For example, certain technologies for renewable energy or energy efficiency could be favored, and performance standards for transferred technologies would be mandated in advance.

However, the CDM will operate in the context of broader market forces affecting energy use and investment. Constraints on project eligibility and laborious project assessment and approval procedures built into the legal structure of the CDM will limit the flexibility of entities trying to form transactions, reduce the number and increase the financial costs of CDM investments. This will have the additional implication of reducing the amount of developing country benefits from the CDM, and possibly aggravating concerns about the distribution of these benefits as investments flow more disproportionately to those countries with large numbers of prime targets for investment. These drawbacks need to be traded off carefully against whatever perceived environmental or other benefits may flow from more constraints on the operation of the CDM.

One alternative—and in my view, superior—perspective for the structure of the CDM starts from the premise that in investor and host countries, the private sector must take a proactive role for the mechanism to be most effective in meeting the needs of developed and developing countries. In particular, private market institutions can bring an irreplaceable expertise in project evaluation, financial, and risk management services. The need for effective market institutions is especially high if, as one would hope from the perspective of encouraging both cost-effective emissions limitation and sustainable development benefits, the scale of CDM activity grows to be substantial. A vibrant and competitive market for CERs encourages efficient pricing of these credits, so that suppliers of lower-cost opportunities can enjoy a surplus return over that gained from more marginal investments. Thus, a well-functioning market can protect developing countries from "predatory" underpricing of CERs in transactions with Annex I entities, just as it protects buyers from the exercise of seller market power.

Some rules are needed to ensure the integrity of the mechanism. But integrity can be promoted if (a) the CDM Executive Board promulgates clear and logical criteria for project evaluation to promote the consistency of projects with the aims of the CDM and (b) public authorities of host countries carry out their responsibilities to mitigate undesired side effects and to facilitate projects with the most favorable development outcomes given their self-interest. In this model, operational entities serve as needed auditors in the enforcement of project performance standards but do not have to be involved in market

transactions. Participation by multilateral organizations in CDM project development and financing to assist the least developed countries can be pursued, but it does not become the norm for CDM transactions. Choices of investment types, terms, and conditions are determined largely by negotiation and competition, not by one-size-fits-all rules. It is also important in establishing the CDM's legal structure that rules for national participation and project eligibility not become tools for obstructing legitimate transactions as a form of rent-seeking or limiting the entry of new sellers.

Evaluating Projects and CERs

A key issue in balancing environmental additionality and cost-effectiveness is establishing how many emission credits a particular CDM activity generates and over what time frame the credits are created. Establishing these characteristics of the CER flow requires a counterfactual baseline and an estimate of the emissions change relative to that baseline (discussed later). Because a counterfactual baseline is inherently subjective and because participants will have incentives to convince less-well-informed monitoring bodies that emission reductions are larger than in actuality, this element of project-based credits is controversial.

Gauging the number of CERs from a project requires a counterfactual assessment of what net emissions would have been in the absence of the project—for example, what technology and fuel might otherwise have been used for electricity generation. Broadly speaking, this assessment can be done project by project or through the use of more-or-less standardized baselines that may vary across sectors, technologies, countries, and geographical regions. One important distinction to maintain in evaluating ways to address these options is the interplay between uncertainty and dynamic change. There is inherent uncertainty before the fact, at the time a project is being designed, about what the "right" baseline might be. In addition, views about the appropriate baseline may change over time as technological and market circumstances evolve.

Much of the policy discussion concerning project baselines has focused on the trade-offs between using relatively generic benchmarks for gauging reduced emissions versus project-specific assessment. Using generic benchmarks that average over heterogeneous technologies and sectors runs the risk of crediting some "paper tons" in the absence of a project-specific assessment. Critics also argue that experience to date with voluntary pilot projects has revealed many instances of excessive optimism in assessing emission reductions because the project baselines did not build in trends toward lower emissions under business-as-usual conditions. But reported experiences also demonstrate that project-specific assessment does not necessarily rule out risks of fictitious abatement. A benchmark approach may make dubious claims about incremental abatement easier to identify and control statistically and provide a crucial benefit in keeping transaction costs (all the overhead costs of setting up and implementing projects) affordable. These points are important not only for Annex I countries but also for developing countries that seek to expand project opportunities—especially small projects in smaller economies that cannot afford high overhead costs.

Expected energy efficiency improvements (if any) over time once a project is in place could be built into specifications of baselines and benchmarks for project performance in an effort to limit the over-crediting of projects. Retroactive revisions in past baselines and CERs for existing projects (as opposed to periodic and expected revisions of baseline conditions for new and ongoing projects) should be avoided because they create unproductive investment uncertainty beyond the control of project investors that hampers project finance and implementation. Over time, judgements about appropriate baselines could be different from what was expected when earlier projects were implemented because of the uncertainties mentioned above. The appropriate response to this phenomenon would be to revise baselines for future projects. Baselines also could be defined over shorter periods (less than the life of the project) and subsequently revised based on new information. (The process of revision presumably would be symmetric—if conditions had evolved such that baseline emissions appeared larger than had been thought when the project was launched, a more favorable baseline for calculating CERs would subsequently be provided.) Investors would recognize these uncertainties up front in their decisions and build them into their project planning.

Whether baselines are set project by project or through more generic benchmarks, they need to be realistic given the circumstances of the investment. Just as the assumption of no technical advance in a developing country's energy system could overstate CDM credits (a new energy-efficient plant might have been built anyway in a few years), the use of an "internationally best available" standard for evaluating efficiency improvements will understate CERs if the prevailing norm is less-advanced technology. Such standards not only are too stringent in terms of assessing environmental additionality; they also can hinder, rather than promote, the spread of advanced technology to developing countries. Faced with higher costs and fewer CERs as a result of such technology standards, investors will reduce and redirect their investment.

CDM Project Implementation and Crediting

In addition to an established baseline, calculating CERs requires an assessment of what reductions the project generates relative to the baseline. Here there is concern about the difference between after-the-fact and before-the-fact project performance because of considerations that include unpredicted deviations in technical performance and issues related to project implementation (for example, a power plant achieves the promised energy efficiency but operates only half as much as expected).

As in defining baselines, there is tension from conflicting needs—to ensure greater environmental quality with a desired distribution of benefits, and to ensure that market transactions occur in a flexible milieu without onerous transaction costs. Even if there is no uncertainty about the net emissions reduction of a project if it is completed, the fact that both host and investor entities possess more information about project performance than their governments (or the CDM authority and its operational entities) creates potentially significant incentives and opportunities for misrepresenting the amount of emission reductions generated by CDM projects.

Many observers have argued that in the interest of environmental integrity, only credits that have resulted from direct observation of project performance after the fact should count toward compliance with Annex I emission targets. Annex I governments might even require their own legal entities to use only such "prime quality" credits for compliance, to ensure that the Annex I governments would not be embarrassed by unanticipated national noncompliance due to disallowed CERs. Speculative forward sales of precertified credits could be allowed, but the buyer would need to beware the risk of nonperformance after the fact.

A certain amount of after-the-fact certification based on direct monitoring and verification by independent bodies under the control of the CDM Executive Board would be necessary to ensure the integrity of the mechanism. However, requiring direct monitoring and verification of all reductions before certification would imply larger monitoring costs and delays in project returns. These results, in turn, would discourage project development—especially for small projects—and limit the realization of some projects that would be cost-effective and beneficial to the host country in certain regions and domains.

One alternative is to rely on a degree of self-certification for compliance coupled with spot monitoring and the threat of significant sanctions if project performance is less than claimed. If the monitoring system is active enough and the sanctions for nonperformance are substantial enough, this approach has the potential to provide adequate performance. In particular, buyers can be held liable for nonperformance of purchased credits so that they have an incentive to police seller conduct and buy insurance in the form of extra credits (see Chapter 22, Allocating Liability in International Greenhouse Gas Emissions Trading and the Clean Development Mechanism). It would also be possible to build in incentives for improved monitoring and reporting by allowing relatively more before-the-fact certification for projects supported by higher-quality monitoring and reporting capacity.

CDM projects could involve either direct reductions in GHG emissions (particularly through energy-related projects) or projects to enhance carbon sequestration (such as reforestation initiatives). Carbon sequestration projects give rise to several special considerations, including those related to measuring carbon storage from the project. These considerations are beyond the scope of this chapter and are addressed in Chapter 13, Carbon Sinks in the Post-Kyoto World.

Providing Development Benefits

Many experts from developing countries have expressed the concern that CDM projects may not meet (or even could contradict) their development priorities. A related but distinct concern is that some developing countries could be left at a serious disadvantage given their own limited capacities to generate and negotiate projects that would meet eligibility requirements and attract investment.

The Kyoto Protocol provides that the decision to participate in the CDM rests with governments; hence, host country public authorities will have the authority to decide whether particular CDM projects are consistent with their development and environmental priorities. Because this capacity may be undermined in some countries by financial pressures and risks of corruption, some observers have suggested that a set of internationally implemented indicators might be used to gauge the benefits to host countries and to ensure that the interests of these countries are protected. An even stronger position would vest with the CDM authority the power to judge whether projects provided sustainable development benefits. However, the set of potential indicators is vast, addressing various economic and social issues, and the assessment of any CDM project depends very much on the specific circumstances of the project. Moreover, imposing international standards would restrict the sovereign discretion of countries. For these reasons, I believe that attention should be focused instead on ensuring that the institutions of the CDM itself provide the best possible opportunities for developing countries consistent with a market mechanism.

Concern continues that the CDM will simply result in a reallocation of official development aid. However, some degree of crowding out probably is inevitable, given the continued downward trends in this aid worldwide. It may be more relevant to focus on how the CDM can provide tangible sustainable development benefits that otherwise might not be reaped at all. There also may be synergies between CDM and official development aid that need to be better developed. For example, spending such aid in part to help develop capacity for CDM in developing countries might be useful—even though it is a nominal form of crowding out—because it may have positive impacts in terms of general capacity building for development policies.

Concern over the development benefits of CDM also is part of a larger and longer-standing debate over the nature of benefits from private foreign direct investment (FDI). Developing countries may not effectively capture a share of benefits from FDI in CDM projects, and the projects selected will not necessarily advance the development interests of the host country. Such concerns about sustainable development vis-à-vis a market mechanism can be addressed in several ways.

Non-Annex I countries can generate CERs on their own or through some kind of international financial pooling (a sort of mutual fund, like the World Bank's Prototype Carbon Fund) and sell them in the international GHG credit market, rather than simply participating in individual joint ventures. Such approaches could allow developing countries to compete effectively as CER suppliers, rather than being subject to the exercise of market power by a single partner or credit buyer. This advantage can be traded off against the possible efficiencies of joint ventures with technologically advanced investment partners, and different kinds of transactional arrangements (for example, service contracts with technologically advanced partners) can be structured to meet the interests of project proponents. It is also important that developing countries be allowed to bank CERs they create, so that they can compete over time for the best contract terms rather than being in a "sell or lose" position with undeveloped credits.

Proposals have been made for the CDM to include rules for the sharing of CERs between the international investor and the host country over time.

In particular, CERs could revert to the host country after a fixed period or if and when the host country assumes its own national GHG control obligations. However, there are inherent trade-offs between project returns in cash and CERs, and insistence on a greater share of CERs will reduce the financial benefits an international investor is prepared to offer. Moreover, fixed rules for sharing credits over time will not replicate the circumstances that best suit the interests of the transacting entities and the overall cost-effectiveness of the mechanism except by accident. These terms should be left to the entities involved to negotiate.

Another possibility is using the CDM to provide leverage in encouraging public infrastructure policies or domestic policy reforms that would trigger GHG-reducing investments even if they were to reflect primarily policy aims other than climate objectives. Examples include public-sector improvements in energy efficiency and investments in transportation as well as financial inducements to reduce energy subsidies. Making those carbon-abating investments that are in synergy with development targets more attractive would broaden the potential for CDM projects, whose attractiveness would in turn be reinforced by the revenue from CERs.

Including these options would require further work on how to assess their environmental additionality. It involves not only the technical assessment of potential changes in emissions but also the politically controversial determination of whether to view such changes as already occurring in a business-as-usual situation. One does not want to create incentives for decisionmakers in developing countries to retard beneficial reforms in the search for financial benefits from the CDM. But the limited real-world success of policies to promote better energy pricing and more efficient public infrastructure in some countries would support a more generous view about allowing CDM credits for such shifts where improved practice otherwise seems unlikely.

The CDM cannot be expected to rectify all the problems of either international equity or capacity shortfalls. Broadening opportunities for host countries to take a more active role in project identification and implementation would also respond in part to the concerns about the potential for an uneven geographical distribution of benefits. Smaller economies would be more able to attract investments consistent with their development priorities by lowering the search and transaction costs for foreign (and domestic) investors.

Attempting to more directly mitigate distributional concerns through project eligibility criteria will reduce total investment and development benefits, even if it does transfer some benefits to the countries with the least CDM potential. A similar concern arises with a plan to (in effect) tax CDM transactions to raise money for a fund to redistribute project investments, as well as to cover CDM administration costs. This plan may lead to some perverse outcomes distributionally, in that Annex I investors will seek to protect their project returns by shifting part of the tax to their host country partners. Consequently, the financing of the redistribution fund will involve a shift of resources from some developing countries to others. Moreover, the greater the CDM tax, the greater the advantage of Annex I emissions trading or joint implementation over the CDM. Some of this tax shifting could be ameliorated if other Kyoto Protocol flexibility mechanisms for Annex I also were taxed, but it is not clear whether and how such an extension of project excise taxes could be justified.

Other revenue-raising options need to be explored to address these concerns. One possibility would be a set of national-level contributions to a CDM fund by Annex I countries that do not inherently affect the returns to specific projects. Such contributions could be thought of as payments for the option to participate in CDM, and they could be scaled in many ways. Although this approach would be more economically efficient than taxing specific projects, it also would be politically challenging. And the thorny questions of who controls the fund and how it is used still would have to be addressed. What particular sustainable development criteria would be used to govern the allocation of the fund? Would it be used to support additional CDM projects, presumably those that did not pass the test of the CDM market, or distributed in some other way such as grants? How could the allocations be made

in a way that does not invite excessive bureaucratic inefficiency, rent-seeking, and corruption? In the end, the least developed countries may benefit as much or more from more institutional kinds of support—the availability of technical assistance in project development and implementation, and the creation of general project eligibility criteria that increase their opportunities for participation in a market-oriented CDM.

CDM and Longer-Term Climate Agreements

Experience suggests that any project-based mechanism will have higher transaction costs than a system of trading in homogeneously defined emissions permits, as in the sulfur dioxide trading program used in the U.S. power sector. The use of project-based credits also raises questions about what is actually happening to aggregate emissions and carbon leakage. One alternative to project-based crediting is the voluntary assumption of national emission caps by non-Annex I countries. These countries then could engage in emissions trading like the Annex I countries. The Kyoto Protocol provides for such actions by non-Annex I countries; however, the conditions under which accession could take place have not been fully spelled out.

Assuming a reasonably accurate national inventory of GHG (or at least CO_2) emissions—a reasonably big "if" for some developing countries—a voluntary but binding national baseline could help increase the accuracy of carbon accounting, diminish concerns about the environmental additionality of any particular GHG reduction activity, and lower transaction costs by making the commodity exchanged (emission permits) more homogeneous. For this approach to succeed, however, the non-Annex I country must perceive more benefit than cost from the assumption of a national baseline. The baseline must not constrain economic growth and associated increases in energy use by more than the value of increased revenues from expanded emissions trade under the more efficient emissions trading system, assuming some constraints on CDM projects in the absence of the baseline out of concern for environmental additionality. One possibility to this end is a "growth baseline" that seeks to limit the carbon intensity of economic activity without limiting the total volume of carbon emissions.

Just as implementing the CDM requires some kind of project-level or sectoral agreements about what is normal and above-normal practice in matters such as energy efficiency, implementing a voluntary national baseline would require agreement on a grander scale about what is an appropriate business-as-usual standard for the GHG emissions of the developing country. Some developing countries fear that commitment to the CDM—and even more so the additional commitment to a voluntary set of aggregate emissions baselines—could prejudice their position in future negotiations. Ultimately, successful international climate policy to limit GHG emissions would require global participation. The distribution of burdens in a long-term global agreement remains very contentious, with developing countries understandably arguing that their long-term development opportunities should not be constrained by the developed world's historic levels of GHG emissions. Developing countries need concrete reassurances that their options in future negotiations are not foreclosed by the project, sectoral, or national baselines established under the CDM.

There also is concern that developing countries could end up supplying too many "low-hanging fruit" projects early in the CDM. These are projects with low GHG abatement costs that developing countries might prefer to retain for later, when they are negotiating requirements for GHG reduction. However, such disadvantages are not preordained. A well-functioning competitive market for CERs, in which suppliers of credits can bank CERs for future sale, can engender an efficient spread of CDM investment over time. (To this end, the negotiation of terms for a second commitment period beyond the Kyoto Protocol would provide greater clarity for the development of future price expectations for CERs.)

Host countries also can exercise their sovereign authority to limit the approval or restrict the term of projects they believe should be retained for future domestic use. However, the costs to such constraints must be considered. Economic and environmental

returns on CER-generating investments can be lost if the investments are delayed. For example, a delay in upgrading industrial or power plant facilities whose energy efficiency and pollution performance are poor implies continued pollution and higher energy expenditures; the same investment made in the future may not satisfy future additionality tests if improved technology has become the norm. Moreover, the supply of such projects is not necessarily exhausted by early investments, as long as there remains a continuing differential in the state of technology between developed and developing countries. More work is needed to reassure developing countries that with appropriate structuring of the CDM, tangible and environmentally responsible development benefits are possible.

Developing countries also have an interest in the evolving debate over the operation of the Annex I flexibility mechanisms. It encompasses both short-term economic impacts and long-term institutional structure. Large volumes of Annex I trading may reduce the demand for CDM projects. Therefore, developing countries have an interest in pushing to ensure that the mechanisms are credible (that is, not just loopholes).

On the other hand, actions that weaken the efficiency of the mechanisms themselves, such as "supplementarity" constraints, may not serve the interests of developing countries as suppliers of emission credits. This issue arises because the Kyoto Protocol refers to the use of international emissions trading (and, by extension, the CDM) as being supplemental to domestic actions. Various countries have advocated a variety of specific constraints on the use of international flexibility mechanisms as a way of limiting the use of these mechanisms. The most commonly advocated constraint is a ceiling on the amount of emissions permits or credits that a country can purchase through the mechanisms.

These proposals reflect a concern by some countries that excessive importation of emissions permits or credits by other countries will limit the scope and stringency of domestic policies, thus retarding the long-term development of technology and improved energy efficiency needed to achieve and surpass the goals of the Kyoto Protocol. However, limits on trading and CDM are blunt instruments to improve the credibility of a nation's commitment to GHG limits. And by increasing the overall cost of compliance with GHG control targets, the restrictions also contribute to a lack of willingness to achieve target reductions. It is not even clear how the CDM and other flexibility mechanisms would effectively operate under supplementarity constraints, because all transactions would be contingent on whether the constraints were binding at a particular moment in time in the credit buyers' countries.

Finally, attacks on "hot air" need to be carefully considered. This pejorative term has been applied to the likely surplus of emission allowances in Russia under the Kyoto Protocol. Clearly, the sale of such a surplus to other Annex I countries implies less aggregate emissions control than if such a surplus did not exist. However, the same mechanism—otherwise known as "providing headroom for growth"—could figure prominently in longer-term negotiations to increase developing country participation in emissions targets. Future climate negotiations doubtless would benefit from more explicit consideration of economic, environmental, and distributional trade-offs, but the allocation of tradable surpluses to some countries should not be precluded.

Concluding Remarks

The CDM is an inherently imperfect mechanism. Even the most ardent CDM proponents concede that it will have higher transaction costs, greater monitoring difficulties, and lower overall efficiency than options such as a full-blown permit trading system. But it is a good place to start. CDM projects can provide

- acceptably tangible GHG reductions,
- concrete information about GHG control opportunities and costs in developing countries,
- cost-saving benefits in developed countries, and
- economic and local environmental benefits in developing countries.

Specific institutional measures, such as a liability system that holds international investors at least

partly responsible for project shortfalls, can improve credibility. The leakage problem with a project-based approach is vexing, but it can be addressed by defining rules of thumb for different project categories that attempt to adjust for leakage and that can be updated over time, as well as by encouraging the voluntary adoption of credible growth baselines in developing countries. To protect the long-term bargaining positions of developing countries, it must be absolutely clear that such interim baselines are for defined periods, with no prejudice as to future divisions of GHG responsibilities.

Acknowledgements

Much of the material in this chapter draws on joint work with Marina Cazorla and a workshop summary prepared with Jean-Charles Hourcade. I am grateful to Urvashi Narain and Klaas van't Veld for sharing with me the results of their ongoing work on CDM benefits. However, I am solely responsible for the points made here.

Suggested Reading

CCAP (Center for Clean Air Policy). 1998. Growth Baselines: Reducing Emissions and Increasing Investment in Developing Countries. Washington, DC: CCAP.

Grubb, Michael, with Christiaan Vrolijk and Duncan Brack. 1999. *The Kyoto Protocol: A Guide and Assessment.* London, U.K.: Royal Institute of International Affairs.

Hagem, Cathrine. 1996. Joint Implementation under Asymmetric Information and Strategic Behavior. *Environmental and Resource Economics* 8(4): 431–47.

Hahn, Robert W., and Robert N. Stavins. 1999. What Has Kyoto Wrought? The Real Architecture of International Tradable Markets. Discussion paper 99-30. March. Washington DC: Resources for the Future. http://www.rff.org.

Haites, Erik, and Farhana Yamin. 2000. The Clean Development Mechanism: Proposals for Its Operation and Governance. *Global Environmental Change* 10(1): 27–45.

Jacoby, Henry, Ronald Prinn, and Richard Schmalensee. 1998. Kyoto's Unfinished Business. *Foreign Affairs* 77(4): 54–66.

Joshua, Frank, and others. 1998. *Greenhouse Gas Emissions Trading: Defining the Principles, Modalities, Rules and Guidelines for Verification, Reporting, and Accountability.* August. Geneva, Switzerland: U.N. Conference on Trade and Development. http://www.weathervane.rff.org/negtable/unctad.html.

Tata Energy Research Institute. 1999. *The CDM Maze: The Way Out.* New Delhi, India: Tata Energy Research Institute. http://www.teriin.org.

Wiener, Jonathan B. 1999. Global Environmental Regulation: Instrument Choice in Legal Context. *Yale Law Journal* 108(4): 677–800. (This article is a broad overview of the Kyoto Protocol flexibility mechanisms.)

22 Allocating Liability in International Greenhouse Gas Emissions Trading and the Clean Development Mechanism

Suzi Kerr

The possibility of international trade in credits for greenhouse gas (GHG) emission reductions is a key "flexibility mechanism" built into the December 1997 Kyoto Protocol for GHG reduction worldwide. The protocol allows entities in Annex I countries (the industrialized countries that agreed to cap their total emissions) to trade emission reductions. Through the Clean Development Mechanism (CDM), investors in Annex I countries also can secure GHG reduction credits for emission-reducing activities in non-Annex I developing countries that have not accepted national emission caps.

For these forms of international emissions trading to be seen as credible forms of real emissions reductions, legal responsibility (or liability) must be assigned for the failure of promised emission reductions embodied in the credits to materialize. Although a well-functioning compliance system is crucial for the integrity of trading, excessive restrictions on trading to enforce responsibility could stifle emission credit markets and raise international compliance costs to unacceptable levels. The desirable allocation of liability trades off these two concerns.

Liability for the "quality" of an emission reduction credit when created could rest with buyer, seller, or both parties; it also could stay with whomever originally is assigned the liability, or it could be transferred as credits are resold. A very high level of compliance by sellers could always be ensured by

"gold plating" credits or permits. Before credits were sold, we could require they be certified by an independent agent. Buyers and sellers would then have to decide how often to bring in the certifiers, trading off the costs of more frequent quality control against the advantage of a more continuous flow of certified credits or permits. Because this approach probably would be quite expensive—because of certification costs or illiquidity—in this chapter, I focus on systems that allow trading of emission permits or credits prior to certification with post-trade liability rules that aim to enhance the credibility of trading.

Designing good compliance systems would be easy if everybody—traders and governments—had lots of information about the emission-reducing activities of different entities and if there were strong legal sanctions within every participating country for nonperformance. In practice, information is scarce and not evenly shared, and both domestic and international enforcement mechanisms are limited in what they can accomplish. Starting with these two points, we first consider some of the general institutional background for international emissions trading. We then consider the assignment of liability in an international GHG trading system for the Annex I developed countries, focusing on the assignment of liability for "bad" emission permits when the seller country is not in compliance with its Kyoto targets, known as "assigned amounts." We turn then to address issues of credibility and liability in the context of CDM joint ventures.

The Institutional Backdrop for Emissions Trading and Liability

As noted above, the Kyoto Protocol envisages two different kinds of international emissions transactions. Exchange among actors in Annex I countries could involve international trade in homogeneous emissions permits created by individual countries in pursuit of their domestic implementation of the Kyoto Protocol targets, or the exchange of project-specific emission reduction credits. In either case, the goal of buyers of emission permits or credits would be to obtain credits at a lower cost than their own domestic compliance efforts while still achieving the Kyoto Protocol targets. The Kyoto Protocol also allows countries to pool and redistribute their assigned amounts, as the members of the European Union have done. We see this effort primarily as an extension of the international negotiation process, distinct from the market-oriented activities involving private-sector entities that are the focus here.

The CDM transactions involve specific joint ventures between actors in Annex I countries and actors in non-Annex I developing countries, wherein the former invest in the latter to obtain emission reduction credits that are less costly for the investor than other forms of GHG control. These projects were referred to previously as *Joint Implementation*, but that term now is reserved for project-specific credit trades among Annex I countries.

Under the Kyoto Protocol, ultimate responsibility for meeting the numerical emission control obligations rests with the national governments of Annex I countries. The protocol contains provisions for the calculation of national emission inventories and the deployment of sanctions in the event that the targets are breached. However, these sanctions have not yet been specified, and monitoring is imprecise. Moreover, because participation in the Kyoto Protocol and the underlying U.N. Framework Convention on Climate Change (UNFCCC) are voluntary decisions by sovereign nations, the ability to exact sanctions is inherently limited; participation must be in the self-

interest of every signatory. Whereas sanctions are difficult to apply internationally, they could be adopted by mutual consent to apply when national compliance falls short. The sanctions could include a country receiving a lower national assigned amount in subsequent commitment periods (to "make up its environmental debt"). Within a trading program, sanctions on selling countries that are out of compliance also could include strictures on future trading opportunities (future permits or credits might be sold at a previously agreed discount, or the capacity to trade might be suspended entirely). When governments delegate emission trading privileges to subnational parties ("legal entities" in the parlance of the Kyoto Protocol), they will apply domestic enforcement pressure, because if their companies cheat and are caught, the government is ultimately responsible.

Figure 1 illustrates the various combinations of trading and enforcement activities. Efficiency of trading probably will be greatest when both buyer and seller are subnational private actors with incentives to make the best deal given the specific information about trading opportunities they possess. However, it is certainly possible that some buying and selling will be done by governments on behalf of or in lieu of their private sectors.

In CDM transactions, seller countries do not have emission obligations and thus do not have national baselines against which to measure emission reductions. Project-level reductions, referred to in the protocol as certified emission reductions (CERs), are measured by assessing the additional reduction achieved relative to some notion of business as usual. This process is controversial and costly. Business as usual is inherently difficult to define, and burdensome project-by-project review will chill incentives for cost-effective CDM investments; yet a generous definition of "additionality" may involve creating CERs that do not reflect real reductions.

Parties to the Kyoto Protocol could minimize the costs of assessing CDM emissions reductions and reduce manipulation of baselines by determining a set of generic criteria for evaluating the additional GHG reductions associated with different classes of investments (for example, enhancements in the power sec-

Figure 1. International and Domestic Enforcement and Potential Trades.

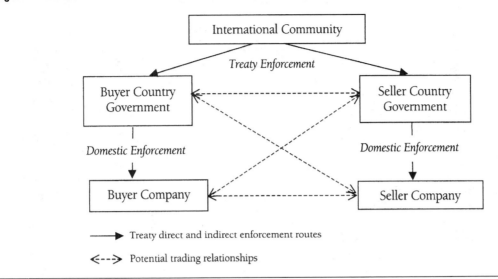

tor, improvements in end-use efficiency, and the use of low-GHG renewable energy sources). To a significant extent, these criteria would be arbitrary and would require revision periodically with experience; however, they would provide a way to avoid the arduous and equally arbitrary project-specific determination of additionality.

In both Annex I and CDM trading—even where noncompliance is clearly defined and measurable and sanctions are used, because monitoring is not perfect and sanctions are limited—parties may misreport reductions or claim inability to meet targets after the fact. In particular, sellers may sell more credits than they really create. Assignment of liability for the validity of credits after the fact is key for controlling these behaviors in a permit market. The question we address is, Which country should bear liability for the validity of traded credits—the buyer, the seller, or both? The fact that Annex I and CDM transactions involve different actors and operate under different domestic regulatory structures leads us to different conclusions about how responsibility for the integrity of the trading system should be imposed.

Before turning to these issues, we briefly address the general question of whether international trading increases the general risk of noncompliance with the Kyoto Protocol. Some observers correctly indicate that if entities in some countries can create and sell credits not backed by real emission reductions, the resulting "export" of noncompliance can magnify the compliance problem. On the other hand, the capacity to enforce national commitments internationally is inherently limited, as already noted; compliance will not be perfect in the absence of trading either. Moreover, the same monitoring problems that complicate implementation of a credible trading system also bedevil the assessment of compliance without trading. Finally, but perhaps of greatest importance, because a successful trading system will reduce compliance costs, it should strengthen national incentives to meet the obligations of the Kyoto Protocol.

Performance Risks and Remedies

Annex I Trading
As already noted, Annex I trading involves either the trading of homogeneous permits based on national emissions, or bilateral contracts for the creation and exchange of project-related credits. In either case, the fact that all the actors trading are in Annex I coun-

tries means that their respective governments are ultimately responsible for their actions, and all trading entities should be governed by their domestic compliance mechanisms. The Kyoto Protocol requires each Annex I country to develop an operational capacity to keep track of their aggregate net emissions as well as a capacity to develop and implement policies and measures to credibly ensure that those emissions remain within the negotiated targets.

If these mechanisms worked as they were supposed to, then effectuating international GHG trading would require only a relatively minor increase in bookkeeping capacity. Whether the domestic systems of both the buying and selling countries are permit-based, or one or both countries allows only project-based trades at the subnational level, regulators in the buying and selling countries need to keep track of the volumes of the international flows. Then, they need to adjust domestic emissions accordingly for the purpose of evaluating domestic compliance. Although it may be useful to keep track of where the permits flow, there is no need to keep detailed regulatory records of individual buyer and seller transactions in this case. If, for example, a U.S. permit holder sold some U.S. permits to a Japanese entity, it would suffice for the seller's permit balance to be debited by U.S. regulators and for Japanese regulators to note that the buyer has obtained a greater capacity to emit. Aggregate flows among countries would need to be recorded as a cross check on each country's report of permit transactions.

In practice, things could go wrong. Sellers could simply sell more permits than they create and openly not comply with their commitments under the protocol. Alternatively, subnational sellers could hide their noncompliance by inflating the number of credits created in a project in their reports to national regulators. The national regulators in seller countries could report higher levels of permits sold than emissions they really reduced. These misleading reports would be given to the international bodies charged with overseeing compliance under the UNFCCC and the Kyoto Protocol. Even if buyers observe misreporting or know that the seller will not be able to comply, they may collude with the seller to get cheaper credits. For example, Russia could sell more credits than it really created (the assigned amount minus actual emissions), and a firm in the United States could buy them cheaply, despite knowing that they are probably bad permits.

In response to concerns about this potential threat to the credibility of Annex I trading, several proposals have been offered for assigning liability for bad permits or credits (that is, those coming from a noncompliant seller country). Within Annex I, seller countries are always liable for the sum of net emissions plus permit sales being below their assigned amount. The question is whether the buying side also should be liable. If buyer countries are also liable, the buyer country government would be responsible for covering their ex post "carbon deficits" if part of their obligations had been met with the purchase of permits or credits from a country found later to be noncompliant. It could be required to purchase additional "secured" permits or credits from other sources in addition to being sanctioned. In all likelihood, the finding of noncompliance in the seller country would be based primarily on an accounting of aggregate emissions plus sales versus assigned amounts; however, it is possible that some spot checking of individual project-based transactions also could occur.

Buyer liability could put all transactions with that country at risk. Alternatively, only those permit or credit sales beyond some target level of seller country emissions might be subjected to this caveat emptor test. For example, Russia could sell credits equal to the difference between its emission quota under the Kyoto Protocol and its 1998 emissions or some other target level—the so-called hot air—without triggering buyer liability, but sales beyond that mark would be at the risk of the buyer ("caveat emptor"). Another alternative would be the "last in, first out" system, whereby permits would be made invalid in reverse order on the basis of the date of trade until the seller was back in compliance.

Introducing buyer liability to Annex I trading in addition to seller liability will increase seller compliance, but it may do so at high cost relative to the gains and to the costs of increasing direct compliance pressure on sellers. I believe the costs of this approach will outweigh the benefits.

For buyer liability to be beneficial, buyers must have a comparative advantage relative to the international community either in monitoring or responding to seller noncompliance. In addition, liable parties will bear the residual risk of performance failure that is beyond their control and thus the risk of penalties for noncompliance. For buyer liability to be beneficial, buyers must not be so risk-averse that bearing these risks would drive large numbers of potential trades from the market.

If these conditions hold, it could be efficient for the international community to "employ" the buyer as an enforcer, with the threat of sanctions as the incentive, rather than increasing efforts to enforce seller compliance. In this system, buyers would affect seller behavior by monitoring seller conduct and then responding to the information they obtain either by changing their choice of trading partners, offering less for more risky credits, or sanctioning noncomplying sellers directly. If buyer countries were liable for seller noncompliance, then buyer countries and their subnational actors—to whom they pass the liability—would undertake various risk management measures. Expert assessments of noncompliance risk would be developed to help buyers gauge the probabilities of having their permits or credits devalued, and these probabilities would be used by buyers to discount the price of purchased credits from less secure sources accordingly. Buyers would refuse to trade with especially risky sellers. In response, reliable sellers would try to provide transparent information on their compliance. Buyers also could directly punish noncompliant sellers through standard international contract law.

However, early on in the advent of international GHG trading, neither buyers nor intermediaries are likely to have a strong comparative advantage in assessing the prospective conduct of sellers, particularly when permit validity depends on national, not project-level, compliance. Anything an individual buyer could do to monitor another country's GHG production or to subcontract monitoring to specialists also could be done (perhaps better and more cheaply, with less duplication) by international bodies charged with overseeing the protocol. Market

reputations about seller conduct also will take some time to develop.

Buyers' options for directly policing or controlling seller behavior are inherently limited if defectiveness of credits is based solely on national aggregate emissions of the seller country rather than project-specific findings. In particular, with homogeneous permits, buyer liability is a form of "joint and several" liability that imposes the burden of noncompliant conduct by one seller (national or subnational) on all buyers. No buyer—and no subnational seller—can control the risk of national noncompliance.

Even with a careful choice of sellers, buyers will still face residual risks beyond their direct control to which they will have to respond by holding additional credits as a reserve margin or buying insurance from a broker. These are direct costs to buyers and to society as a whole that could be avoided if the international community put in place more direct sanctions against noncompliant seller countries. These costs are borne for good trades as well as bad ones and hence limit trading by reliable sellers as well as those who are unreliable. Risk-averse buyers will tend to be overly cautious and discount or exclude many good trades. The resulting loss of legitimate gains to trade could be significant.

In considering the trade-off between this loss and the goal of increased reliability of trading, it is also important to remember that shared liability between buyers and sellers could reduce the incentives for credible trading that sellers may face under the protocol. Although the financial sanctions against Annex I sellers have not yet been articulated, they may be stronger than those available against noncompliant buyers. Overall incentives for compliance thus could actually fall under shared buyer and seller liability.

I believe that on balance an appropriately defined seller liability program is the best way to allocate liability for performance failures in trades by most Annex I countries. The sanctions would have to be serious enough to motivate seller country governments to comply and to deter cheating by their private actors (for example, forfeiture of future assigned amounts, limits on future access to the market). These sanctions are reasonably credible and potentially serious.

An effective seller liability program in Annex I must distinguish sellers with good reputations and domestic enforcement from those whose weaker enforcement engenders more uncertainty about the credibility of their permits. To do this, we propose coupling qualifying requirements for Annex I countries to sell permits in the Annex I trading market with whatever after-the-fact sanctions are devised by the conference for noncompliant Annex I sellers. The qualifying requirements would distinguish high-risk sellers in advance and prevent them from creating large numbers of invalid permits.

The details of the qualifying conditions would themselves be subject to international negotiation. The national qualification process presumably would look at countries' domestic monitoring and enforcement capabilities (including the general strengths of the legal systems for regulatory and contract enforcement), past treaty compliance, and perhaps at other factors that bear on their commitment and capacity to meet their obligations. Countries that fail to meet the qualifications would still be able to sell credits, but only after national emissions during the commitment period have been certified (that is, at the end of the commitment period) or through the project-level CDM described later, with buyer liability. A more complex variant of this approach would impose various discounts on credits of countries that failed to fully meet all the qualifying requirements. Whereas determining a more fully articulated set of qualification criteria and applying them in practice would be inherently subjective, the problem does not seem that much greater than determining countries' eligibility to participate in the International Monetary Fund or the World Trade Organization.

Ultimately, for the trading program to work well and help achieve the goals of the UNFCCC, the combination of preconditions and the threat of future sanctions must deter cheating by sellers in the permit market. Countries may simply walk away from their UNFCCC obligations, but this is a problem with or without trading. As already noted, trading may enhance incentives for compliance by reducing the costs. What is important, especially in the early stages, is to do as much as possible to develop a liquid and flourishing international trading system

among Annex I countries subject to some basic credibility tests. We may have to live with the system we create for a considerable period. We want a system that has low transaction costs, many trades, and low national compliance costs and that consequently strengthens national interests in reducing GHG emissions.

CDM Trading

The situation with the CDM is fundamentally different from Annex I trading in at least three important respects. First, developing country hosts of Joint Implementation/CDM projects are not currently obligated to cap aggregate emissions. Thus, these governments do not have the same obligations for policing the performance of domestic GHG abatement as their Annex I counterparts. Second, several non-Annex I countries clearly have less technical and institutional infrastructure for overseeing Joint Implementation/CDM project results and enforcing contractual performance obligations than their Annex I counterparts. Third, Annex I actors obtaining the emissions credits also have a direct involvement in financing and possibly managing the investment project producing the credits. Their comparative advantage in monitoring and enforcement and the prospects for collusion to misrepresent project results do not arise with more arms-length permit or credit transactions.

Consider the problem of a seller inflating the results of a CDM project to increase the number of certified credits that can be sold. If the buyer observes this and is not liable for the validity of the credits, it has no incentive to report accurately. The buyer can strategically collude with the seller to increase the payment to the seller while reducing the compliance burden of the buyer. However, if the buyer is liable and there is some risk of detection and subsequent sanction, it may act as a more effective monitor and enforcer.

In contrast to the situation in Annex I trading, buyers in CDM can be good monitors and enforcers with respect to their capacities in monitoring and enforcement and their capacities to manage risks. As partners in CDM joint ventures, Annex I buyers have the ability to directly observe the number of credits created. They, as well as the sellers, can employ third-party certifiers to monitor the specific project and refuse to pay for invalid credits or sue to improve contract enforcement.

Even where liability could be shared, it might be better to hold the buyer liable instead of the seller. Because international law is weak, it may be better to hold one party primarily liable rather than risk the dilution of sanctions through ambiguity about liability. If both actors can observe and control the outcome, then the best way to deter such behavior is to assign liability for misrepresentation, once detected, to that actor with the greatest vulnerability to punishment.

Annex I buyer countries have more to lose through potential damage to their to reputations than the non-Annex I sellers, so they have stronger incentives to enforce domestic regulation. Combining this strong incentive with strong domestic regulatory capability in Annex I countries, Annex 1 subnational buyers of CERs will face a wider array of credible domestic sanctions for noncompliance (including the possibility of citizen suits as well as formal regulatory punishments). In contrast, although seller countries can be punished through restraints on future trading, this potential repercussion may not have much impact before the fact on the behavior of subnational actors. This conclusion is strengthened by the observation that the buyer is better able than the seller to bear the increased economic risk implied by this penalty system.

Buyers will face high costs when they are held liable in CDM transactions, and this liability will limit CDM trading. However, I believe that the higher levels of compliance to be achieved would justify the higher costs. As a project participant, the buyer is able to control a significantly large part of the risk it faces, and many of the trades that will be inhibited would have involved bad credits. By holding buyer countries liable—and through them, the domestic actors engaged in CDM transactions—the incentive for being honest about credits is increased.

Concluding Remarks

The credibility of traded permits is key in international emission permits markets. Permit markets can exacerbate or alleviate international compliance problems. In trading among Annex I countries, the potential compliance benefits from adding buyer liability to seller liability are outweighed by the high transaction costs created and the level of risk buyers would still have to bear. In contrast, for trading with non-Annex I countries through the CDM, the primary form of liability should be buyer liability. Buyers have considerably stronger incentives to comply than sellers do because buyers are more vulnerable to punishment and their governments are able to enforce domestic regulation. Because trades are project-based, subnational buyers are able to observe and control accurate reporting.

Suggested Reading

General

Baron, Richard. 1999. *An Assessment of Liability Rules for International GHG Emissions Trading.* Unpublished report prepared for OECD and IEA Project for the Annex I Expert Group on the UNFCC. Paris, France: International Energy Agency.

Corfee-Morlot, J. 1998. Monitoring, Reporting and Review of National Performance under the UNFCCC and the Kyoto Protocol. OECD Information Paper. Paris, France: Organisation for Economic Co-operation and Development. http://www.oecd.org/env/docs/cc/MON_98.pdf.

Fisher, B., S. Barrett, P. Bohm, M. Kuroda, J.K.E. Mubazi, A. Shah, and R. Stavins. 1996. An Economic Assessment of Policy Instruments to Combat Climate Change. In *Change 1995: Economic and Social Dimensions of Climate Change*, edited by J. Bruce, H. Lee, and E. Haites. Contribution of Working Group III to the Second Assessment Report of the Intergovernmental Panel on Climate Change. Cambridge, U.K.: Cambridge University Press.

Goldemberg, Jose. 1998. *Issues and Options: The Clean Development Mechanism.* New York: U.N. Development Programme.

Hahn, Robert, and Robert N. Stavins. 1995. Trading in Greenhouse Permits: A Critical Examination of Design and Implementation Issues. In *Shaping National Responses to Climate Change*, edited by Henry Lee. Washington, DC: Island Press.

Haites, E. 1998. International Emissions Trading and Compliance with Greenhouse Gas Emissions Limitation Commitments. Working paper for International

Academy of the Environment Policy Dialogue on International Emissions Trading under the Kyoto Protocol: Rules, Procedures and the Participation of Domestic Entities. September 6–7, 1998, Geneva, Switzerland.

Haites, Erik, and Farhana Yamin. 2000 The Clean Development Mechanism: Proposals for Its Operation and Governance. *Global Environmental Change* 10(1): 27–45.

Hargrave, Tim, Suzi Kerr, Ned Helme, and Tim Denne. 2000. Treaty Compliance as Background for an Effective Trading Program. In *Global Emissions Trading: Key Issues for Industrialized Countries,* edited by Suzi Kerr. Gloucestershire, U.K.: Edward Elgar.

Kerr, Suzi. Forthcoming. Additional Compliance Issues Arising from Trading. In *Global Emissions Trading: Key Issues for Industrialized Countries,* edited by Suzi Kerr. Gloucestershire, U.K.: Edward Elgar.

Mitchell R., and A. Chayes. 1995. Improving Compliance with the Climate Change Treaty. In *Shaping National Responses to Climate Change*, edited by H. Lee. Washington, DC: Island Press.

OECD (Organisation for Economic Co-operation and Development). 1992. *Climate Change: Designing a Tradable Permit System.* Paris, France: OECD.

Tietenberg, Tom, Michael Grubb, Byron Swift, Axel Michaelowa, Zhong Xiang Zhang, and Frank T. Joshua. 1998. *Greenhouse Gas Emissions Trading: Defining the Principles, Modalities, Rules, and Guidelines for Verification, Reporting, and Accountability.* Geneva, Switzerland: U.N. Conference on Trade and Development.

Tietenberg, Tom, and David G. Victor. 1994. Possible Administrative Structures and Procedures. In *Combating Global Warming: Possible Rules, Regulations and Administrative Arrangements for a Global Market in* CO_2 *Emissions Entitlements.* New York: United Nations, U.N. Conference on Trade and Development.

Werksman, J. 1996. Designing a Compliance System for the U.N. Framework Convention. In *Improving Compliance with International Environmental Law,* edited by J. Cameron, J. Werksman, and P. Roderick. London, U.K.: Earthscan Publications Ltd.

Wiener, Jonathan. 1999. Prices vs. Quantities: The Impact of the Legal System. *Yale Law Journal* 108(4): 677–800.

Yamin, Farhana. 1996. The Use of Joint Implementation to Increase Compliance with the Climate Change Convention. In *Improving Compliance with International Environmental Law,* edited by J. Cameron, J. Werksman, and P. Roderick. London, U.K.: Earthscan Publications Ltd.

Technical

Becker, Gary S. 1968. Crime and Punishment: An Economic Analysis. *Journal of Political Economy* 76(2): 169–217.

Chayes, Abram, and Antonia Handler Chayes. 1993. On Compliance. *International Organization* 47(Spring): 175–205.

Hagem, C. 1996. Joint Implementation Under Asymmetric Information and Strategic Behavior. *Environmental and Resource Economics* 8: 431–47.

Harrington, Winston. 1988. Enforcement Leverage When Penalties are Restricted. *Journal of Public Economics* 37(1): 29–53.

Holmström, B. 1982. Moral Hazard in Teams. *Bell Journal of Economics* 13: 324–40.

Holmström, B., and P. Milgrom. 1990. Regulating Trade Among Agents. *Journal of Institutional and Theoretical Economics* 146: 85–105.

Jackson, Tim. 1995. Joint Implementation and Cost-Effectiveness under the Framework Convention on Climate Change. *Energy Policy* 23(2): 117–38.

Keohane, Robert O., and Marc A. Levy (eds.). 1995. *Institutions for Environmental Aid.* Cambridge, MA: MIT Press.

Loske, R., and S. Oberthür. 1994. Joint Implementation under the Climate Change Convention. *International Environmental Affairs* 6(1): 45–58.

Martin, Lisa L. 1993. Credibility, Costs, and Institutions: Cooperation on Economic Sanctions. *World Politics* 45(3): 406–32.

Milgrom, P., D. North, and B. Weingast. 1990. The Role of Institutions in the Revival of Trade. *Economics and Politics* 2(1): 1–23.

Mitchell, Ronald B. 1996. Compliance Theory: An Overview. In *Improving Compliance with International Environmental Law,* edited by J. Cameron, J. Werksman, and P. Roderick. London, U.K.: Earthscan Publications Ltd.

Mitchell, Ronald B. 1994. Regime Design Matters: Intentional Oil Pollution and Treaty Compliance/ *International Organization* 48(3): 425–58.

Tirole, Jean. 1992. Collusion and the Theory of Organizations. In *Advances in Economic Theory: Sixth World Congress of the Econometric Society.* Vol. II, edited by Jean-Jacques Laffont. Cambridge, U.K.: Cambridge University Press.

Varian, Hal R. 1990. Monitoring Agents with Other Agents. *Journal of Institutional and Theoretical Economics* 146: 153–74.

23 International Equity and Climate Change Policy

Marina V. Cazorla and Michael A. Toman

Defining (or divining) an internationally equitable distribution of the burdens of reducing climate change risks has been a core concern for as long as greenhouse gas (GHG) emissions policies have been debated. Countries clearly differ greatly in their vulnerability to climate change, their historical and projected contributions to global GHG emissions, and their ability to bear the costs of mitigating GHG emissions. And neither history nor philosophy provides a definitive guide to what would constitute a fair distribution of burden.

In this chapter, we first review briefly the history and background of the international climate equity debate. Next, we put the equity debate in the broader context of the challenge of achieving international environmental (or other) agreements in general. Then, we review the many alternative equity principles that have been advanced for defining common but differentiated responsibilities; our key conclusion is that no single principle can be expected to govern resolution of this issue. However, approaches that involve adjusting responsibilities over time on the basis of more than one criterion may offer more promise for successful negotiation over the longer term. We conclude with some observations concerning constructive first steps for advancing international agreement on sharing the burden of risk associated with climate change.

Background

Article 3 of the 1992 U.N. Framework Convention on Climate Change (UNFCCC) contains the following language: "The Parties should protect the climate system for the benefit of present and future generations of humankind, on the basis of equity and in accordance with their common but differentiated responsibilities and respective capabilities. Accordingly, the developed country Parties should take the lead in combating climate change and the adverse effect thereof."

The phrase "common but differentiated responsibilities" has become a touchstone for the international climate equity debate. During the first Conference of Parties to the UNFCCC (COP-1; held in 1995 in Berlin, Germany), negotiators debated whether developing countries, in addition to developed countries, would commit to binding reductions for GHG emissions. The developing countries ultimately rejected binding commitments, asserting that the historical responsibility for climate change was not theirs, that they had less financial ability to pay for reductions, and that they had more urgent priorities for their limited resources.

In the Kyoto Protocol negotiated at the third Conference of Parties (COP-3) in December 1997, the Annex I developed countries (referred to as Annex B countries in the protocol document) agreed to legally binding commitments to reduce their collective GHG emissions by an average of 5% compared with 1990 levels during the first commitment period (2008–12). Consistent with the Berlin Mandate at COP-1, developing countries did not agree to any targets at Kyoto. Developing countries share with developed countries some common obligations for emissions monitoring and reporting, and under the UNFCCC, all countries are generally exhorted to take steps to enhance sustainable development that would limit the growth of GHG emissions. However, developing countries at Kyoto did not accept language that would have required them to pursue "best efforts" to reduce GHG emissions through such measures as energy market reforms and also rejected language that would have set future limits on their emissions as a negotiating target.

It is worth noting that different percentage emissions limits were agreed to by Annex I countries in the Kyoto Protocol. The differentiated targets among Annex I countries were not based on a standardized formula but rather on complex negotiations on multiple issues. In addition, Article 4 of the protocol allows groups of countries to negotiate a redistribution of their collective GHG emissions limits. This flexibility has been used by the European Union to set widely varying emissions targets for GHGs among member countries. These complex procedures may foreshadow even more complex future negotiations among developed and developing countries over obligations for GHG emissions control.

Debate over a long-term equitable sharing of the burden has continued since the negotiation of the Kyoto Protocol, while the details of the protocol's implementation are still being negotiated. Several developing country representatives continue to press the claim that the Annex I countries should do even more to reduce GHG emissions given their historical responsibility for GHG emissions. Developed countries accept the concept of differentiated responsibilities but want greater assurance on when and how developing countries would start assuming greater responsibilities, because developing countries are likely to be the primary industrial emitters in the future. This concern was underscored in the Byrd–Hagel resolution, passed 95–0 by the U.S.

Senate prior to COP-3. That resolution opposed ratification of a Kyoto treaty until developing countries committed to binding emission limits in the same time frame as the United States. Developing countries counter that developed countries generally have done little to implement their previous voluntary reduction commitments, even while some developing countries have slowed their emissions growth through reforms in the economic and energy sectors. This ongoing debate about the adequacy of commitments illustrates how differently countries interpret what constitutes fairness or equity in mitigating the risks of global climate change.

In the midst of this debate, one set of observations is widely accepted: Developed countries are responsible for the largest share of cumulative past GHG emissions by far. Moreover, current emissions intensities vary dramatically between rich and poor countries (see Table 1). In addition, developing countries need to increase their emissions over some period in the future as a necessary consequence of critically needed economic growth and improved living standards. Ultimately, however, the sources of climate change risk (GHG emissions and land use changes) are globally distributed. Therefore, responsibility for resolving the problem also must ultimately be widely shared. This point is vividly illus-

Table 1. Comparison of Emissions Intensities among Countries.

	Per capita GDP[a] (1995$)	Per capita CO_2 emissions (metric tons, 1995)	CO_2/GDP (kg/$)
United States	26,980	20.5	0.76
European Union[b]	19,050[c]	7.9[c]	0.41
Russia	4,820	12.2	2.53
China	2,970	2.7	0.91
Africa	1,760[c]	1.1[c]	0.63
India	1,420	1.0	0.70

[a] Based on estimated purchasing power parities rather than exchange rates.

[b] Luxembourg is not included.

[c] Population figures are from 1998.

Source: World Resources Institute (see Suggested Reading).

trated by the calculations of future changes in atmospheric GHG emissions presented by Jacoby and others (see Suggested Reading), who analyze the consequences of having continued and strengthened GHG emissions controls within the Annex I industrialized countries while allowing rapid business-as-usual emissions growth in the developing world. Their projections indicate that even if the current developed world drives its net GHG emissions to zero by the end of this century, the effects on the atmosphere would be small given the strong expected future growth of emissions in developing countries.

It is obvious from Table 1 that emission patterns and the ability to absorb the costs of GHG emissions mitigation (as indicated by per capita income) vary widely across countries. One other point worthy of mention is the differential vulnerability of developing and developed countries to climate change. The IPCC Second Assessment Report presented results of studies showing that the damaging effects of a doubling of concentrations of GHG emissions in the atmosphere could cost on the order of 1.0–1.5% of gross domestic product (GDP) for developed countries and 2.0–9.0% of GDP for developing countries, with some developing countries (for example, low-lying island states) much more vulnerable still. Although controversy surrounds the scale of these estimates, the increased vulnerability of developing countries is subject to little debate. Their high vulnerability reflects both a greater dependence on natural systems (such as agriculture), which could be affected by climate change, and a more limited capacity to adapt to climate change given limited resources.

International Negotiation of GHG Mitigation

The problem of achieving effective and lasting international agreements can be stated simply: A self-enforcing deal is easiest to strike when the stakes are relatively small, or when no other option exists (a clear and present risk). No global police organization exists to enforce an international climate agreement. As such, an agreement must be voluntary and self-enforcing—all sovereign parties must have no incentive to deviate unilaterally from the terms of the

Explaining Trends in CO_2 Emissions

Trends in CO_2 emissions can be explained by changes in population, the GDP-to-population ratio (per capita GDP), the energy-to-GDP ratio (energy intensity ratio), and the CO_2-to-energy ratio (the carbon intensity ratio). They are illustrated by the following equation:

$$CO_2 \text{ emissions} = (\text{Population}) \times (\text{GDP/Person}) \\ \times (\text{Energy consumption/Unit GDP}) \\ \times (CO_2 \text{ emissions/Unit energy consumption})$$

Other things being equal, slower population growth means less growth in CO_2 release, whereas higher GDP per capita signifies a greater volume of CO_2 emitted. Energy/GDP is a measure of an economy's aggregate energy intensity that reflects the structural, technological, and energy-use characteristics of the society. All else unchanged, a falling energy-to-GDP ratio means less CO_2 emitted. The forces that can contribute to a falling energy-to-GDP ratio—such as efficiency improvements through use of combined-cycle technology in generating electricity—are key factors in considering CO_2 abatement strategies. Finally, the CO_2/Energy element spotlights the effect of a changing mix of energy sources and forms with varying carbon characteristics.

Clearly, an important issue in the determination of CO_2 mitigation possibilities and costs is the ease or difficulty of altering that mix away from carbon-intensive components (such as coal) toward ones lean in, or devoid of, carbon (for example, natural gas, solar power, and nuclear energy).

agreement. As part of this negotiation process, countries can negotiate various enforcement provisions, such as trade sanctions for noncompliance. This does not change the basic point that the agreement as a whole, including enforcement provisions, must satisfy the self-interest of the participants. (Sanctions also must be credible to be meaningful; if sanctions will cost more than the benefit of obtaining greater compliance, they are not credible.)

Nations have a common interest in responding to the risk of climate change, yet many are reluctant to reduce GHG emissions voluntarily or unilaterally. They hesitate because climate change is a global public good—all nations can enjoy protection against the risks of climate change, regardless of whether they participate in a treaty intended to mitigate those risks. Each nation's incentive to reduce emissions is thus limited because it cannot be prevented from enjoying the fruits of other nations' efforts. This incentive to free ride reflects the divergence between national actions and global interests.

A self-enforcing agreement is hardest to achieve in the gray area between low and infinite stakes. By free riding, some nations can be better off refusing an agreement. The greater the global net benefits of cooperation, the stronger the incentive to free ride; therefore, a self-enforcing agreement is harder to maintain. A self-enforcing agreement is most easily maintained when the global net benefits are not much bigger than those in the absence of an agreement. This is a basic paradox of international agreement.

To achieve significant reductions in GHG emissions, it is not necessary that every country in the world participate. The Organisation for Economic Co-operation and Development (OECD) countries, eastern Europe, Russia, and Ukraine, along with a few large developing countries (China, India, and Brazil) account for most of the current volume of GHG emissions and a significant fraction of future emissions as well. However, even a self-enforcing agreement that involved only a subset of the world's emitters would probably lead to total emissions in excess of desirable global targets if several small nations refused to agree, and achieving a self-enforcing agreement even among a few major emitting countries or regions is not easy, as the negotiations among the Annex I countries reveal.

Many decisionmakers in industrialized countries worry about the consequences to their economies of reducing emissions while developing countries face no limits. This situation could adversely affect comparative advantages in the industrialized world, whereas the "leakage" of emissions from controlled to uncontrolled countries would limit the environmental effectiveness of a partial agreement. Estimates of carbon leakage vary from a few percent to more than one-third of the Annex I reductions, depending on model assumptions regarding substitutability of different countries' outputs and other factors. On the other hand, many developing countries see more immediately pressing national needs, such as clean water and a stable food supply. Given their limited resources, developing countries will be less inclined to join an agreement that they see as imposing unacceptable costs on them, even if the costs are manageable and acceptable for developed countries.

All of these considerations must be incorporated when addressing equity issues related to GHG emissions agreements. Equity may be one motivation for countries to pursue GHG emissions policies. However, equity principles will not override other elements of national self-interest. Moreover, differences in perceptions about what constitutes equitable distributions of effort complicate any agreement. The international policy objective is obvious but elusive: finding incentives to motivate nations with strong and diverse self-interests to move voluntarily toward a collective goal of reduced carbon emissions.

Equity Principles for Burden Sharing

The concept of equity can be interpreted in many ways, and it often has been the root of misunderstandings and conflict between developed and developing countries in past negotiations. However, any criteria that might be used to distribute current and future burdens of GHG mitigation must be based, explicitly or otherwise, on some concept of equity. Table 2 summarizes some of the many equity principles that have been applied to the issue of sharing the burden of climate risk, with illustrations of how each could be translated into rules for the distribution of GHG emissions abatement costs. Obviously, some of the categories in the table overlap. Moreover, most of the principles would yield more than one possible distribution of mitigation burden and cost, depending on how key parameters are set.

The *egalitarian* principle is popular with many developing countries, because it would shift most abatement responsibility to developed countries

Table 2. Alternative Equity Criteria for Climate Change Policy.

Equity principle	Interpretation	Implied burden-sharing rule
Egalitarian	People have equal rights to use atmospheric resources.	Reduce emissions in proportion to population or equal per capita emission.
Ability to pay	Equalize abatement costs across nations relative to economic circumstances.	Net cost proportions are inversely correlated with per capita GDP.
Sovereignty	Current rate of emissions constitutes a status quo right now.	Reduce emissions proportionally across all countries to maintain relative emission levels between them ("grandfathering").
Maxi-min	Maximize the net benefit to the poorest nations.	Distribute the majority of abatement costs to wealthier nations.
Horizontal	Similar economic circumstances have similar emission rights and burden sharing responsibilities.	Equalize net welfare change across countries so that net cost of abatement as a proportion of GDP is the same for each country.
Vertical	The greater the ability to pay, the greater the economic burden.	Set each country's emissions reduction so that net cost of abatement grows relative to GDP.
Compensation (Pareto rule)	"Winners" should compensate "losers" so that both are better off.	Share abatement costs so that no nation suffers a net loss of welfare.
Market justice	Make greater use of markets.	Create tradable permits to achieve lowest net world cost for emissions abatement.
Consensus	Seek a political solution that promotes stability.	Distribute abatement costs (power weighted) so the majority of nations are satisfied.
Sovereign bargaining	Principles of fairness emerge endogenously as a result of multistage negotiations.	Distribute abatement costs according to equity principles that result from international bargaining and negotiation over time.
Polluter pays	Allocate abatement burden corresponding to emissions (may include historical emissions).	Share abatement costs across countries in proportion to emission levels.
Kantian allocation rule	Each country chooses an abatement level at least as large as the uniform abatement level it would like all countries to undertake.	Differentiate by country's preferred world abatement, possibly in tiers or groups.

Source: Adapted from Burtraw and Toman, Ringius and others, and Rose 1992 (see Suggested Reading).

where per capita emissions are much higher than in developing countries. The *sovereignty* principle in its strict form implies symmetrical cost burdens. The *ability to pay* approach would allocate the majority of costs to wealthy nations; developing nations would pay more as they became wealthier over time and increasingly were able to pay for mitigation. The *maxi-min* principle would maximize the welfare of the least-well-off countries by distributing the majority of abatement costs to wealthier nations.

The next three approaches are based on allocation of net benefit, not only cost. *Horizontal* and *vertical* equity principles seek to distribute costs relative to GDP. The meaning of these terms in this context is similar to their meaning in debates about tax policy: Horizontal equity seeks to treat equals equally, whereas vertical equity seeks to increase the burden as the ability to pay grows. The *compensation* approach would entail payments through income transfers or investment to avoid loss by any nation.

The next three principles are process-based and assume that the correct process of cost allocation will result in an equitable outcome. The principle of *market justice* is based on the theory that a free market mechanism for cost allocation is the fairest because its efficiency results in greater overall welfare. *Consensus* equity simply asserts that the process of political negotiation itself results in an equitable outcome. A variation on the consensus principle is the *sovereign bargaining* principle, in which principles of fairness emerge endogenously as a result of international multistage negotiations.

The *polluter pays* principle is self-explanatory and relatively simple. Although developed countries are responsible for the majority of past and present pollution, developing countries will be the primary emitters within a few decades. Thus, this principle implies a reallocation of burdens over time (as do other rules that depend on cost or income shares). Under the *Kantian allocation rule*, each country would choose an abatement level at least as large as the uniform abatement level it would like all countries to undertake, ensuring that countries' true preferences for abatement are revealed so that each country takes on a relative cost burden equal to what it expects of other countries.

The criteria in Table 2 are presented in a relatively static context. Interest is now growing in the possibility of long-term differentiation and graduation formulas for burden sharing. These formulas would define the terms under which developing countries would accept commitments and how the commitments of developed countries would be strengthened over time.

To illustrate this approach, Nordhaus proposed a climate allowances protocol that "would embed the price and quantity [of] targets within a framework that considers both environmental and economic objectives and sets the policies to maximize net benefits" (see Suggested Reading). The climate allowances protocol would issue tradable allowances or permits with carbon price ceilings and floors that are derived from a dynamic cost–benefit analysis of slowing climate change. Participant countries would be allocated emissions permits for each budget period according to what are in essence rolling targets

The Montreal Protocol Model

The 1987 Montreal Protocol on Substances that Deplete the Ozone Layer is one of the best examples of successful implementation of equity principles in relation to an international environmental issue. The participation of developing countries was one of the reasons for the success of the Montreal Protocol, and their participation was primarily due to the protocol's inclusion of developing country concerns about equity, economic constraints, and flexibility. The IPCC Second Assessment Report cites the following elements as important in encouraging developing country participation (see Banuri and others in Suggested Reading):

- differentiated standards for developed and developing country parties,
- additional financial assistance to developing country parties,
- technology transfer facilitated by protocol financial resources if necessary, and
- acknowledgement that developing country compliance was contingent on effective implementation of financial assistance and technology transfer obligations.

The Montreal Protocol was frequently discussed as a model prior to the Kyoto Protocol. However, the Montreal Protocol addressed a problem that was in several ways less complex to resolve than the climate change problem (see Chapter 2, How the Kyoto Protocol Developed). That point notwithstanding, the questions of financial assistance and technology transfer continue to be important topics in climate change negotiations.

rather than targets based on an arbitrary historical benchmark. Developing countries would begin to participate as their economies grew, making small emissions reductions as their per capita incomes reached entry levels and making greater emissions reductions when their incomes reached full participation levels. The allocation of permits would be based on a formula that reflects each country's uncontrolled emissions and its ability to pay. The climate allowances protocol would provide an incentive to participate through the imposition of duties on the carbon content of imports from nonparticipants into participating countries. The aim of this proposal is thus to promote forward-looking rather than backward-looking formulas for the allocation of emissions reduction costs and burdens. How well a forward-looking approach (with essentially endogenous targets) could work remains unclear.

The idea of making per capita emissions the basis for equitable burden sharing is a much-discussed option that is favored by many developing countries. Such formulas are often referred to as convergence measures. However, different formulas have different bases for convergence and thus different consequences.

A dynamic example of this approach from the Global Commons Institute is "contraction and convergence" (see Suggested Reading). Under this option, developed countries would reduce emissions over time in proportion to their population, and developing countries would increase emissions according to their population. Eventually, developed and developing countries would converge to the same per capita emissions ratio. For the environmental goals of the UNFCCC to be met, the ratio and length of expected of time until convergence would have to be calculated to ensure the necessary amount of GHG emissions reductions. Another example of this approach is a Brazilian proposal made during the Kyoto Protocol negotiations to allocate permits according to relative historical contribution to global warming. Various criteria for contributing to cumulative climatic impacts could be used to determine which countries have reached a threshold for sharing the burden of emission reduction.

Edmonds and Wise attempt to sidestep direct debate over equity rules by proposing a technology-based graduation framework (see Suggested Reading). In their proposal, any new fossil fuel electric power capacity in Annex I nations installed after the year 2020 must scrub and dispose of the carbon from its exhaust stream, any new synthetic fuels capacity must capture and dispose of carbon released in the conversion process, and non–Annex I nations that participate must undertake the same obligations that Annex I nations undertake when their per capita income equals the average for Annex I nations in 2020. The authors note that in their technology graduation scenario, China is the first to join the protocol in approximately 2040, and countries in South Asia and East Asia join by 2055. They acknowledge that a technology-based protocol would be economically inefficient but state that its graduation mechanism would allow developing nations to achieve an economic status comparable to that of developed nations before being required to undertake binding obligations.

Several authors also have suggested that by broadening the number of elements included in the terms of agreement among negotiating parties, multiple-criteria formulas may make political agreement among negotiating entities with very disparate interests more likely. For example, Ridgley proposed that several equity criteria could be used together (see Suggested Reading). The factors he considers include population (as an indicator of entitlement), baseline emissions (as an indication of responsibility), and ability to pay. Grubb adds to this list factors such as basic trends in energy efficiency, competitiveness, and institutional capacities in arguing for a sectorally differentiated approach (see Suggested Reading).

Analyzing Impacts of Different Burden-Sharing Rules

Proposals for per capita contract-and-converge options face several challenges in reaching agreement. China and India would be able to increase their national emissions significantly before reaching the population-to-emissions ratio limit because of their large and rapidly growing populations and rapid economic growth. For this increase to not endanger environmental goals, stringent long-term environmental limits would have to be placed on the industrialized countries, or the emissions ceiling for the developed world would have to grow less than proportionally with future population increases. In the absence of tighter constraints on the growth of developing countries' ceilings, the convergence limits would in turn imply a level of either domestic compliance cost (without international emissions trading) or wealth transfer (with trading) that is currently unacceptable to developed countries. By the same token, however, allocating emission limits based on historical emissions conveys a huge advantage to the developed world, given their historically overwhelming share of emissions. Simply extending this approach to developing countries limits their incentives to pursue modest "win-win" emissions controls in the short term, for fear that doing so would put them on a lower emissions baseline and jeopardize their position in future negotiations.

It is also unlikely that a majority of nations would accept a multiple-criteria approach simply because it includes a lot of equity principles. Countries will still look to see how a specific formula affects their self-interest. Contentious negotiation on the criteria and weights would still be required, even in determining how to divide burdens among developing countries. For example, within Latin America, Brazil and Honduras have large differences with regard to wealth, past and current emissions, and vulnerability to climate change impacts.

Several modelers have attempted to assess the potential outcomes of different burden-sharing criteria. Unfortunately, it is difficult to compare results because each model uses different time frames, parameters, variables, and methods. One analysis by Rose and others attempted to quantify the minimum cost of the different equity principles for each of nine world regions: United States, Canada and western Europe, other OECD countries, eastern Europe and the former Soviet Union, China, the Middle East, Africa, Latin America, and Southeast Asia (see Suggested Reading). The analysis is framed in terms of a global allocation of a fixed number of tradable emission permits. Under this scenario, global allocation of abatement activity is cost-effective in all cases, and the initial permit allocation affects only the distribution of costs. The study used a nonlinear programming model based on an abatement cost function that corresponded to each equity principle to calculate the net cost of implementing each principle. The analysis of the cost allocation from each principle was performed for three future years (2005, 2020, and 2035), and this cost was discounted to a 1990 present value. Table 3 illustrates the findings of the study.

The analysis of the distributional implications of the different principles in Table 3 illustrates the large range of net cost outcomes for a given country or region. Under the egalitarian principle, for exam-

Table 3. Cost of Different Allocation Rules in 2020 (present value, in billions of $1990).

Country or area	Sovereignty	Egalitarian	Horizontal	Vertical	Consensus
United States	44.1	354.5	52.2	95.7	121.1
Canada and western Europe	29.9	156.2	38.1	19.2	17.8
Other OECD countries	8.0	65.3	21.6	50.9	−25.7
Eastern Europe and former Soviet Union	37.2	337.6	24.2	7.5	272.0
China	23.3	−109.1	8.3	0.1	43.2
Middle East	6.3	1.1	8.4	9.2	−11.0
Africa	8.1	−226.3	5.2	0.1	99.6
Latin America	7.4	56.6	8.1	0.5	−21.4
Southeast Asia	13.0	345.5	10.9	0.1	−119.3

Notes: The operational rules for the equity principles in this analysis are as follows: Sovereignty distributes permits in proportion to emissions; egalitarian distributes permits in proportion to population; horizontal distributes permits to equalize net welfare change (net gain or loss as proportion of GDP equal for each nation); vertical progressively distributes permit (net gain/loss) proportions inversely with per capita GDP; consensus distributes permits in a manner that satisfies the (power weighted) majority of nations.

Source: Rose and others (see Suggested Reading).

ple, large amounts of wealth are transferred from the United States and the former Soviet Union to China, Africa, and Southeast Asia. The vertical principle results in large burdens for OECD countries and minute burdens for developing countries, but no wealth transfer. The sovereignty and horizontal equity principles result in roughly similar outcomes.

The use of per capita emissions as the criteria for emissions reduction was also the focus of a study by Manne and Richels (see Suggested Reading). They calculate the welfare effects on five regions (United States, OECD countries, the former Soviet Union, China, and "rest of world" [ROW]) of a transition from allocation of emissions permits based on 1990 baseline emissions levels to allocation based on per capita emissions. In a relatively rapid transition completed by 2030, all cooperating regions benefit from reduced climate change damages compared with a business-as-usual scenario. However, under this scenario, the majority of the burden falls on the United States, OECD countries, and the former Soviet Union, whereas China and ROW benefit from reduced damages and the sale of emissions permits. The distribution of overall burdens is very different in a scenario of slower transition from permit allocation according to 1990 baselines to permit allocation according to per capita emissions with the transition completed by 2200 (170 years later than in the rapid transition scenario). In this scenario, damage and nonmarket costs are the same as in the rapid transition scenario, but the OECD countries receive positive net benefits and the rest of the world no longer benefits due to smaller wealth transfers. China still benefits, but not as much as under the rapid transition scenario.

Reiner and Jacoby argue that it would be most equitable to allocate differentiated burdens based on a starting point of zero for all nations instead of historical emissions levels (such as 1990 emissions; see Suggested Reading). Starting from this premise, their Emissions Prediction and Policy Assessment (EPPA) model calculates a percentage allocation of emissions permits according to either per capita emissions or the carbon intensity of GDP. The resulting allocations are shown in Table 4.

As shown in Table 4, zero-based allocation according to a per capita emissions rule would require draconian decreases in allowed emissions from OECD countries and allow large increases for populous less-developed countries such as China and India. Zero-based allocation according to carbon intensity of GDP, however, results in a distribution of allowed emissions that is closer to the status quo and therefore would provide no headroom for growth by developing countries. Using the carbon intensity ratio, the United States would receive 25% of all permits, the European Union would receive 16%, and energy-exporting countries (such as OPEC contains) would receive 14%.

In 2000, Rose and Stevens simulated net benefits for different countries or areas in various scenarios that involved participation in emissions limits and international emissions permit trading (see Suggested Reading). Net benefits incorporate both avoided climate change damages and the costs of abating GHG emissions, plus potential revenue flows from permit sales and purchases. Scenario 1 is a base case with fixed Kyoto emission targets for de-

Table 4. Percentage Allocation of Emissions Permits According to the Emissions Prediction and Policy Assessment (EPPA) Model.

	Allocation according to C/POP (%)	Allocation according to C/GDP (%)
United States	4	25
Japan	2	10
European Union	5	16
Other OECD countries	2	6
Former Soviet Union	5	6
Central and eastern Europe	2	2
Energy-exporting countries	16	14
India	16	3
Brazil	3	2
China	20	8
Dynamic Asian economies	3	3
Rest of the world	22	5

Note: C = carbon; POP = population.

Source: Reiner and Jacoby (see Suggested Reading).

veloped countries but no interperiod or interregional permit trading; in Scenario 2, all developing countries must constrain their emissions starting in 2020 based on 2020 emissions levels; and in Scenario 3, developing country accession to a GHG emissions protocol is based on per capita income (PCI). The authors divided developing countries into two groups: Group A, with greater than $1,000 PCI (Middle East, rapidly developing "Asian Tigers," and Latin America), and Group B, with less than $1,000 PCI (China, South Asia, India, and Africa). Group A developing country limits begin in 2010, and Group B developing country limits begin in 2020, with Kyoto target levels still in effect for Annex I countries. Their findings are summarized in Table 5.

In Scenario 1, the present discounted value of the global net benefits of achieving the Kyoto targets by Annex I countries is actually a loss of $12.6 billion. Net benefits are also negative for Annex I (that is, industrialized) countries but positive for developing countries, particularly China and South Asia. In Scenario 2, all regions except for Africa, India, and the Asian Tigers are better off than in the baseline Kyoto case. In Scenario 3, global net benefits are less relative to Scenario 2 because the increases in mitigation costs for Group A countries greatly outweigh the benefits not only to themselves but also to all other countries; Group A countries are worse off and all Group B and Annex I countries are better off than in the baseline Kyoto case. In each scenario, some countries are "losers," and losing countries could require compensation in return for their participation in the climate change mitigation regime.

Conclusions

Attempts to limit global climate change by mitigating GHG emissions would be futile without relatively broad international participation. Global participation also is supported by a powerful economic argument for cost-effectiveness. Poorer countries tend to have less efficient technology and more rapid growth rates for population and per capita income than richer countries do. Consequently, poorer countries are likely to have a comparative advantage in carrying out mitigation, and richer countries are

Table 5. Summary of Net Benefits from Different Participation and Permit Trading Simulations.

Country or area	Scenario		
	1	2	3
Africa	24.2	16.6	11.2
Australia/New Zealand	−2.1	−1.3	−1.3
Canada	−6.9	−3.7	−4.4
China	81.0	88.8	123.9
Eastern Europe	−2.8	1.1	3.3
Former Soviet Union	−0.1	41.9	58.8
India	27.5	26.8	30.2
Japan	−36.1	−13.0	−15.9
Latin America	23.2	33.7	3.5
Middle East	9.6	20.2	−16.2
Asian Tigers	8.1	2.3	−17.8
South Asia	62.2	91.9	98.1
United States	−84.6	−50.3	−58.4
Western Europe	−115.8	−46.4	−56.2
Global net benefits	−12.6	208.6	158.8
Cumulative net benefit increase (relative to Kyoto base case)		221.2	171.4

Notes: Results are given in billions of present discounted 2010$ summed over the entire time horizon 2010–35 (4% discount rate). Scenario 1 = Kyoto quotas for Annex I countries; Scenario 2 = Kyoto, interperiod and interregional permit trading, with 2020 quotas for developing countries; Scenario 3 = Kyoto, interperiod and interregional permit trading, with Group A quotas in 2010 and Group B quotas in 2020. See text for additional discussion.

Source: Rose and Stevens 2001 (see Suggested Reading).

likely to have a comparative advantage in developing and supplying the necessary technical opportunities, at least over the short to medium term. But to achieve a successful international agreement for GHG emissions abatement, this cost-effectiveness argument must be supplemented with policies for acceptably distributing mitigation costs across countries as well as over time.

The application of simple, fixed principles of equity will most often result in winners and losers. For example, the net costs to major emitters such as the United States and China differ significantly depending on which equity principles or criteria are applied. As principles of equity are discussed and analyzed in the international arena, it is sometimes difficult to distinguish between a country's concern with equitable burden sharing under a climate change regime and its informed calculation of how use of a given principle for cost allocation might affect its welfare.

Our review of equity analyses highlights the difficulty of constructing a single formula—even a simple composite formula—that is in the basic self-interest of enough developed and developing countries to provide a foundation for organizing international climate change policy. Efforts to find a magic solution to equity disputes are likely to be in vain, and the question of which international climate policies will be equitable over the long term will require a great deal of additional time and effort to resolve.

However, it appears that dynamic graduation formulas offer a critical degree of flexibility for balancing the short-term concerns of developing countries (bearing excessive costs of mitigating GHG emissions) against the long-term concerns of developed countries (expanding participation in GHG emissions mitigation and reducing leakage). Although this kind of approach still requires difficult negotiation, it is broadly consistent with the general notion of "common but differentiated" responsibilities in the UNFCCC; it moves the debate from *whether* developing countries should act to *how* and *when*.

The long-term equity debate over climate change mitigation often has been framed as a property rights problem—how rights to use the global atmospheric commons for releases of GHG emissions might be allocated. Because different answers to this question have profoundly different implications for global wealth distribution, the solution to this problem is obviously quite difficult.

Critics claim that there is no hope for solving this problem. They endorse other solutions, such as individually administered national carbon taxes. However, this approach is not a panacea for distributional concerns. An initial allocation of rights and responsibilities is implicit in any international control agreement, including taxes. Moreover, the argument for taxes rests on the willingness of the developing world to implement substantially higher energy taxes than exist today. Although developing countries theoretically would reap some advantages of increased energy taxes (for example, more reliable revenue than from income taxes), it is unclear whether the advantages would be so compelling in practice. Without broad participation, the tax approach becomes an inefficient partial agreement, like the Kyoto Protocol. It also is difficult to monitor changes in domestic tax provisions that would nullify the effects of carbon taxes in the developing world and, to an extent, in the developed world. Several efficiency and political economy arguments favor a quantity-based over a tax-based approach; however, negotiating a global distribution of emission rights and responsibilities is a formidable task.

If the long-term climate equity problem is this difficult, then one important policy challenge is to keep this difficulty from impeding useful short-term progress toward cooperative mitigation of GHG emissions. Richer and poorer countries need to continue to develop relationships that will support long-term commitment to pursuing shared benefits in a mutually agreeable fashion. Given the uncertainties that surround future economic growth, climate change impacts, and willingness to pay to ameliorate those risks, this process needs to be adaptive rather than immediate. Initial steps to cooperate in climate policy and share benefits can be pursued without precluding options in the future.

The Clean Development Mechanism (CDM) offers one avenue toward short-term policy cooperation with shared benefits (see Chapter 21, Establishing and Operating the Clean Development Mechanism). It involves efforts to provide GHG emissions credits for developed countries through offsetting GHG emissions reductions in developing countries, thereby lowering the cost of mitigating GHG emissions in Annex I countries and providing local environmental, economic, and technological benefits in developing countries.

For this policy tool to be successful, developing countries need assurance that participation in the CDM meets their near-term needs and does not compromise their long-term interests. As for meeting near-term needs, analysis and experience should help to demonstrate the immediate economic and environmental benefits of CDM participation. Concrete steps in CDM design also can be useful—

notably, allowing developing countries to create and bank their own CDM emission credits so they are not subject to profit-taking market power by CDM credit buyers. As for the protection of long-term interest, Annex I countries need to establish a firm policy precedent that future binding limits for developing countries will not be based on some historical base year, which would penalize earlier efforts to reduce GHG emissions in those countries. This discussion could be the beginning of a broader negotiation of mutually acceptable graduation protocols.

Suggested Reading

Agarwal, A., and S. Narain. 1991. *Global Warming in an Unequal World: A Case of Environmental Colonialism*. New Delhi, India: Centre for Science and Environment.

Bac, Mehnet. 1996. Incomplete Information and Incentives to Free Ride on International Environmental Resources. *Journal of Environmental Economics and Management* 30(3): 301–15.

Banuri, T., K. Göran-Mäler, M. Grubb, H. K. Jacobson, and F. Yamin. 1996. Equity and Social Considerations. In *Climate Change 1995: Economic and Social Dimensions of Climate Change*, edited by James P. Bruce, Horsang Lee, and Erik F. Haites. Contribution of Working Group III to the Second Assessment Report of the Intergovernmental Panel on Climate Change. New York: Cambridge University Press, 83–124.

Barrett, Scott. 1994. Self-Enforcing International Environmental Agreements. *Oxford Economic Papers* 46: 878–94.

Benestad, Olav. 1994. Energy Needs and CO_2 Emissions: Constructing a Formula for Just Distributions. *Energy Policy* 22(9): 725–34.

Burtraw, Dallas, and Michael Toman. 1992. Equity and International Agreements for CO_2 Containment. *Journal of Energy Engineering* 118(2): 122–35.

Carraro, Carlo, and Dominico Siniscalco. 1993. Strategies for the International Protection of the Environment. *Journal of Public Economics* 52(3): 309–28.

Cooper, Richard. 1998. Toward a Real Global Warming Treaty. *Foreign Affairs* 77(2): 66–79.

den Elzen, Michel, Marco Janssen, Jan Rotmans, Rob Swart, and Bert de Vries. 1992. Allocating Constrained Global Carbon Budgets. *International Journal of Global Energy Issues* 4(4): 287–301. (Special Issue on Energy and Sustainable Development)

Edmonds, Jae, and Marshall Wise. 1999. Exploring a Technology Strategy for Stabilizing Atmospheric CO_2. In *International Environmental Agreements on Climate Change*, edited by Carlo Carraro. Dordrecht, the Netherlands: Kluwer Academic Publishers.

Frankhauser, Samuel, and others. 1998. Extensions and Alternatives to Climate Change Impact Valuation: On the Critique of IPCC Working Group III's Impact Estimates. *Environment and Development Issues* 3(1): 59–81.

Global Commons Institute. 1997. *Contraction and Convergence: A Global Solution to a Global Problem*. July. London, U.K.: Global Commons Institute. http://www.gn.apc.org/gci/contconv/cc.html (accessed October 17, 2000).

Grubb, Michael. 1989. *The Greenhouse Effect: Negotiating Targets*. London, U.K.: Royal Institute of International Affairs, Energy and Environmental Affairs Programme.

Hoel, Michael. 1992. International Environment Conventions: The Case of Uniform Reductions of Emissions. *Environmental and Resource Economics* 2(2): 141–59.

———. 1996. *Climate Change 1995: Economic and Social Dimensions of Climate Change*, edited by James P. Bruce, Horsang Lee, and Erik F. Haites. Contribution of Working Group III to the Second Assessment Report of the Intergovernmental Panel on Climate Change. New York: Cambridge University Press.

Jacoby, Henry, Ronald Prinn, and Richard Schmalensee. 1998. Kyoto's Unfinished Business. *Foreign Affairs* 77(4): 54–66.

Jacoby, Henry D., Richard Schmalensee, and Ian Sue Wing. 1998. *Toward a Useful Architecture for Climate Change Negotiations*. Cambridge, MA: MIT Joint Program on the Science and Policy of Global Change. http://web.mit.edu/globalchange/www/rpt49.html (accessed October 17, 2000).

Jacoby, Henry D., Richard S. Eckaus, A. Denny Ellerman, Ronald G. Prinn, David M. Reiner, and Zili Yang. 1997. *CO_2 Emissions Limits: Economic Adjustments and the Distribution of Burdens*. Report 9. July. Cambridge, MA: MIT Joint Program on the Science and Policy of Global Change. http://web.mit.edu/globalchange/www/rpt9.html (accessed October 17, 2000).

Manne, Alan, and Richard Richels. 1995. The Greenhouse Debate: Economic Efficiency, Burden Sharing and Hedging Strategies. *The Energy Journal* 16(4): 1–37.

McKibbon, Warwick J., and Peter J. Wilcoxen. 2000. *Beyond the Kyoto Protocol*. Presented at the International Conference on the Sustainable Future of the Global System, United Nations University, May 24–25, Tokyo, Japan. http://www.msgpl.com.au/msgpl/download/unu300.pdf (accessed October 17, 2000).

Morgan, Granger. 2000. Climate Change: Managing Carbon From the Ground Up. *Science* 289(5488): 2285.

Nordhaus, William. 1997. *Climate Allowances Protocol (CAP): Comparison of Alternative Global Tradable Emissions Regimes.* Presented at the NBER Workshop on Design on International Emissions Trading Systems, June, Snowmass, CO.

Nordhaus, William D., and Zili Yang. 1996. A Regional Dynamic General-Equilibrium Model of Alternative Climate Change Strategies. *American Economic Review* 86(4): 741–65.

Paterson, Matthew, and Michael Grubb (eds). 1996. *Sharing the Effort: Options for Differentiating Commitments on Climate Change.* London, U.K.: The Royal Institute of International Affairs.

Pearce, David, and others. 1996. The Social Costs of Climate Change: Greenhouse Damage and the Benefits of Control. In *Climate Change 1995: Economic and Social Dimensions of Climate Change,* edited by James P. Bruce, Horsang Lee, and Erik F. Haites. Contribution of Working Group III to the Second Assessment Report of the Intergovernmental Panel on Climate Change. New York: Cambridge University Press.

Reiner, David M., and Henry D. Jacoby. 1997. *Annex I Differentiation Proposals: Implications for Welfare, Equity and Policy.* Report 27. October. Cambridge, MA: MIT Joint Program on the Science and Policy of Global Change. http://web.mit.edu/globalchange/www/rpt27.html (accessed October 17, 2000).

Ridgley, Mark. 1996. Fair Sharing of Greenhouse Gas Burdens. *Energy Policy.* 24(6): 517–29.

Ringius, Lasse. 1997. *Differentiation, Leaders and Fairness: Negotiating Climate Commitments in the European Community.* Report 1997-08. Oslo, Norway: Center for International Climate and Environment Research.

Ringius, Lasse, Asbjørn Torvanger, and Bjart Holtsmark. 1998. Can Multi-Criteria Rules Fairly Distribute Climate Burdens? *Energy Policy* 26(10): 777–93.

Rose, Adam. 1990. Reducing Conflict in Global Warming Policy. *Energy Policy* 18(10): 927.

———. 1992. Equity Considerations of Tradable Carbon Emission Entitlements. *Combating Global Warming: Study on a Global System of Tradable Carbon Emission Entitlements.* New York: U.N. Conference on Trade and Development.

Rose, Adam, and Brandt Stevens. 1993. The Efficiency and Equity of Marketable Permits for CO_2 Emissions. *Resource and Energy Economics* 15(1): 117–46.

———. Forthcoming. An Economic Analysis of Flexible Permit Trading in the Kyoto Protocol. *International Environmental Agreements* 1(2): 219–42.

Rose, Adam, Brandt Stevens, Jae Edmonds, and Marshall Wise. 1998. International Equity and Differentiation in Global Warming Policy: An Application to Tradeable Emission Permits. *Environmental and Resource Economics.* 12(1): 25–51.

Tol, Richard S. J. 1995. The Damage Costs of Climate Change toward More Comprehensive Calculations. *Environment and Development Economics* 5: 353–74.

Toman, Michael. 2000. *Establishing and Operating the Clean Development Mechanism.* RFF Climate Issue Brief 22. Washington, DC: Resources for the Future. http://www.rff.org/issue_briefs/PDF_files/ccbrf22_Toman.pdf (accessed October 17, 2000).

Torvanger, Asbjørn, and Odd Godal. 1999. *A Survey of Differentiation Methods for National Greenhouse Gas Reduction Targets.* Report 1999-05. Oslo, Norway: Center for International Climate and Environmental Research (CICERO). http://www.cicero.uio.no/Publications (accessed October 17, 2000).

Welsch, Heinz. 1993. A CO_2 Agreement Proposal with Flexible Quotas. *Energy Policy* 21(7, July): 748.

Weyant, John, and Jennifer Hill. 1999. Introduction and Overview. *The Energy Journal,* Special Issue (The Costs of the Kyoto Protocol: A Multi-Model Evaluation): vi–xiiv.

Wiener, Jonathan. 1999. Global Environmental Regulation: Instrument Choice in Legal Context. *Yale Law Journal* 108(4): 677–800

———. 1997. Policy Design for International Greenhouse Gas Control. RFF Climate Issue Brief 6. Revised July 2000. Washington, DC: Resources for the Future. http://www.rff.org/issue_briefs/PDF_files/ccBrf6_rev.pdf (accessed October 17, 2000).

WRI (World Resources Institute). 1998. *World Resources 1998–99: A Guide to the Global Environment.* New York: Oxford University Press.

Yang, Zili. 1999. Should the North Make Unilateral Technology Transfers to the South? North-South Cooperation and Conflicts in Responses to Global Climate Change. *Resource and Energy Economics* 21(1): 67–87.

24 The Economics of Climate-Friendly Technology Diffusion in Developing Countries

Allen Blackman

Recent efforts to forge some consensus on the role that developing countries should play in reducing global greenhouse gas (GHG) emissions have focused attention on climate-friendly technologies (CFTs). Developing countries are expected to supersede industrialized countries as the leading source of GHG emissions in the next 30 years. Yet, their ability and willingness to contribute to abatement efforts is constrained by limited financial resources, weak regulatory institutions, and the perception that they should not have to bear the costs of mitigating a problem created primarily by industrialized countries. CFTs are seen by many as a means of surmounting these obstacles.

Many types of CFTs—notably, innovations in energy efficiency—not only reduce emissions of GHGs but also cut production costs. As a result, such technologies could conceivably diffuse spontaneously throughout developing countries, obviating the need for government financing and regulation. Climate strategies that focus on promoting the diffusion of CFTs probably will garner widespread support because

- to developing countries, they represent opportunities to enhance productivity and abate local pollution and

- to industrialized countries, they represent opportunities to boost exports of equipment and expertise.

But how likely is it that technology-based strategies will have a significant impact on GHG emissions in the near to medium term? In part, the answer depends on whether, once introduced, CFTs would diffuse at a reasonably rapid pace and whether policymakers will be able to speed the rate of diffusion. In this chapter, I summarize some of the key findings of the extensive economics literature on the diffusion of new technologies and assess the implications of these findings for the ongoing debate about technology-based strategies for dealing with climate change. In short, the literature suggests that a wide variety of policies are likely to speed the diffusion of CFTs, including rationalizing energy prices, improving information, and investing in energy infrastructure.

The Economics of Technology Diffusion

Even though not all the evidence on technology diffusion is conclusive, there is broad agreement on two points. First, new technologies are never adopted by all potential users at the same time. The widespread diffusion of new technologies can take anywhere from 5 to 50 years. Second, countless studies have confirmed that the diffusion of new technologies follows a predictable intertemporal pattern—technologies are adopted rather slowly at first, then more rapidly, and then slowly again as a technology specific "adoption ceiling" is reached. These stylized facts have prompted researchers to focus on two related questions:

- Why do some firms adopt a given innovation before others?
- Why do some innovations diffuse more quickly than others?

Researchers have addressed these questions using various theoretical constructs that emphasize different aspects of the diffusion process. *Epidemic models* focus on the dissemination of information about new technologies via day-to-day contact among firms, likening the process to the spread of a disease. These models imply that some firms adopt before others because they happen to become "infected" first and that some innovations diffuse faster than others because they are more "contagious" than others by virtue of their profitability and limited riskiness.

Rank models are premised on the observation that, given differences in capital vintage, size, access to technical information, labor productivity, and environmental regulatory costs, some firms will get a higher return from a new technology than others. Hence, one may rank all potential adopters on the basis of their expected returns. Only firms with a sufficiently high ranking will adopt when an innovation first becomes available. However, over time—as sector-wide production and information costs fall, the new technology is refined, and the existing capital depreciates—low-ranking firms will adopt as well.

Order models are applicable when there is a fixed critical input into production, such as a pool of specially trained labor or a scarce natural resource. In such situations, the order of adoption clearly matters—initially, only first movers who secure access to the critical input will find it profitable to adopt.

Finally, *stock models* are also premised on the idea that early movers obtain higher returns on the new technology. However, they attribute this phenomenon to the fact that as the stock of firms that have adopted a cost-saving innovation grows, average production costs fall, and eventually output prices fall as well. Thus, initially, it will only be profitable for a limited number of firms to adopt.

It is important to note that these four kinds of theoretical models are not mutually exclusive. Indeed, the diffusion of any specific technology probably will be influenced by some combination of the factors emphasized by the models: information and learning, the characteristics of potential adopters, the characteristics of technology, the scarcity of critical inputs, and the sensitivity of output prices to technological change.

Most empirical (versus theoretical) investigations of technology diffusion have sought to understand exactly how the characteristics of new innovations and potential adopters influence diffusion (thus, they essentially constitute tests of the rank model). Not surprisingly, they have found that relatively profitable, small-scale, and simple innovations are adopted fastest. In addition, new technologies are adopted fastest by firms that are large, have well-trained staff, incur high regulatory costs when using an existing technology, have infrastructure complementary to the new technology, are in fast-growing industries, invest more in R&D, pay relatively low prices for inputs used intensively by the new technology, and have relatively old existing capital. Despite considerable research, the evidence regarding the impact of market structure (that is, the degree to which the market is competitive or controlled by a small number of firms) on the timing diffusion is inconclusive.

Policy Prescriptions

What does the current state of knowledge suggest for technology-based climate change policy? Two implications are immediately obvious. First, even if CFTs that significantly lower production costs can be transferred to developing countries, diffusion will not be immediate. Second, firms in developing countries will not necessarily adopt CFTs rapidly simply because they reduce production costs in industrialized countries. A broad range of firm-level, sector-level, and country-level characteristics determine whether and how quickly new technologies are adopted, and systematic differences are likely to exist between developing countries and industrialized countries in nearly all of these characteristics. For example, labor is generally much less costly relative to capital in developing countries. Therefore, labor-saving technology that is profitable in industrialized countries will not necessarily be profitable in developing countries.

In addition, it is important to note that fast technology diffusion is not necessarily beneficial. Diffusion may be "too fast" if firms adopt a technology before it is profitable to do so, or if firms adopt a new technology today that effectively preempts the adoption of a superior technology in the future.

Given the theory and evidence presented here, there appear to be seven kinds of policy levers available to influence the speed of diffusion of CFTs in developing countries. They concern information; input prices; regulation; credit; subsidies; human capital, infrastructure, and research and development (R&D); and intellectual property restrictions.

Information

Economic theory suggests that the dissemination of information about a new technology is likely to be a critical determinant of diffusion. (Limited empirical support for the importance of information probably stems only from the difficulty of measuring information flows.) Government intervention to enhance the dissemination of technical information is likely to be justified because almost every means by which firms acquire information about new technologies is imperfect.

When firms acquire technical information through day-to-day contact with other firms, early adopters supply information about the new technology to later adopters, but the former do not capture any of the benefits from this information transfer themselves. As a result, they do not have proper incentives to make this information available to their rivals. When firms acquire technical information through active search, firms operating independently will inefficiently duplicate search efforts. And finally,

when firms acquire technical information from advertising, technology suppliers who are concerned about market share, not the diffusion of the technology, have incentives to oversupply technical information, which may lead to too-rapid diffusion of intermediate technology.

Policy options for enhancing the flow of information about new technologies include demonstration projects, advertising campaigns, the testing and certification of new technologies, and subsidies to technological consulting services. Have such mechanisms had a verifiable impact on diffusion? Demonstration projects have received wide application in the context of agriculture in developing countries, and many industrialized countries have set up regional information clearinghouses to provide consulting services to small- and medium-sized businesses that presumably can least afford search costs associated with adoption. For example, the U.S. Department of Commerce sponsors a network of Manufacturing Technology Centers. Similar networks have been established in the United Kingdom (the Advanced Information Technology Programme and the Regional Office Technology Transfer Programme), in France (Centres Régionaux d'Innovation et de Transfert de Technologie), and in the European Union (the SPRINT Programme). As yet, however, evaluations of these programs have been limited, and we know little about their effectiveness.

Finally, information-based polices may actually retard diffusion by fostering the expectation that improved technologies are forthcoming and therefore creating incentives to defer adoption. A similar problem arises when public provision of information about new technologies crowds out private information.

Input Prices

Both theory and evidence attest to the important role that input prices play in the diffusion of new technologies. In particular, energy prices clearly have a critical impact on the adoption of energy-saving technologies. In several developing and reforming economies, energy prices are still subsidized. Removing or significantly scaling back energy subsidies in these countries would create strong incentives to adopt energy-saving technologies. In countries where energy prices are not subsidized, taxing energy to raise its effective price above the market price would have the same effect as removing subsidies elsewhere.

Regulation

As noted above, empirical research supports the hypothesis that firms subjected to stricter environmental regulation are more likely to adopt clean technologies, including CFTs. The opportunities for and barriers to effective regulation in developing countries have received considerable attention, especially during the past 10 years. Two points deserve mention. First, the use of market-based incentives such as pollution taxes and marketable permit systems (including a credible international GHG emissions trading system) is analogous to raising the price of a critical factor of production—namely, environmental services. Therefore, the same arguments about the link between factor prices and technology adoption are applicable. Second, even when institutional and financial constraints make formal public sector–led regulation problematic, private sector–led informal initiatives (such as grassroots efforts to deter polluters by stigmatizing them) can be an effective substitute.

Credit

Lack of access to credit may be a critical barrier to adoption. Subsidizing credit for specific kinds of investments has been a common policy response. However, thus far these programs—both public and private—have had very mixed results. Chronic problems include the diversion of loans by borrowers to nontargeted activities, low repayment rates, the creation of financially unsustainable lending institutions, the politicization of lending decisions, and the undermining of existing credit markets. Because there is growing support for the view that the costs of "targeted" credit outweigh the benefits, a wiser approach to overcoming financial barriers to technological innovation is to focus on improving banking, which, in developing countries, is often hamstrung by unstable monetary policy, interest rate restrictions, and weak property rights.

Subsidies

An obvious mechanism for speeding the diffusion of a new technology is for governments to subsidize it. But in addition to being quite expensive, subsidies are likely to be subject to many of the same problems as targeted credit—namely, the politicization and distortion of input markets.

Human Capital, Infrastructure, and R&D

Empirical research on the links between the early adoption of innovations on one hand and human capital, infrastructure, and R&D on the other suggests that there is an argument for subsidizing education, technical training, infrastructure, and R&D. It need not be a broad-brush strategy if the subsidies are focused on GHG-intensive sectors such as energy.

Intellectual Property Restrictions

Intellectual property restrictions such as patents and licenses have countervailing effects on technology diffusion. On one hand, they stimulate R&D, which in turn stimulates technology diffusion. Perhaps more important for developing countries, they also are likely to encourage foreign investment, which can be a significant source of new technologies. But on the other hand, intellectual property restrictions attach significant costs to the adoption of new technologies that can retard diffusion. In many developing countries, the adaptation of existing technologies, rather than the creation of substantially new ones, accounts for the bulk of productivity growth. Therefore, there is reason to suspect that the negative impact of intellectual property restrictions on diffusion in developing countries could be substantial.

Practical First Steps

Of the broad range of policy options presented here, which are likely to be politically practical? Technology-based strategies will generate political support to the extent that they represent obvious "win-win" opportunities for the parties involved. Many policies fit this description to some degree. Information, human capital, and infrastructure policies will enhance productivity; rationalizing energy prices will boost allocative efficiency; improving banking should stimulate saving and investment; and strengthening regulation should produce environmental benefits.

However, some of these policies involve up-front economic costs that are more immediate and payoffs that are more delayed than others, making them unattractive to decisionmakers with short time horizons. For example, although investments in banking, human capital (broadly defined), and environmental regulation may have tremendous benefits in the long run, they involve substantial up-front costs. Thus, the most practical policy options discussed here would seem to be rationalizing energy prices, improving information, and investing in energy infrastructure.

Suggested Reading

General

Stoneman, P. 1991. Technological Diffusion: The Viewpoint of Economic Theory. In *Innovation and Technology in Europe: From the Eighteenth Century to the Present Day*, edited by P. Mathias and J. Davis. The Nature of Industrialization series. Oxford, U.K.: Blackwell, 162–84.

World Bank. 1992. *World Development Report: Development and the Environment, 1992*. Oxford, U.K.: Oxford University Press.

Technical

Blackman, A. 1997. Economic Research on Technology Diffusion: Implications for Developing Country Climate Change Policy. Mimeo. Washington, DC: Resources for the Future.

Ecchia, G., and M. Mariotti. 1994. A Survey on Environmental Policy: Technological Innovation and Strategic Issues. Working paper 44.94. Milan, Italy: Fondasione Eni Enrico Mattei EEE.

Karshenas, M., and P. Stoneman. 1993. Rank, Stock, Order and Epidemic Effects in the Diffusion of New Process Technology. *Rand Journal of Economics* 24(4): 503–27.

Kemp, René. 1997. *Environmental Policy and Technical Change*. Brookfield, VT: Edward Elgar.

Mansfield, E. 1968. *Industrial Research and Technological Innovation*. New York: W.W. Norton.

Pargal, S., and D. Wheeler. 1996. Informal Regulation of Industrial Pollution in Developing Countries: Evidence from Indonesia. *Journal of Political Economy* 104(6): 1314–27.

Stoneman, P. 1983. *The Economic Analysis of Technological Change*. New York: Oxford University Press.

Stoneman, P., and P. Diederen. 1994. Technology Diffusion and Public Policy. *Economic Journal* 104(425): 918–30.

Including Developing Countries in Global Efforts for Greenhouse Gas Reduction

Ramón López

The United States and other countries that agreed to binding greenhouse gas (GHG) emissions reductions in the Kyoto Protocol (collectively known as the Annex I countries in the U.N. Framework Convention on Climate Change) have increasingly called upon developing countries to "participate meaningfully" in the implementation of the Kyoto Protocol. In particular, recent discussion has focused on the various ways developing countries could participate voluntarily in international carbon reduction efforts and eventually accede to the protocol by entering the nascent international carbon trading system.

The scope for reducing carbon emissions in developing countries via "win-win" policies or mechanisms with low net costs is broad. First, the removal of energy subsidies could enable developing countries to reduce their total emissions of carbon while enhancing overall economic efficiency and freeing up scarce funds for pressing spending needs in other areas. Second, reductions in biomass and forest burning could result in potentially large revenues from the sale of carbon credits, if these credits were legitimate under the Kyoto Protocol. The principal economic benefit of burning forests is expansion of agricultural frontiers, but this benefit may be less than the potential economic value of carbon sequestration in standing forests after an international carbon-trading regime is established.

I begin by examining energy subsidies in developing countries. Next, I discuss biomass as a source of carbon emissions and the role that developing countries could play in limiting these emissions and then address implementation issues. Finally, I offer some recommendations for initiating the participation of developing countries in implementing the Kyoto Protocol.

Energy Subsidies

Energy subsidies that promote coal and oil consumption cause significant local environmental damage as well as increased GHG emissions. One attempt to model the effect of a phase-out of coal subsidies on GHG emissions in both Organisation for Economic Co-operation and Development (OECD) and non-OECD countries indicates that in western Europe and Japan, global carbon dioxide emissions would be lowered by 5% in 2005 relative to what otherwise would have been emitted under a business-as-usual scenario (see Anderson and McKibbin in Suggested Reading). If western Europe, Japan, and major non-OECD or developing countries all removed subsidies together, then global carbon dioxide emissions would fall by 8% in 2005. These figures highlight the potential for GHG savings in developing countries, but they also underscore the potential in developed countries. They do not account for the potential for reduced mining-related releases of coal-bed methane (a potent GHG in its own right) due to reduced coal demand. Eliminating subsidies also increases the gross domestic product (GDP) in many countries as a result of efficiency gains—resources are freed up from distorted sectors and reallocated through the global economy, yielding higher rates of return.

Although energy subsidies have been reduced in several developing countries (see Table 1), they are still widespread in many others. They take the form of *direct subsidies* to consumers through underpriced energy services and *implicit subsidies* to producers through trade barriers that limit the availability of technologies that are more energy efficient.

Countries that currently subsidize energy could be induced to cut their subsidies if they were allowed to sell carbon emissions reduction credits to developed countries that reflected the avoided emissions. These countries could obtain large gains in terms of carbon permit revenues and higher long-term economic efficiency, without compromising future growth. They could gradually reduce their permit sales over time if desired to accumulate a stock of credits for domestic use if and when the quantitative emission commitments for developing countries are negotiated.

Biomass Burning from Land Clearing as a Source of Carbon Emissions

The technical capacity to measure carbon emissions from biomass burning has not been perfected, and any figures given are rough estimates. Nevertheless, it appears that biomass burning for land clearing in developing countries contributes a significant amount of GHGs to the atmosphere annually. The most common forms of biomass burning are tropical deforestation, burning of savanna, and combustion of agricultural waste. Biomass burning has been estimated by some sources to account for about one-fifth of all emissions, and tropical forest burning accounts for more than half of the total contribution. The primary motivation for burning is to clear land for agricultural or other purposes, and it has been estimated that only 20% of the total biomass burning (mostly of

Table 1. Energy Price Reform in Developing Countries.

Energy source	Real price change (final price as % of original)	Time interval
Coal		
India	106	1990–94
South Africa	119	1991–94
Petroleum		
Brazil	122	1991–95
China	221	1991–94
India	102	1990–95
Mexico	138	1990–95

Source: Reid and Goldemberg (see Suggested Reading).

fuelwood and charcoal) in developing countries is actually used to generate energy (see the papers by Andreae and Levine in Suggested Reading).

In many regions, much of the land clearing and associated biomass burning is caused by organized interests—such as cattle ranching, timber companies, or mining companies—as opposed to small landholders or subsistence farmers. Given current distortions in land use policies and property rights institutions, at least some of this land clearing is not justified on economic grounds, without consideration of GHG issues. Even where land clearing might be justified apart from carbon considerations, developing countries may be better off with less land clearing and more carbon sequestration, if these avoided carbon releases could be packaged and sold to developed countries as carbon credits. I return to this point later.

The possibilities are illustrated in an analysis of estimated land values of settlements in which the market value of forestland for agriculture in eight settlements in the Brazilian Amazon was found to be $2–300/hectare, depending on the attributes of the land (remoteness, infrastructure availability, and the like) (see Schneider in Suggested Reading). Because these figures include the value of investments attached to the land as well as the value of the public infrastructure required to bring the land into production, the actual value of undeveloped forestland for agriculture may be even lower.

These values can be roughly compared to the value of retaining carbon in biomass by combining figures on the amount of carbon stored per hectare of Amazon forest with figures for actual or proposed carbon taxes in various countries. This calculation implies a value of $700–5,000/hectare. This range illustrates what emitters in developed countries might be willing to pay to maintain carbon stored in the forest biomass and use it to offset their own emissions. The large difference in the carbon storage and agricultural land values indicates an enormous potential for mutually beneficial carbon trade between developed and developing countries.

Implementing Biomass Carbon Retention Credits

In principle, implementing controls over groups of organized interests in land clearing would be easier than doing so with small landholders. Nevertheless, such reductions would be difficult to achieve in practice. Incorporating biomass carbon sources into international GHG mitigation efforts would be complicated given the diffuse and geographically diverse sources of the emissions, the complicated land use dynamics and economics of deforestation and agricultural burning, the technical difficulty of measuring changes in biomass carbon emissions, and the sociopolitical difficulties of negotiating acceptable baselines.

Credits usable in an international carbon trading system could be given to countries that institute and enforce national policies to reduce biomass carbon sources through the elimination of government subsidies for land colonization, or "transmigration"; improvement of land tenure and improved assignment of land property rights; joint forest management programs (involving co-management of forest resources with local communities to provide incentives for conservation by, for example, improving land tenure); restrictions on foreign and local logging industries; and programs to reduce household fuelwood usage, such as the adoption of new cooking technologies and the creation of household woodlots.

To tap these large potential gains and to provide adequate incentives for developing country participation in these GHG reduction efforts, three scien-

tific, legal, and policy requirements would need to be implemented:

- **A system of monitoring based on remote sensing information would need to be established.** This system would be used to monitor the status of forests and other biomass and to establish net rates of carbon sequestration or emissions. There are already international efforts to coordinate such a system, such as the U.N. Environment Programme's Integrated Global Observing Strategy program.

- **Internationally tradable permits based on reduced biomass burning would have to be legitimated by the Conference of Parties to the United Nations Framework Convention on Climate Change.** The Kyoto Protocol mandated the establishment of an international system of permit trading and the creation of a clean development mechanism for project-based permit creation through emissions reductions, but the details of these mechanisms remain to be established. The technical capability to count emission reductions from reduced biomass burning is not clearly established and would have to be improved. Proposed caps on trading due to supplementarity concerns would significantly reduce potential gains for developing countries, thus reducing their incentives to cooperate in trading.

- **Deforestation permits in settled areas could be allocated in a decentralized way to communities, individuals, nongovernmental organizations, firms, and local governments that are in actual or potential control of the forests.** One interesting idea is the establishment of tradable development rights. Once a value is placed on tradable emissions permits and governments have economic incentives to conserve forest resources, governments could commit to a moratorium in road and infrastructure construction in core forest areas and allocate tradable development permits to local interests and authorities. Local communities would lease their permits to domestic firms or to firms located elsewhere, meaning that they commit to preventing forest fires and other forms of carbon emissions within their jurisdictional geographic areas. Rural communities thus could benefit directly from carbon trade, and the infusion of income into the rural communities of developing countries could help reduce rural poverty. Subsistence farmers, who cause a significant portion of forest fires, could obtain higher incomes by devoting part of their time to preventing further deforestation.

Conclusion

Governments in many developing countries are beginning the difficult political process of reducing energy subsidies and finally are beginning to implement serious efforts to curtail forest burning. Awareness in these countries is growing of the negative socioeconomic consequences of energy price distortions and the often relatively small benefits of forest clearing compared with resulting environmental and ecological damage. Moreover, the high cost burden of energy subsidies in certain developing countries creates fiscal imbalances that increasingly concern their populations. Similarly, forest burning is becoming an urban problem that affects millions of people in large urban centers, a development that has contributed to increasing awareness of the forest fire problem among politically influential urban populations.

Reducing energy subsidies and forest burning, however, are slow processes subject to significant political difficulties. Powerful economic interests would be hurt by the removal of energy subsidies, and the control of forest fires requires political clout to enact and enforce laws and regulations that affect forests.

A system of international carbon trading that provided credits for these activities could provide significant economic benefits that would increase the political weight of those pushing for the elimination of subsidies and for forest protection in developing countries. Revenues from carbon permit sales could also greatly increase the funding for the enforcement of forestry laws. In addition, technical assistance programs for developing countries could help support the process of subsidy removal and the development of institutions and technical capacity to protect forests. Developing countries thus could gain a great deal by choosing to participate in the international emissions trading regime that is being created.

Developed countries also could gain from reduced carbon abatement costs due to international trading if they take care to design a trading system that takes into account developing countries' concerns and interests.

Suggested Reading

Anderson, K., and W. McKibbin. 1997. Reducing Coal Subsidies and Trade Barriers: Their Contribution to Greenhouse Gas Abatement. Discussion Paper 1698. October. London, U.K.: Centre for Economic Policy Research.

Andreae, M. 1991. Biomass Burning: Its History, Use, and Distribution and Its Impact on Environmental Quality and Global Climate. In *Global Biomass Burning*, edited by J. Levine. Cambridge, MA: MIT Press.

Fischer, C., and M. Toman. 1998. Environmentally and Economically Damaging Subsidies: Concepts and Illustrations. RFF Climate Issues Brief 14. October. Washington, DC: Resources for the Future.

International Energy Agency. 1999. *World Energy Outlook, 1999 Insights, Looking at Energy Subsidies: Getting the Prices Right*. Paris, France: International Energy Agency.

Levine, J. 1990. Global Biomass Burning: Atmospheric, Climatic and Biospheric Implications. *EOS* 71(September): 1075–77.

Mattos, M., C. Uhl, and D. Goncalvez. 1992. Economic and Ecological Perspective on Ranching in the Eastern Amazon in the 1990s. Brasilia, Brazil: Instituto do Homen e Medio Ambiente de Amazonia, EMBRAPA (Empresa Brasileira de Pesqkuisa Agropecuária).

Ozorio de Almeida, A. 1992. Deforestation and Turnover in Amazon Colonization. Unpublished discussion paper. Washington, DC: The World Bank.

Panayotou, T. 1994. Financing Mechanisms for Environmental Investments and Sustainable Development. Environmental Economics Series Paper 15. Nairobi, Kenya: U.N. Environment Programme, Environment and Economics Unit.

Reid, W., and J. Goldemberg. 1998. Developing Countries are Combating Climate Change. *Energy Policy* 26(3): 233–37.

Schneider, R. 1994. Government and the Economy on the Amazon Frontier. Regional Studies Program, Report 34. May. Washington, DC: The World Bank, Latin America and the Caribbean Technical Department.

World Resources Institute. 1998. *World Resources 1998–99*. New York: Oxford University Press.

26 Moving Ahead with Climate Policy

Michael A. Toman

After a decade of international meetings and negotiations, more than 160 nations signaled their commitment to address the problem of climate change by initialing the Kyoto Protocol in December 1997. During that decade and subsequently, climate negotiators made a series of key policy decisions with far-reaching consequences that were not fully appreciated at the time. Negotiators often passed up options that might have lowered the costs of achieving the long-term objectives of the U.N. Framework Convention on Climate Change (UNFCCC). Based on the rich content of the economic analysis provided in previous chapters of this volume—analysis based on more than a decade of work by scholars all over the world—I believe that (re)consideration of these options in the international negotiations will be crucial for long-term success in limiting greenhouse gas (GHG) emissions.

The material in this chapter was presented at the Global Climate Change Conference held at the James A. Baker III Institute for Public Policy, Rice University, on September 8, 2000. An earlier version of the same ideas was presented at the Arco Forum of Public Affairs held at Harvard University on March 15, 2000. It draws on several ideas developed by or with colleagues, including Jason Shogren, Dick Morgenstern, Billy Pizer, Marina Cazorla, and John Anderson. Responsibility for the content of the paper, however, is mine alone.

Several kinds of flexibility were introduced in the protocol. The first is *where flexibility,* through provisions for international trade in carbon emission rights to promote the most cost-effective and lowest-cost abatement opportunities; "where" flexibility includes the participation of developing countries in emissions abatement through the Clean Development Mechanism (CDM). Also included is *what flexibility,* in that targets can be met by controlling several different gases as well as by increasing long-term uptake of atmospheric CO of trading in multiple gases. However, the protocol focuses on year-to-year emissions of GHGs in this treaty, rather than the concentration of those gases in the atmosphere affecting the climate over the long term. The protocol further emphasizes a short-term timetable (2008–12) rather than the century-long schedule required to effectively reduce GHG concentrations. This last point is important because many paths can be taken to reach a specific long-term concentration; those with greater *when flexibility* will be less costly than others. Specifically, approaches with a more gradual beginning that gain momentum over time will cost less, whereas the more front-loaded the targets, the higher the costs to reach them.

In addition, the fact that the Kyoto Protocol negotiators found no agreement on more specific policy targets for developing countries simply postponed an inevitable day of reckoning on how these countries are to be incorporated in limiting global emissions, assuming that the international community agrees to seek limits on GHGs. Just as the Kyoto Protocol's quantitative targets for Annex I countries in themselves delay by only a few years the inexorable growth of global GHG concentrations, so the omission of developing countries from quantitative emissions limits implies only a modest slowing of GHG growth, no matter how draconian the long-term policies implemented by Annex I countries.

Four points seem central in assessing how the climate policy debate can move forward:

- Efforts are warranted to strengthen the flexibility mechanisms in the Kyoto Protocol, but we must be realistic about their performance.
- A more gradual but accelerating trajectory of GHG abatement would be a better policy in practice on economic and environmental grounds.
- It is worthwhile to take modest but effective domestic actions to abate GHG emissions in the United States today and to prepare for additional actions in the future.
- Continued constructive efforts to involve the cooperation of developing countries to reduce global emissions is crucial.

Some of these points can be seen as natural extensions of current climate policy negotiations and policy debates. On the other hand, the second point represents a conspicuous departure from business as usual in the climate policy process. This and other less dramatic gaps between the current state of play and my propositions reveal to some extent how the economic analysis of the past decade has not been taken up in the formulation of climate policy.

Strengthening the Kyoto Protocol Mechanisms

Economic tools help cut the costs of achieving a GHG emissions target because they generate a market price for GHG emissions, which are otherwise treated as a free good. With either carbon taxes or emissions trading, consumers respond to the resulting price signals in various ways: switching to fuels that are less carbon-intensive (for example, from coal to natural gas); increasing energy efficiency per unit of output by using technologies that are less energy-intensive; adopting technologies to reduce the emissions of other GHGs (assuming that they are covered in the program); reducing the production of what become high-cost, carbon-intensive goods; increasing the sequestration of carbon through reforestation; and developing and refining new technologies (for example, renewable energy resources) for avoiding GHG emissions.

The Kyoto Protocol allows for both formal GHG emissions trading among the Annex I developed countries and activities with developing countries through the CDM. Annex I trading could involve tying together domestic emissions trading programs or a project-level approach in which participants can generate emission credits from emission-reducing actions in other Annex I countries (so-called joint implementation). These various endeavors could be organized and financed by Annex I investors, the developing countries themselves, and international third parties.

Hahn and Stavins describe the practical difficulties of operating a transaction-specific, credit-based joint implementation program internationally with heterogeneous domestic GHG emissions measures (see Suggested Reading). They point out the trade-off between international cost-effectiveness and domestic policy sovereignty that result from such an undertaking. Operational challenges also beset the CDM. However, I think the CDM could generate both low-cost emissions reductions for developed countries and tangible benefits to the host country through the transfer of efficient low–carbon-emitting technology. One key immediate question is how to design a credible monitoring and enforcement system that does not impose such high transaction costs that it chokes off CDM trades. People will not start a project if the time, effort, and financial outlays needed to search out, negotiate, and obtain governmental approvals are too onerous.

The United States has been a strong advocate in international negotiations for the option of broad international trading to meet Kyoto Protocol commitments. Other countries, notably in western Europe and in some parts of the developing world, have been cooler toward decentralized private-sector emission trading. Some nations like trading, but only if strict rules are imposed, which in a sense may ultimately be a self-defeating approach. European negotiators have advocated trading limits that restrict the degree to which Kyoto Protocol targets could be met through international flexibility mechanisms. These countries may have many motivations for taking this view: less concern for the cost of GHG emissions control if economic growth is slower in Europe than in the United States, a desire to increase comparative advantage by limiting U.S. access to low-cost abatement opportunities, and concern that the United States not use harder-to-verify international reductions to displace long-lasting domestic action. Such supplementarity constraints (as they are termed in the debate) have been stoutly resisted by the United States for fear that they would unduly restrict opportunities for cost-effective emissions control and delay the evolution of effective GHG emissions permit markets.

There is every reason to believe that the flexibility provided by the Kyoto Protocol mechanisms can lower Annex I countries' costs of meeting their Kyoto Protocol targets. Moreover, the CDM provides a valuable potential for the productive engagement of developing countries in activities that promote sustainable economic development while slowing growth trends for GHG emissions. For these reasons, the United States should continue to promote the implementation of these mechanisms and resist calls for undue supplementarity or other limitations on their use (for example, limits on the projects and technologies eligible for inclusion in the CDM).

But these mechanisms are never going to operate in a textbook fashion. They will inevitably be subject to transaction costs that limit their efficiency, and concerns about the flows of funds out of the United States to acquire foreign-supplied emissions credits will have to be overcome if the mechanisms are to be widely used. It is therefore misleading to be too optimistic about the potentially low costs of meeting the Kyoto Protocol targets, as I believe the Clinton administration's 1998 estimate of compliance costs was (see CEA in Suggested Reading). In practice, the

costs may be significantly larger, especially for meeting policy targets as ambitious as the Kyoto Protocol (which likely will require an emissions reduction of one-third or more relative to business as usual). This expectation strengthens the case for explicitly considering costs in evaluating compliance and for considering greater flexibility in the timing of emissions reductions.

Increasing the Flexibility of International Climate Targets

The Issue of Timing*

The UNFCCC declares in Article 3, Section 3, that "policies and measures to deal with climate change should be cost-effective so as to ensure global benefits at the lowest possible cost." Although debate continues on this point, conviction is growing among economists that the Kyoto Protocol may seek too much abatement, too soon compared with a slower path that achieves long-term climate protection goals at lower costs and with less economic dislocation. To address this concern requires more gradual but accelerating targets for the control of GHG emissions and flexible compliance standards that take into account compliance costs as well as quantitative emissions targets. Otherwise, political agreement for achieving the desired environmental goals is likely to remain elusive.

Several papers published during the past several years conclude that the lowest-cost path to any targeted concentration of GHGs in the atmosphere would begin gradually and leave the more drastic emissions reductions until the later decades of the program. The sharp emissions reductions by 2008–12 under the Kyoto Protocol provide little "when" flexibility and thus do not lie on the lowest-cost path to any plausible long-term GHG concentration target.

Four reasons for back-loading the deeper cuts in emissions are generally offered in the literature. First, the world's investments in fuel-burning equipment

---

*The discussion in this section draws heavily on Toman, Morgenstern, and Anderson (1999).

are vast, and to replace that equipment before its useful life has ended would be expensive. It is quite true that technologies are available to use energy more efficiently and with lower GHG emissions than much of the present equipment does. But a plant that generates electric power, for example, is often built with an expected lifetime of 40 years. To replace it after only 20 years imposes an added cost on society.

If the Kyoto Protocol is ratified and put into force, the United States would be obligated to reduce its GHG emissions during 2008–12 to a level equal to 7% below 1990 emissions. But with economic expansion, the country's emissions are already 10% above the 1990 level, and U.S. Energy Information Administration projections indicate that under the present policy, by 2010, U.S. emissions will be more than one-third above 1990 levels. Meeting this target provides a time frame hardly longer than the life of an automobile, let alone heavy industrial equipment.

Second, technology steadily finds ways to use energy more efficiently, with fewer emissions. By postponing drastic cuts in emissions for several decades, the world's economy could take advantage of technology that is not yet available. Third, the discount rate argues in favor of delaying the heavy expenditures. A dollar invested in 1998 at a real (after inflation) interest rate of 2% a year would be worth $1.88 in 2030 and thus could buy almost twice as much then as now. More generally, if a dollar is saved today and invested in science, or education, or any of the other contributors to economic development, it will result in a richer society. Each dollar spent on emissions control two or three decades from now would represent a smaller proportion of society's wealth, and would be a smaller burden on society.

Fourth and finally, CO_2 in the atmosphere is constantly absorbed by oceans, forests, soil, and other carbon sinks. Not all of the CO_2 that human activity emits into the air is absorbed, of course—and that is why atmospheric concentrations are rising—but about half of the CO_2 emitted is estimated to disappear that way. If the sharp cuts in emissions are postponed, then some of the CO_2 emitted to the atmosphere in the meantime will have disappeared naturally before the concentration reaches the target.

This point is relevant because total GHG concentrations, not annual GHG emissions, influence climate change.

Each of these four points has generated controversy. Will research and development (R&D), for example, produce significant advances in energy technology without the pressure exerted by mandatory emissions cuts? What exactly are the incentives that will produce the technology the world would need to reduce emissions substantially without crippling industrial production? Goulder and Mathai find that when knowledge is gained through R&D investments, less abatement is needed in the present, thereby supporting the notion of back-loading (see Suggested Reading). However, when knowledge is gained through learning by doing, the impact on the efficient timing of abatement is ambiguous.

Another central argument against the option of beginning gradually is that, in the absence of dramatic and forceful change, people will not take seriously the need to curb emissions. Companies' management and private consumers, the argument goes, will not change their habits and begin to make the investments necessary to carry them into an era of lower emissions. Postponing severe action will merely mean that, a generation from now, the world will face the same need to reduce emissions, but from a much higher level.

However, credibility cuts both ways. If a program turns out to be too expensive and disruptive, it will collapse and discredit the whole idea of controlling emissions. Politicians are left to struggle with the same question they have confronted for more than a decade: What is the size and shape of an emissions cut big enough to persuade people that profound change is coming, but not so big that they dismiss the whole idea as unrealistic? In confronting this question anew, they must take into account the large potential cost savings from increased "when" flexibility.

Building Flexibility into Policy Choices and Compliance Rules

The costs of meeting the Kyoto Protocol targets are uncertain, as illustrated by a comparison of leading computerized models of energy use and economic activity used to study this issue (see Weyant and Hill in Suggested Reading). Optimistic estimates are provided by the Clinton administration's own analyses and by a study of technological potential by several U.S. national laboratories. Other estimates run higher—in some cases, substantially so. For example, even if opportunities for low-cost technology and management improvements that reduce GHG emissions exist at present, will these opportunities be ongoing as GHG targets get stricter, or will they be used up quickly? The rules for international emissions trading under the Kyoto Protocol also have yet to be worked out. Because broad and substantial GHG emissions reductions have no precedent, it is not possible to know what assumptions and projections are more accurate or realistic.

A substantial body of economic analysis suggests that debating different strict quantitative targets is debating the wrong issue. Instead, we should be trying to develop policies that lead to gradual but accelerating limits on GHG emissions over time. Achieving agreement on such policy goals requires addressing the costs associated with meeting different policy targets.

One way to address the issue of cost uncertainty in the context of the Kyoto Protocol targets is to negotiate agreement on compliance rules for Annex I countries that would limit the potential increase in energy costs over the short term while still requiring meaningful abatement measures be taken. The stringency of abatement then could be gradually increased over time, as businesses and consumers adjust to pressures for increased energy efficiency and new technologies for avoiding GHG emissions emerge. If emissions control is as cheap as advocates of the Kyoto Protocol suggest, then the limits on abatement activities to contain costs would never be triggered; however, if the advocates are wrong, then insurance against unacceptable cost increases in the short term is critical to expanding the base of political support for meaningful action.

Kopp and others (2000) and McKibbon and Wilcoxen describe ways for providing this flexibility (see Suggested Reading), proposing what is sometimes known as a safety-valve option. This option would allow Annex I governments to relax their strict quantitative limits after the market price of GHG emissions permits (and thus the domestic cost of additional abatement) reaches a certain common ceiling. The ceiling would be maintained by governments standing ready to supply permits on demand at the ceiling price.

The approach suggested by McKibbon and Wilcoxen would allow for national emissions trading within Annex I countries subject to a common ceiling price in lieu of international GHG trading, which in their view could lead to costly and counterproductive international income and capital flows. The proposal from Kopp and others would retain the option of international GHG trading and would have Annex I governments that used the safety valve paying into a compliance fund an amount equal to their national shortfall in abatement (relative to the Kyoto Protocol targets) times the ceiling price (presumably, this payment would be accomplished by charging their regulated entities who exceeded whatever national emission standards were in force). This fund could be used in various ways, for example, for investment in additional emissions control in developing countries through the CDM (with resources allocated to specific projects via a competitive auction).

One objection to these kinds of approaches is that they would not guarantee to produce a specific reduction on an agreed timetable. One then returns to the question, What guarantee of emissions reductions exists in the current situation? If the Kyoto Protocol is not ratified in a timely way, few if any emission reductions will materialize. Moreover, by starting with more modest policy targets and gaining experience with GHG control while also providing more time for scientific knowledge to accumulate, the world gives up little in the way of options to act more decisively in the future as warranted to limit GHGs.

Early Reduction of GHG Emissions

As pointed out in Chapter 19 (Parry and Toman), much debate in the United States has surrounded different policies for encouraging early reductions of GHG emissions prior to the date when the Kyoto Protocol targets would go into force (if the protocol is ratified, and even if it is not ratified in the near future or at all). Many of these proposals involve programs for granting early reduction credits—a promise of more generous regulatory allotments of allowed emissions in the future—to those good actors who voluntarily reduce emissions beyond business as usual.

A common drawback of these proposals is that their effectiveness is stymied by uncertainty over the imposition of mandatory emissions limits in the future. Moreover, awarding early credits requires many subjective judgements about what constitutes "additional reductions" versus reductions that would have occurred anyway over time. Depending on how the programs are constructed, they may induce too much or too little early reduction.

With several other colleagues from Resources for the Future, I have advocated the imposition of a mandatory formal emissions trading program for controlling CO_2 (and gradually other GHGs) within the next three years (see Kopp and others 1999 in Suggested Reading). This program would apply comprehensively to fossil fuel supplies (so there is no need to measure or impute the emissions of myriad sources of emissions). Fossil fuel supply permits would be auctioned off, and the resulting revenues would be used to offset other burdensome taxes, with transitional assistance to those most adversely affected by the policy (especially displaced workers and affected businesses). To limit the economic risk, the government would supply supplemental permits as needed at a fixed price that would rise over time (a domestic application of the safety valve idea mentioned earlier).

This kind of policy deserves continued consideration as the climate policy debate moves ahead. It prepares the way for future action if and when the Kyoto Protocol is ratified while allowing a more explicit balancing of environmental goals against economic costs. Because the program has fixed targets and clearly established claims for emissions permits, it avoids the disadvantages of other early reduction programs. Finally, several other Annex I countries already are moving ahead with various moderate policies for GHG emissions control. By joining this group with an extremely cost-effective policy design, the United States has a chance to regain a real leadership position in the international climate policy debate.

Engaging Developing Countries in Cooperative GHG Control

Ultimately, achieving international agreement on global responsibility for reducing GHG emissions is tantamount to establishing international agreement on how to divide up rights to emit GHGs over the long term. This task is enormously difficult. The difficulty is reflected in the real-world controversies that surround international equity in climate negotiations, in which developing countries vehemently oppose proposals from the United States and other developed countries for a flexible approach to mitigating GHG emissions on the grounds that such flexibility may create economic or environmental disadvantages for the developing world.

Advocates for developing countries in this debate have proposed several formulas for allowing developing countries to increase their emissions concurrent with needed economic growth before beginning to rein in their emissions. It follows almost arithmetically that given any particular long-term target for GHG concentrations in the atmosphere, the longer and higher the emissions grow in developing countries, the faster and lower the reductions in emissions must be from industrialized countries. Many of the formulas reflect a basic premise of equal rights to emit GHGs per capita, coupled with an assignment of primary responsibility for GHG emissions control to developed countries, given their greater affluence and responsibility for historic emissions. Developed countries could expand emissions beyond their allotment, but only by purchasing additional emission allowances from developing countries. Developed countries, in turn, have staunchly resisted such proposals as involving an after-the-fact liability for emissions and a degree of income redistribution that are politically unacceptable (not to mention inadequate provisions for ensuring that resources transferred would generate international environmental benefits).

Even a cursory review of the growing literature on international equity in GHG emissions policy casts grave doubt on the notion that some simple formula will resolve this dilemma. Although it is possible to find combinations of allocation rules that generate less lopsided distributional consequences than simple approaches such as equal GHG emissions allocations per capita or per unit of GDP, it is still questionable whether such ad hoc approaches can command broad political support.

In Chapter 20, Wiener provides a strong argument for international transfers to developing countries for expanding participation in GHG emissions control and stabilizing the atmosphere. He notes that such transfers are much more effectively accomplished by the international allocation of GHG emissions rights and their sale through market channels than through intergovernmental redistribution of carbon tax revenues. In the midst of long-term uncertainty about how to structure long-term global agreement (again, given the premise of an important long-term environmental threat from climate change), it may be useful to look for a process that can gradually increase the engagement of developing countries in GHG emissions control while increasing understanding of the options available and building international confidence in international mechanisms as they evolve. The CDM is a first key step in pursuing this approach. Even the most ardent proponents of the CDM concede that it will have higher transaction costs, greater monitoring difficulties, and lower overall efficiency than other options—such as a full-blown permit trading system—but it is a good place to start. CDM projects can provide concrete information on GHG emissions control opportunities and costs in developing countries as well as relatively tangible GHG emissions reductions, cost-saving benefits in developed countries, and economic benefits in developing countries.

Developing countries have expressed concern about an uneven playing field for CDM negotiations. A proposed solution to this problem has been to embed CDM in a multilateral institution, in which developing countries individually and collectively can influence which CDM projects are undertaken and how. The problem with this approach is that it risks drowning an imperfect but potentially valuable international market mechanism in a more inflexible international bureaucracy. Developed countries should stand fast to the idea of promoting the CDM as a market mechanism with maximum operational participation by the private sector while acknowledg-

ing the need for auditing to ensure environmental credibility and the right of host developing countries to veto projects they find unacceptable.

Another concern of developing countries has been the risk of "giving up the low fruit"—having low-cost CDM projects undertaken early with international partners and then not having these options available later, when or if binding commitments are undertaken by developing countries. However, developing countries can negotiate terms and conditions for CDM projects that reflect the best trade-off for them between short-term economic and environmental benefits and the long-term value of retaining low-cost options for domestic control of GHG emissions. Given the scarcity of capital in many developing countries, the prospects for local environmental benefits from CDM projects and the possibility that today's low-cost GHG emissions abatement opportunities may be tomorrow's "lost fruit" because of technical progress, CDM projects could yield substantial benefits to developing countries as well as their Annex I partners.

Although the CDM can help start international climate cooperation, it is not a substitute for broadening formal international commitments and developing more cost-effective mechanisms. Eventually, some kind of bargaining will be needed over the distribution of national responsibilities. It is unlikely to be accomplished in a one-shot deal. One way to approach the problem is through a graduation formula of the type discussed in Chapter 23 (Cazorla and Toman). The essence of this idea is that developing countries gradually assume more responsibility, in a relatively predictable way, as their economic circumstances and thus their ability to pay improves. Although negotiating the terms of such an approach is not easy either, it may offer enough flexibility to provide ways to accommodate the interests of both developing and developed countries.

Concluding Remarks

It is a gigantic undertaking to organize a cooperative worldwide program that limits GHG emissions. Some 180 governments are now involved, and all realize that any useful attempt must affect their economies deeply. But amidst all the gaping uncertainties that surround this subject, one thing can be stated with assurance: A program that offers great flexibility and relatively low economic costs will have a better chance of adoption than an inflexible, expensive one. In the range of all possible approaches, the Kyoto Protocol lacks significant elements of "when" and "how" flexibility.

At first glance, the present process of post-Kyoto negotiations looks like nothing more than a tedious process of jockeying over minor technical points. However, many governments have begun to realize the extent to which the Kyoto Protocol poses difficult questions that require additional thought. They are now grappling not with final details of the treaty itself but its enormous economic implications. Progress toward completing the protocol is turning out to be very slow, and its fate remains uncertain.

If this slow progress continues, governments will eventually have to consider alternative policies—in economic terms—as a way to move ahead. These policies include not only the development of the Kyoto Protocol mechanisms but also increased "when" and "how" flexibility through emissions control targets that are early and modest but accelerate over time and with price-based policies to limit compliance cost shocks. For developed and developing countries alike, the challenge is to find ways to use the CDM concept to achieve the best possible outcome and to continue discussions on how all parties' interests can be served by a gradual broadening of commitments to the control of GHG emissions. This evolutionary approach to policy is consistent with the still-evolving science of climate change risks and the legitimate aspirations of developing countries for increased development.

Suggested Reading

Burtraw, D., and M. Toman. 1992. Equity and International Agreements for CO_2 Containment. *Journal of Energy Engineering* 118(2): 122–35.

CEA (Council of Economic Advisers). 1998. *The Kyoto Protocol and the President's Policies to Address Climate Change: Administration Economic Analysis.* July. Washington, DC: Executive Office of the President. http://www.weathervane.rff.org/refdocs/wh_analysis.pdf.

Cooper, R.N. 1998. Toward a Real Global Warming Treaty. *Foreign Affairs* 77(2): 66–79.

Goulder, Lawrence H., and Koshi Mathai. 2000. Optimal CO_2 Abatement in the Presence of Induced Technological Change. *Journal of Environmental Economics and Management* 39(1): 1–38.

Grubb, M., T. Chapuis, and M. Ha-Duong. 1995. The Economics of Changing Course: Implications of Adaptability and Inertia for Optimal Climate Policy. *Energy Policy* 23(4/5): 417–32.

Ha-Duong, M., M.J. Grubb, and J-C. Hourcade. 1997. Influence of Socioeconomic Inertia and Uncertainty on Optimal CO_2-Emission Abatement. *Nature* 390: 270–3.

Hahn, Robert W., and Robert N. Stavins. 1999. What Has Kyoto Wrought? The Real Architecture of International Tradable Permit Markets. RFF Discussion Paper 99-30. March. Washington, DC: Resources for the Future. http://www.rff.org/disc_papers/PDF_files/9930.pdf (accessed October 18, 2000).

Interlaboratory Working Group (IWG). 1997. Scenarios of U.S. Carbon Reductions: Potential Impacts of Energy Technologies by 2010 and Beyond. Report LBNL-40533 and ORNL-444. September. Berkeley, CA, and Oak Ridge, TN: Lawrence Berkeley National Laboratory and Oak Ridge National Laboratory.

Jacoby, H., R. Prinn, and R. Schmalensee. 1998. Kyoto's Unfinished Business. *Foreign Affairs* 77(4): 54–66.

Kolstad, C. A. 1996. Learning and Stock Effects in Environmental Regulation: The Case of Greenhouse Gas Emissions. *Journal of Environmental Economics and Management* 31(1): 1–18.

Kopp, Raymond, Richard Morgenstern, and William Pizer. 2000. *Limiting Cost, Assuring Effort, and Encouraging Ratification: Compliance under the Kyoto Protocol.* Prepared for the RFF/CIRED Workshop, June 26–27. Washington, DC: Resources for the Future. http://www.weathervane.rff.org/features/parisconf0721/KMP-RFF-CIRED.pdf (accessed October 18, 2000).

Kopp, Raymond, Richard Morgenstern, William Pizer, and Michael Toman. 1999. *A Proposal for Credible Early Action in U.S. Climate Policy.* February 16. Washington, DC: Resources for the Future. http://www.weathervane.rff.org/features/feature060.html (accessed October 18, 2000).

Manne, A.S., and R. Richels. 1997. On Stabilizing CO_2 Concentrations—Cost-Effective Emission Reduction Strategies. *Environmental Modeling and Assessment* 2(4): 251–65.

McKibbon, Warwick J., and Peter J. Wilcoxen. 1997. A Better Way to Slow Climate Change. Brookings Policy

Brief 17. June. Washington, DC, Brookings Institution. http://www.brookings.edu/comm/PolicyBriefs/pb017/pb17.htm (accessed October 18, 2000).

National Research Council (NRC). 2000. *Reconciling Observations of Global Temperature Change.* Washington, DC: National Academy Press.

Parry, Ian W.H., and Michael Toman. 2000. Early Emissions Reduction Programs: An Application to CO_2 Policy. RFF Discussion Paper 00-26. June. Washington, DC: Resources for the Future. http://www.rff.org/disc_papers/PDF_files/0026.pdf (accessed October 18, 2000).

Richels, R., and J. Edmonds. 1995. The Economics of Stabilizing Atmospheric CO_2 Concentrations. *Energy Policy* 23(4/5): 373–78.

Rose, A., and B. Stevens. 1998. A Dynamic Analysis of Fairness in Global Warming Policy: Kyoto, Buenos Aires, and Beyond. *Journal of Applied Economics* 1(2): 329–62.

Rose, A., B. Stevens, J. Edmonds, and M. Wise. 1998. International Equity and Differentiation in Global Warming Policy: An Application to Tradeable Emission Permits. *Environmental and Resource Economics* 12: 25–51.

Shogren, J., and M. Toman. 2000. Climate Change Policy. In *Public Policies for Environmental Protection* (Second Edition), edited by Paul Portney and Robert Stavins. Washington DC: Resources for the Future.

Toman, M., and J.-C. Hourcade. 1999. Policies for the Design and Operation of the Clean Development Mechanism. Summary of discussions at a multinational workshop organized by the Centre international de recherche sur l'environnement et le développement (CIRED) and Resources for the Future (RFF). September 24–25. Washington, DC: Resources for the Future. http://www.weathervane.rff.org/research/toman%5Fpolicies%5Fcdm.htm (accessed October 18, 2000).

Toman, M., R. Morgenstern, and J. Anderson. 1999. The Economics of "When" Flexibility in the Design of Greenhouse Gas Abatement Policies. *Annual Review of Energy and the Environment* 24: 431–60.

Weyant, John P., and Jennifer N. Hill. 1999. Introduction and Overview. *The Energy Journal*, Special Issue (The Costs of the Kyoto Protocol: A Multi-Model Evaluation): vi–xiiv.

Wiener, Jonathan B. 1999. Global Environmental Regulation: Instrument Choice in Legal Context. *Yale Law Journal* 108(4): 677–800.

Wigley, T.M.L., R. Richels, and J.A. Edmonds. 1996. Economic and Environmental Choices in the Stabilization of Atmospheric CO_2 Concentrations. *Nature* 379 (6562): 240–43.

Glossary

This glossary is adapted from Resources for the Future's *Weathervane* online climate glossary (http://www.weathervane.rff.org/glossary/index.html). Some definitions are taken from the texts of the Kyoto Protocol and the U.N. Federation Convention on Climate Change (UNFCCC); both documents are accessible via the UNFCCC website (www.unfccc.de).

additionality – Any improvements that would not have otherwise occurred without the existence of a project. In the context of a *joint implementation* or a *Clean Development Mechanism* project, any reduced emissions of *greenhouse gas* or increased *carbon sequestration* that occurs beyond the baseline projections are additionalities.

Annex B parties – Industrialized countries and economies in transition that are listed in Annex B of the *Kyoto Protocol*. Their responsibilities under the Kyoto Protocol would include legally binding national emissions ceilings during the period 2008–12 as per Article 3 of the protocol. The ceilings for Annex B parties range from an 8% decrease to a 10% increase relative to 1990 levels. (Note that Belarus and Turkey are listed in Annex I but not Annex B; Croatia, Liechtenstein, Monaco, and Slovenia are listed in Annex B but not Annex I; and the Czech Republic and Slovakia in Annex B replaced Czechoslovakia in Annex 1.) See also *Annex I parties*.

Annex I parties – Industrialized countries and economies in transition that are listed in Annex I of the *United Nations Framework Convention of Climate Change* (UNFCCC). Their responsibilities under the UNFCCC included a nonbinding commitment to return their *greenhouse gas* emissions to 1990 levels by the year 2000 as per Article 4.2(a) and (b) of the convention. See also *Annex B parties*.

anthropogenic – Directly or indirectly related to (or caused by) human influence on nature (for example, anthropogenic emissions).

assigned amounts – Allowed greenhouse gas emissions based on the emissions reductions for *Annex B* countries in the *Kyoto Protocol,* relative to 1990 baseline emissions.

auction – A sales method whereby the items are sold to the highest bidder. In a domestic *emissions trading* regime, permits or allowances for emitting *greenhouse gases* could be allocated by an auction.

banking – Saving emissions permits or *certified emission reduction units* for future use.

benefit–cost analysis (or cost–benefit analysis) – An economic technique applied to decisionmaking that attempts to quantify in dollar terms the advantages (benefits) and disadvantages (costs) associated with a particular policy.

biomass – Plant material that could be used as a renewable fuel source.

bubble – A collection of emission sources that are required to meet a single overall emission limit, rather than having separate limits for each source. Can apply to countries as well as facilities (see *E.U. bubble*).

cap-and-trade policy – A system of emissions control in which a target level of aggregate emissions is distributed among sources, and these sources can achieve more or less control by selling or buying emission permits from other sources (see *emissions trading*).

carbon sequestration – The collection or absorption of carbon in *sinks* such as oceans, forests, or soils that reduces the total amount of carbon in the atmosphere. See also *carbon sink*.

carbon sink – Any process, activity, or mechanism that removes carbon from the atmosphere. See also *carbon sequestration*, *sink*.

carbon taxes – Surcharges on the carbon content of oil, coal, and gas meant to discourage the use of fossil fuels and reduce carbon dioxide emissions. Can also apply to other greenhouse gases based on their *global warming potentials*.

certified emission reductions (CERs) (or certified emission reduction units) – Verified and authenticated units of *greenhouse gas* reductions from abatement or *carbon sequestration* projects that are certified by the *Clean Development Mechanism*.

Clean Development Mechanism (CDM) – A modified version of *Joint Implementation* that was included in the *Kyoto Protocol* for project-based activities in developing countries. The CDM was established to help developing countries achieve sustainable development and help *Annex I* parties meet their obligations for emissions limitations and reductions.

climate change – The global effect on the world's climate of an increase in the atmospheric concentration of *greenhouse gases,* attributed directly or indirectly to human activity (for example, deforestation; the burning of fossil fuels such as gasoline, oil, coal, and natural gas; and the release of chlorofluorocarbons [CFCs] from refrigerators, air conditioners, and other sources), that inhibit the transmission of some of the Sun's energy from the Earth's surface to outer space.

command-and-control regulation – In emissions reduction, requirements for polluters to install and use specific types of equipment to reduce emissions.

commitment period – A range of years within which parties to the *Kyoto Protocol* are required to meet their *greenhouse gas* emissions reduction target, which is averaged over the years of the commitment period. The first commitment period will be 2008–12.

Conference of Parties (COP) – The supreme body of the *United Nations Framework Convention on Climate Change* (UNFCCC), comprising more than 170 nations that have ratified the convention. Its role is to promote and review the implementation of the UNFCCC. It will periodically review existing commitments in light of the convention's objective, new scientific findings, and the effectiveness of national climate change programs. The first COP session was held in Berlin, Germany, in 1995, and the parties have continued to meet annually.

discounting – A method used by economists to determine the dollar value today of costs and benefits in the future. Future money values are weighted by a value <1, or "discounted." This reflects a commonly observed preference of individuals for reaping benefits sooner than later while delaying costs. It also reflects the fact that deferred consumption—investment—leads to increased future consumption opportunities.

distortionary taxes – Taxes that reduce the economic efficiency of production or consumption decisions. In practice, all taxes cause economic distortions; income taxes reduce overall economic activity, and specific commodity taxes inefficiently shift input or purchasing decisions.

double dividend – The notion that environmental taxes can reduce both pollution (the first dividend) and the overall economic costs associated with the tax system (the second dividend) by using the revenue generated to reduce *distortionary taxes*.

downstream – In the U.S. fossil fuel economy, it is commonly interpreted to be industrial boilers, electric utilities, and other major energy users but also applies, in theory, to all consumers of gasoline, coal, electricity, and other fuels. See also *upstream*.

emissions leakage – Emissions abatement achieved in one location that is offset by increased emissions in unregulated locations. In the short term, leakage can arise as emissions abaters reduce energy demand or timber supply, influencing world prices for these commodities and increasing the quantity emitted elsewhere; in the long term, it can arise as industries relocate to avoid controls.

emissions tax – Charge imposed on emissions to provide incentives for firms and households to reduce their emissions. The greater the level of the emissions tax, the greater the incentive to reduce emissions.

emissions trading – An economic incentive–based alternative to *command-and-control regulation*. In an emissions trading program, sources of a particular pollutant are given permits to release a specified volume of the pollutant. The government issues only a limited number of permits consistent with the desired level of emissions. The owners of the permits may keep them and release the pollutants, or reduce their emissions and sell the permits. The fact that the permits have value as an item to be sold or traded gives the owner an incentive to reduce emissions.

E.U. bubble – The policy and reallocation of the *assigned amounts* of individual European Union members.

externalities – The inadvertent impacts on the well-being of one person caused by the activity of another. Many aspects of environmental degradation, such as air pollution, *global warming*, loss of wilderness, and contamination of water bodies are viewed as externalities of economic transactions.

flexibility mechanisms – As established by the *Kyoto Protocol*, methods that seek to increase options and thereby reduce the costs of reducing emissions. The three primary mechanisms contained in the protocol are the *Clean Development Mechanism*, *emissions trading*, and *Joint Implementation*.

fossil fuels – Nonrenewable energy sources that include coal, petroleum products, and natural gas.

global warming – The progressive gradual increase of the Earth's surface temperature, thought to be caused by the *greenhouse effect* and responsible for changes in global climate patterns. See *climate change*.

global warming potential (GWP) – An index that allows for equal comparison of the various *greenhouse gases* due their varying power to accelerate *global warming* and/or the duration of their presence in the atmosphere.

grandfathering – In an *emissions trading* regime, a method for distributing permits for *greenhouse gas* emissions among emitters according to their historical emissions.

greenhouse effect – The progressive, gradual warming of Earth's atmospheric temperature, caused by the insulating effect of carbon dioxide and other *greenhouse gases* that have proportionately increased in the atmosphere. This effect disturbs the way the Earth's climate maintains the balance between incoming and outgoing energy by allowing short-wave radiation from the Sun to penetrate through to warm the Earth but preventing the resulting long-wave radiation from escaping back into the atmosphere.

greenhouse gases (GHGs) – Gaseous constituents of the atmosphere, both natural and *anthropogenic*, that act to retain heat energy from the sun. GHGs include the common gases of carbon dioxide, water vapor, and methane as well as rarer gases such as chlorofluorocarbons (CFCs). The increase in greenhouse gases in the atmosphere, which contributes to *global warming*, is a result of the burning of *fossil fuels*, the emission of pollutants into the atmosphere, and deforestation. See also *climate change*, *greenhouse effect*.

hot air – In recent *climate change* negotiations, surplus *assigned amounts* from unintentional reductions in emissions of *greenhouse gases* (as opposed to reductions that resulted from intentional efforts), such as those that followed the economic collapse of the former Soviet Union.

Joint Implementation – An arrangement in which industrialized countries meet their obligations for reducing *greenhouse gas* emissions by receiving credits for investing in emissions reductions in developing countries.

Kyoto forest – Forest that complies with the specifications of the Kyoto Protocol. Under Article 3, carbon sequestration will be credited only for forests planted after January 1, 1990, and only for carbon sequestered during the commitment period of 2008–12.

Kyoto Protocol – An international agreement among 159 nations that attended the third *Conference of Parties (COP-3)* of the *United Nations Framework Convention on Climate Change* (held in December 1997 in Kyoto, Japan). Delegates to COP-3 agreed to the following specific provisions.

Developed Countries: Thirty-eight developed countries agreed to reduce their emissions of six *greenhouse gases.* Collectively, developed countries agreed to cut back their emissions by a total of 5.2% between 2008 and 2012 from 1990 levels. The six gases are carbon dioxide, methane, nitrous oxide, and three ozone-damaging fluorocarbons not covered by the Montreal Protocol on Substances that Deplete the Ozone Layer (adopted in 1987), which banned global chlorofluorocarbons (hydrofluorocarbons, perfluorocarbons, and sulfur hexafluoride). The European Union agreed to reduce its total emissions by 8% below 1990 levels; the United States signed on to a 7% reduction; and Japan agreed to a 6% reduction. Some countries, particularly Russia and Ukraine, do not in practice have to make any reductions, given their drop in emissions since 1990. Countries including Australia, Iceland, Norway, and New Zealand are allowed to increase their emissions relative to 1990 levels, though the allowed growth is likely to be less than expected without greenhouse gas emissions limits.

Countries with Economies in Transition: Countries that are undergoing the transition to a market economy but also are classified with the European Union, Japan, and the United States as *Annex I* parties to the convention (the Czech Republic, Hungary, Poland, and others) face smaller reductions.

Developing Countries: Countries that are in the process of becoming industrialized and have limited resources (such as China and India) have no formal binding targets but have the option to set voluntary reduction targets.

leakage – Improvements (abatement) achieved in one location that are offset by worsening conditions (increased emissions) in unregulated locations. See also *emissions leakage.*

nitrogen oxides (NO$_x$) – Often mentioned in discussions of nitrogen-based air pollution as a reference to both nitric oxide (NO) and nitrogen dioxide (NO$_2$). In addition to *particulate matter* and *sulfur dioxide,* NO$_x$ is one of the major local and regional pollutants related to fossil energy combustion. It can transform to nitrates (a harmful fine particulate) in the atmosphere.

OECD (Organisation for Economic Co-operation and Development) – An international organization whose member countries include Australia, Austria, Belgium, Canada, the Czech Republic, Denmark, Finland, France, Germany, Greece, Hungary, Iceland, Ireland, Italy, Korea, Japan, Luxembourg, Mexico, the Netherlands, New Zealand, Norway, Poland, Portugal, Spain, Sweden, Switzerland, Turkey, the United Kingdom, and the United States.

ozone – A form of ground-level air pollution that is produced when *nitrogen oxides* and hydrocarbons react in sunlight. It is not to be confused with stratospheric ozone, which is found 9–18 miles high in Earth's atmosphere and protects people from harmful solar radiation. Ground-level ozone pollution is a problem mainly during hot summer days.

particulate matter – A form of air pollution that includes soot, dust, dirt, and aerosols. It has readily apparent effects on visibility and exposed surfaces. Particulate matter can create or intensify breathing problems and heart problems and can lead to premature death.

renewable resources – Energy sources that are not exhaustible. Sources of renewable energy include water, wind, the Sun (solar energy), the Earth (geothermal energy), and some combustible materials (for example, landfill gas, *biomass,* and municipal solid waste).

revenue-raising instruments – Policies that raise funds as a goal of regulation. In environmental policy, revenue-raising instruments include *emissions taxes,* which are levied against producers of pollution, and auctioned *tradable emissions permits,* which can be bought or sold by coal-burning electric utilities and other industries.

revenue recycling – Using the funds raised by a policy to reduce *distortionary taxes,* pay down government deficits, or pay rebates to consumers.

sink – Any process, activity, or mechanism that removes a *greenhouse gas* from the atmosphere. Sinks often refer to the absorption of atmospheric carbon by a forest. See also *carbon sequestration, carbon sink, source.*

source – Any process, activity, or mechanism that contributes a *greenhouse gas* to the atmosphere; the opposite of a *sink.*

stock – A cumulative amount or volume of something durable or persistent, as in the atmospheric stock of carbon or the capital stock of the electricity sector.

sulfur dioxide (SO_2) – A gaseous form of air pollution. SO_2 is a by-product of the combustion of fuels that contain sulfur and is most prevalent in the combustion of coal.

supplementarity – In the context of the *Kyoto Protocol,* limits on use of the *flexibility mechanisms* such as *emissions trading* to lower *greenhouse gas* mitigation costs, in order to increase the force of domestic energy and other policies for greenhouse gas reduction.

sustainable development – A broad concept referring to the need to balance the satisfaction of near-term interests with the protection of the interests of future generations, including their interests in a safe and healthy environment. As expressed by the 1987 U.N. World Commission on Environment and Development (the Brundtland Commission), sustainable development "meets the needs of the present without compromising the ability of future generations to meet their needs."

tradable emissions permits – An environmental regulatory scheme in which firms emitting the pollutant to be regulated are given permits to release a specified volume of the pollutant. The government issues only a limited number of permits consistent with the desired level of emissions. The owners of the permits may keep the permits and release the pollutants, or reduce their emissions and sell the permits. The fact that the permits have value as an item to be sold gives owners an incentive to reduce their emissions. See also *emissions trading.*

United Nations Framework Convention on Climate Change (UNFCCC) – The centerpiece of global negotiation to combat *global warming.* It was adopted in June 1992 at the "Earth Summit" in Rio de Janeiro, Brazil, and entered into force on March 21, 1998. The primary objective of the UNFCCC is the "stabilization of *greenhouse gas* concentrations in the atmosphere at a level that would prevent dangerous *anthropogenic* interference with the climate system. Such a level should be achieved within a time frame sufficient to allow ecosystems to adapt naturally to climate change, to ensure that food production is not threatened, and to enable economic development to proceed in a sustainable manner." The convention also contains several other provisions, including measures to promote increased financial and technical assistance for developing countries and the use of cost-effective policy instruments.

upstream – In the U.S. fossil fuel economy, it is commonly interpreted to mean the input to oil refineries, coal processing plants, and natural gas pipelines. See also *downstream.*

Index

Note: A page entry followed by the letter *f* indicates a figure; a page entry followed by the letter *t* indicates a table.